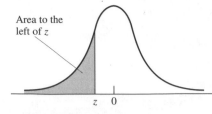

Area to the left of z

Each table value is the cumulative area to the left of the specified z-value.

STANDARD NORMAL CURVE AREAS (NEGATIVE z-VALUES)

The Second Decimal Digit of z

z	.00	.01	.02	.03	.04	.05	.06	.07	.08	.09
−3.4	.0003	.0003	.0003	.0003	.0003	.0003	.0003	.0003	.0003	.0002
−3.3	.0005	.0005	.0005	.0004	.0004	.0004	.0004	.0004	.0004	.0003
−3.2	.0007	.0007	.0006	.0006	.0006	.0006	.0006	.0005	.0005	.0005
−3.1	.0010	.0009	.0009	.0009	.0008	.0008	.0008	.0008	.0007	.0007
−3.0	.0013	.0013	.0013	.0012	.0012	.0011	.0011	.0011	.0010	.0010
−2.9	.0019	.0018	.0018	.0017	.0016	.0016	.0015	.0015	.0014	.0014
−2.8	.0026	.0025	.0024	.0023	.0023	.0022	.0021	.0021	.0020	.0019
−2.7	.0035	.0034	.0033	.0032	.0031	.0030	.0029	.0028	.0027	.0026
−2.6	.0047	.0045	.0044	.0043	.0041	.0040	.0039	.0038	.0037	.0036
−2.5	.0062	.0060	.0059	.0057	.0055	.0054	.0052	.0051	.0049	.0048
−2.4	.0082	.0080	.0078	.0075	.0073	.0071	.0069	.0068	.0066	.0064
−2.3	.0107	.0104	.0102	.0099	.0096	.0094	.0091	.0089	.0087	.0084
−2.2	.0139	.0136	.0132	.0129	.0125	.0122	.0119	.0116	.0113	.0110
−2.1	.0179	.0174	.0170	.0166	.0162	.0158	.0154	.0150	.0146	.0143
−2.0	.0228	.0222	.0217	.0212	.0207	.0202	.0197	.0192	.0188	.0183
−1.9	.0287	.0281	.0274	.0268	.0262	.0256	.0250	.0244	.0239	.0233
−1.8	.0359	.0351	.0344	.0336	.0329	.0322	.0314	.0307	.0301	.0294
−1.7	.0446	.0436	.0427	.0418	.0409	.0401	.0392	.0384	.0375	.0367
−1.6	.0548	.0537	.0526	.0516	.0505	.0495	.0485	.0475	.0465	.0455
−1.5	.0668	.0655	.0643	.0630	.0618	.0606	.0594	.0582	.0571	.0559
−1.4	.0808	.0793	.0778	.0764	.0749	.0735	.0721	.0708	.0694	.0681
−1.3	.0968	.0951	.0934	.0918	.0901	.0885	.0869	.0853	.0838	.0823
−1.2	.1151	.1131	.1112	.1093	.1075	.1056	.1038	.1020	.1003	.0985
−1.1	.1357	.1335	.1314	.1292	.1271	.1251	.1230	.1210	.1190	.1170
−1.0	.1587	.1562	.1539	.1515	.1492	.1469	.1446	.1423	.1401	.1379
−0.9	.1841	.1814	.1788	.1762	.1736	.1711	.1685	.1660	.1635	.1611
−0.8	.2119	.2090	.2061	.2033	.2005	.1977	.1949	.1922	.1894	.1867
−0.7	.2420	.2389	.2358	.2327	.2296	.2266	.2236	.2206	.2177	.2148
−0.6	.2743	.2709	.2676	.2643	.2611	.2578	.2546	.2514	.2483	.2451
−0.5	.3085	.3050	.3015	.2981	.2946	.2912	.2877	.2843	.2810	.2776
−0.4	.3446	.3409	.3372	.3336	.3300	.3264	.3228	.3192	.3156	.3121
−0.3	.3821	.3783	.3745	.3707	.3669	.3632	.3594	.3557	.3520	.3483
−0.2	.4207	.4168	.4129	.4090	.4052	.4013	.3974	.3936	.3897	.3859
−0.1	.4602	.4562	.4522	.4483	.4443	.4404	.4364	.4325	.4286	.4247
−0.0	.5000	.4960	.4920	.4880	.4840	.4801	.4761	.4721	.4681	.4641

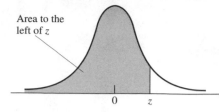

Area to the left of z

Each table value is the cumulative area to the left of the specified z-value.

STANDARD NORMAL CURVE AREAS (POSITIVE z-VALUES)

The Second Decimal Digit of z

z	.00	.01	.02	.03	.04	.05	.06	.07	.08	.09
0.0	.5000	.5040	.5080	.5120	.5160	.5199	.5239	.5279	.5319	.5359
0.1	.5398	.5438	.5478	.5517	.5557	.5596	.5636	.5675	.5714	.5753
0.2	.5793	.5832	.5871	.5910	.5948	.5987	.6026	.6064	.6103	.6141
0.3	.6179	.6217	.6255	.6293	.6331	.6368	.6406	.6443	.6480	.6517
0.4	.6554	.6591	.6628	.6664	.6700	.6736	.6772	.6808	.6844	.6879
0.5	.6915	.6950	.6985	.7019	.7054	.7088	.7123	.7157	.7190	.7224
0.6	.7257	.7291	.7324	.7357	.7389	.7422	.7454	.7486	.7517	.7549
0.7	.7580	.7611	.7642	.7673	.7704	.7734	.7764	.7794	.7823	.7852
0.8	.7881	.7910	.7939	.7967	.7995	.8023	.8051	.8078	.8106	.8133
0.9	.8159	.8186	.8212	.8238	.8264	.8289	.8315	.8340	.8365	.8389
1.0	.8413	.8438	.8461	.8485	.8508	.8531	.8554	.8577	.8599	.8621
1.1	.8643	.8665	.8686	.8708	.8729	.8749	.8770	.8790	.8810	.8830
1.2	.8849	.8869	.8888	.8907	.8925	.8944	.8962	.8980	.8997	.9015
1.3	.9032	.9049	.9066	.9082	.9099	.9115	.9131	.9147	.9162	.9177
1.4	.9192	.9207	.9222	.9236	.9251	.9265	.9279	.9292	.9306	.9319
1.5	.9332	.9345	.9357	.9370	.9382	.9394	.9406	.9418	.9429	.9441
1.6	.9452	.9463	.9474	.9484	.9495	.9505	.9515	.9525	.9535	.9545
1.7	.9554	.9564	.9573	.9582	.9591	.9599	.9608	.9616	.9625	.9633
1.8	.9641	.9649	.9656	.9664	.9671	.9678	.9686	.9693	.9699	.9706
1.9	.9713	.9719	.9726	.9732	.9738	.9744	.9750	.9756	.9761	.9767
2.0	.9772	.9778	.9783	.9788	.9793	.9798	.9803	.9808	.9812	.9817
2.1	.9821	.9826	.9830	.9834	.9838	.9842	.9846	.9850	.9854	.9857
2.2	.9861	.9864	.9868	.9871	.9875	.9878	.9881	.9884	.9887	.9890
2.3	.9893	.9896	.9898	.9901	.9904	.9906	.9909	.9911	.9913	.9916
2.4	.9918	.9920	.9922	.9925	.9927	.9929	.9931	.9932	.9934	.9936
2.5	.9938	.9940	.9941	.9943	.9945	.9946	.9948	.9949	.9951	.9952
2.6	.9953	.9955	.9956	.9957	.9959	.9960	.9961	.9962	.9963	.9964
2.7	.9965	.9966	.9967	.9968	.9969	.9970	.9971	.9972	.9973	.9974
2.8	.9974	.9975	.9976	.9977	.9977	.9978	.9979	.9979	.9980	.9981
2.9	.9981	.9982	.9982	.9983	.9984	.9984	.9985	.9985	.9986	.9986
3.0	.9987	.9987	.9987	.9988	.9988	.9989	.9989	.9989	.9990	.9990
3.1	.9990	.9991	.9991	.9991	.9992	.9992	.9992	.9992	.9993	.9993
3.2	.9993	.9993	.9994	.9994	.9994	.9994	.9994	.9995	.9995	.9995
3.3	.9995	.9995	.9995	.9996	.9996	.9996	.9996	.9996	.9996	.9997
3.4	.9997	.9997	.9997	.9997	.9997	.9997	.9997	.9997	.9997	.9998

STATISTICS IN PRACTICE

 Saunders College Publishing
Harcourt Brace College Publishers

Fort Worth   Philadelphia   San Diego   New York   Orlando   Austin
San Antonio   Toronto   Montreal   London   Sydney   Tokyo

STATISTICS

IN PRACTICE

ERNEST A. BLAISDELL

Elizabethtown College
Elizabethtown, Pennsylvania

Second Edition
Chapters 13–16

Requests for permission to make copies of any part of the work should be mailed to:
Permissions Department, Harcourt Brace & Company, 6277 Sea Harbor Drive, Orlando, Florida 32887-6777.

Some material in this work previously appeared in STATISTICS IN PRACTICE, First Edition, copyright © 1993 by Saunders College Publishing. All rights reserved.

Publisher: Emily Barosse
Product Manager: Nick Agnew
Associate Editor: Alexa Epstein
Project Editor: Nancy Lubars
Production Manager: Alicia Jackson
Art Director: Lisa Caro

Cover Credit: © Bill Brooks/Masterfile

Printed in the United States of America

STATISTICS IN PRACTICE, Second edition
0-03-023499-9

MINITAB is a registered trademark of Minitab, Inc.

Windows is a trademark of Microsoft Corporation.
Macintosh is a registered trademark of Apple Computer, Inc.

Library of Congress Catalog Card Number: 97-80332

8901234567   039   10 987654321

Dedication

To my loving Mother, Thelma

*With fondness, Alf and I recall
those many trips to and fro,
especially when you multiplied
by* n, *more or less.*

Now the final journey home remains.

NEW IN THE SECOND EDITION

Menu Commands:
Stat ➤
Basic Statistics ➤
1-Sample t
In **Variables** *box*
enter **C1**
Select **Test mean**
button and enter **50**
Select OK

- Chapter 1 now includes a discussion of **MINITAB for Windows,** and MINITAB usage is seamlessly integrated throughout the book for both the traditional *Session Commands* and the newer *Windows Menus/Dialog Boxes.* This is accomplished by accompanying each set of session commands in the main body of the text with a marginal box containing optional instructions for using menus and dialog boxes.

- The exercises have been extensively revised using recent information and data from a variety of cited sources. The number of problems is now 1,897, with 283 exercises from the first edition replaced by 547 new problems. The number of MINITAB exercises is increased by nearly 40 percent to 333. Additionally, the exercises reflect a greater emphasis on critical thinking and the need for students to provide explanations with their solutions.

- New sections were added to Chapters 11, 13, and 15. Section 11.6, **Determining Required Sample Sizes,** considers the problem of obtaining the necessary sample sizes for two-sample inferences concerning population means and proportions. Analysis of variance in Chapter 13 has been expanded to include **randomized block designs** and their analyses. The **Friedman F_r-test** for a randomized block design has been added to Chapter 15 on nonparametric statistics.

- Chapter 16, **Multiple Regression Models,** is new and provides a comprehensive introduction to the process of fitting a model to a set of multivariate data, evaluating the model's potential usefulness, and utilizing the model for estimation and prediction. Calculations are de-emphasized through the frequent utilization and interpretation of computer output.

- Chapter 10 now includes exercises on **operating characteristic curves** and **determining β,** the probability of committing a Type II error.

- Several marginal notes of interest have been added or updated to enhance students' awareness and appreciation of the prominent role that the field of statistics plays in our daily events.

- Highlighted boxes are now clearly differentiated as either **Procedure, Key Concept, Note,** or **Tip Boxes.**

- Constructed by the author exclusively for the text, a completely new test bank of 800 distinct multiple-choice exercises is available in both written and computerized form for Windows, DOS, and Macintosh platforms.

- A data disk for 550 exercises, including all MINITAB problems, is now included with the text.

- To reduce the cost to students who will use the text for only a one-semester course, Chapters 13 through 16 are now bound in a separate paperback volume. For use in a second course, this second volume can be purchased for a reasonable price either separately or bundled with the first 12 chapters.

PREFACE

"The time may not be very remote when it will be understood that for complete initiation as an efficient citizen of one of the great complex world states that are now developing, it is necessary to be able to compute, to think in averages and maxima and minima, as it is now to be able to read and to write."

Though written more than 65 years ago, H. G. Wells' passage from *Mankind in the Making* seems particularly relevent in today's electronic age of global communication. Understanding the uses of statistics and its role in assimilating information contained in reports, scientific journals, political coverage, or even the daily newspaper is a necessary part of modern education. I have attempted to bring my training as a statistician and my years of teaching experience to the shaping of a clear and concisely written text. In addition, there is always the difficulty of persuading students, many of whom have been conditioned into math anxiety, that they might actually enjoy statistics, that statistical concepts are worth learning, and that statistics derives from real problems in the real world. My goal has been to provide a presentation that is pedagogically and mathematically sound, yet sufficiently gentle to minimize math anxiety.

Content Features

- To make data as vivid as possible, traditional methods of summarizing data are blended in Chapter 2 with more recently developed **data analysis techniques** such as dotplots, stem-and-leaf displays, 5-number summaries, and boxplots.

- **Regression analysis,** frequently used in many disciplines, can be introduced much earlier than it usually is in the traditional course. Chapter 3, therefore, is a concise introduction to the descriptive aspects of correlation and regression. It is written so that an instructor can vary its placement within the course syllabus. Furthermore, if the instructor wishes, this coverage can be complemented later with a detailed discussion in Chapter 14 of inferential methods in regression analysis. Chapter 16, Multiple Regression Models, allows additional flexibility in the depth of coverage for regression analysis.

- Included in Chapter 4, which introduces probability, is a **concise section on elementary counting techniques.** Students often find this topic difficult because they tend to approach each problem as either "a permutation or a combination." I believe a greater understanding can be achieved by de-emphasizing permutation formulas and stressing the versatility of the multiplication rule.

- The critical topics of **confidence intervals** and **hypothesis testing** merit separate chapters. They are introduced in Chapters 9 and 10, respectively, and in Chapter 11 they are jointly used to discuss two-sample inferences.

- **P-values** are prominent in the research literature of virtually all disciplines. Consequently, after the introduction of hypothesis testing in Chapter 10, the reader is frequently exposed to the use of P-values throughout the remainder of the book. This is done, however, only after the student has had adequate opportunity to comprehend the basic concepts of hypothesis testing and rejection regions.

Exercises and Examples

- There is an abundant quantity of **interesting exercises** (1,897) and illustrations based on real-life situations and cited sources from a wide spectrum of disciplines. They are stated concisely, without burying the reader in verbiage.

- The **exercises** have been **carefully selected** and constructed to ensure that they meaningfully contribute to the learning process and enhance an appreciation of how statistics intermingles with our daily lives. The order of presentation progresses from mastering the basics to practical applications. Data used in previous exercise sets are always reproduced when used in subsequent applications. To serve the needs of instructors, odd-numbered problems are frequently paired with comparable even-numbered problems. Answers are given in the book for all review exercises and for all odd-numbered end-of-section exercises.

- **Worked examples** are set up so that students can "walk through" them step by step. This approach helps the student understand the rationale of each statistical procedure. **Procedure boxes** that recap in a step-by-step manner what students should understand about a given process are liberally provided, as are **Key Concept, Note,** and **Tip boxes.**

Pedagogical Features

- Each chapter opens with a preview, "Looking Ahead." The opening photograph and accompanying caption set the theme for the chapter and give the students a foretaste of what they will explore in the pages that follow.

- To enhance the book's appearance and reader friendliness, liberal use is made of photos, marginal notes of interest, newspaper and magazine excerpts, and historical highlights of prominent mathematicians and statisticians.

- To help students master and retain the concepts in each chapter, the end-of-chapter material includes a summary, "Looking Back," a "Key Words" list, a "MINITAB Commands" list, and "Review Exercises," a set of comprehensive problems.

- Important formulas, concepts, procedures, tips, notes, and computer commands are highlighted so that the reader can give them priority on first reading and, later, during review.

- A removable, detailed formula card and separate tables card have been bound into the book for possible use during examinations.

- For quick reference, the normal, chi-square, and t tables are reproduced on the inside covers.

Use of MINITAB

This book is unique in its abundant use of the statistical computer package MINITAB. Instead of being used as a mere appendage to each chapter, MINITAB is woven throughout the text, with each command explained when it is first used in an application. Each set of commands is accompanied by a marginal box containing optional instructions for using menus and dialog boxes. The integration of MINITAB emphasizes the computer's role as a practical tool for relieving much of the drudgery associated with data sets, allowing the user more time to focus on other aspects of the analysis such as selecting a proper procedure, describing and interpreting data, and displaying the results. MINITAB's use also enhances the comprehension of many statistical concepts and techniques presented in a first course.

Although there are many statistical software packages currently available, I chose MINITAB because of its wide acceptance in educational instruction and its extensive use around the world in business and government. It is currently used at more than 2,000 colleges and universities and by 75% of Fortune 500 companies. Of equal importance, MINITAB can be learned quickly and easily, providing students with a powerful data analysis system that can be used in other courses and in their professional careers.

Because not all students have access to MINITAB, I have incorporated it into the text in a manner that affords an instructor considerable flexibility concerning its usage in the course. An instructor can choose any of the following options, each of which has been used by the author during the class-testing of this book.

- **Active Computer Usage.** Sufficient MINITAB instruction is provided so that students can execute their own commands for statistical analyses. The book contains a total of 333 MINITAB assignments. They are flagged with the symbol **M▶** and are placed at the end of an exercise set. MINITAB coverage is so extensive that the usual MINITAB supplement manual is unnecessary.

- **Passive Computer Usage.** Students can be directed to examine only the output of the MINITAB exhibits and to just look over (or ignore) the instructions used to generate the results.

- **No Computer Usage.** An instructor may prefer to have the class skip entirely the MINITAB exhibits. Implementation of this option is facilitated by the fact that all MINITAB output is prominently highlighted.

Flexibility in Topical Coverage

The book is designed so that an instructor has a great deal of flexibility in topical coverage. The diagram following displays several possibilities.

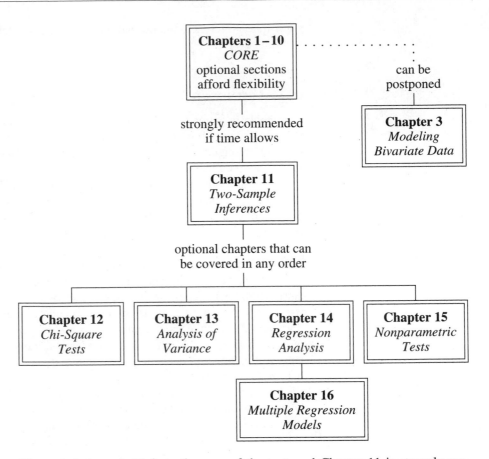

Chapters 1 through 10 form the core of the text, and Chapter 11 is strongly recommended if time allows. For instructors who want to spend less time covering these chapters, optional sections 6.4, The Hypergeometric Probability Distribution; 6.5, The Poisson Probability Distribution; and 11.3, Small-Sample Inferences for Two Means: Independent Samples and Unequal Variances, can be excluded. Additional time can be gained by also excluding sections 9.7, Chi-Square Probability Distributions; 9.8, Confidence Interval for a Variance; 10.5, Hypothesis Test for a Variance; 11.7, F Probability Distributions, and 11.8, Inferences for Two Variances.

Supplements

The following supplements have been prepared to enhance the use of this book. They are available, free of charge, to instructors who adopt the text.

- **Student Solutions Manual.** This supplement was prepared by Ronald L. Shubert of Elizabethtown College and is available to students for purchase. It contains detailed solutions for all review exercises and all odd-numbered end-of-section exercises.

- **Instructor's Manual.** Also prepared by Ronald L. Shubert, this manual contains detailed solutions to all exercises, and a sample course syllabus with helpful suggestions.

- **ExaMaster™ Computerized Test Bank.** Available for PCs (Windows and DOS) and Macintosh computers, this test bank contains 800 multiple-choice exercises, each written especially for this text. A virtually unlimited number of tests can be custom designed by an instructor, and grading keys can be generated. Tests can contain a mixture of multiple-choice and free-response questions added by an instructor. Full documentation and computer software for managing student records accompany the test bank.

- **Printed Test Bank.** A printed version of the Computerized Test Bank is also available. This supplement was prepared by the author of the text.

- **Data Disk.** New to this edition, a computer diskette for DOS and Windows is included with the text and contains the data sets for 550 exercises, including all data sets for the MINITAB problems. The data sets are stored as ASCII files. A Macintosh version of this disk is also available from the publisher.

Saunders College Publishing may provide complimentary instructional aids and supplements or supplement packages to those adopters qualified under our adoption policy. Please contact your sales representative for more information. If as an adopter or potential user you receive supplements you do not need, please return them to your sales representative or send them to

Attn: Returns Department
Troy Warehouse
465 South Lincoln Drive
Troy, MO 63379

ACKNOWLEDGMENTS

Sincere thanks are extended to the many individuals who contributed to the development of this book. In particular, I am grateful for the constructive comments and suggestions from the following reviewers of the second edition.

Patti B. Collings
Brigham Young University

Ethel Muter
Raritan Valley Community College

Byron A. Dyce
Sante Fe Community College

Reider Peterson
Southern Oregon State College

Douglas W. Gosbin
Rowan College of New Jersey

Donald Richards
University of Virginia

Kenneth Grace
Anoka-Ramsey Community College

Robert K. Smidt
California State Polytechnic University

Cheryl R. Groff
Florida Community College

Donal B. Staake, Jr.
Jackson Community College

Joyce K. Hill
Essex Community College

Linn M. Stranak
Union University

William E. Hinds
Midwestern State University

Richard Watkins
Tidewater Community College, Virginia Beach

Lloyd R. Jaisingh
Morehead State University

I also want to acknowledge the contributions from the following who served as reviewers of the manuscript for the first edition.

Graydon Bell
Northern Arizona University

William E. Hinds
Midwestern State University

Patricia M. Buchanan
Pennsylvania State University

Kermit Hutcheson
University of Georgia

Chris Burditt
Napa Valley College

Marlene J. Kovaly
Florida Junior College at Jacksonville

Darrell F. Clevidence
Carl Sandburg College

Mike Orkin
California State University, Hayward

Pat Deamer
Skyline College

Larry Ringer
Texas A & M University

William D. Ergle
Roanoke College

Gerald Rogers
New Mexico State University

Bryan V. Hearsey
Lebanon Valley College

Adele Shapiro
Palm Beach Community College

George Sturm
Grand Valley State University

Douglas A. Zahn
Florida State University

Mary Sue Younger
University of Tennessee

Considerable effort has been devoted to making the text as accurate as possible. I am grateful to David Mathiason at the Rochester Institute of Technology and Donal B. Staake, Jr. at Jackson Community College for serving as accuracy reviewers of all examples and answers at the back of the text. I also thank Beth Foremsky who proofread the Printed Test Bank and checked the accuracy of the answers. Any errors that might remain in the text or its test bank are, however, the sole responsibility of the author.

For their constructive suggestions concerning the first edition, I would like to thank Temple University Professors Richard Heiberger, Burt Holland, Damaraju Raghavarao, and Jagbir Singh. Special thanks are extended to Ronald Shubert at Elizabethtown College who wrote the Student Solutions Manual and the Instructor's Manual. Joyce Curry-Daly and Jim Daly of California Polytechnic State University — San Luis Obispo and Sudhir Goel of Vadosta State University proofread the Instructor's Manual. Their work is greatly appreciated.

I am grateful to Minitab, Inc. for their technical assistance and for providing a copy of the latest release of MINITAB available at the time of this writing — Release 11.12 for Windows. Information about MINITAB can be obtained by contacting

Minitab, Inc.
3081 Enterprise Drive
State College, PA 16801-3008
Phone: (814) 238-3280
Fax: (814) 238-4383
E-mail: info@minitab.com
URL: http://www.minitab.com

I am greatly indebted to the staff at Saunders College Publishing. I wish to thank Bill Hoffman (Executive Editor), Terri Ward (Developmental Editor), Alexa Epstein (Sr. Associate Editor), Nancy Lubars (Sr. Project Editor), Linda Boyle (Project Editor), Alicia Jackson (Production Manager), Lisa Caro (Sr. Art Director), Jane Sanders (Photo Developmental Editor), George Semple (Photo/Permissions Manager), Monika Ruzyc (Editorial Assistant), Nick Agnew (Product Manager) and Adrienne Krysiuk (Marketing Assistant).

Finally, I wish to extend special thanks to my wife, Judy, for her support and many hours of assistance during the preparation of this work.

E.A.B.
Lebanon, Pennsylvania

CONTENTS

INDEX OF APPLICATIONS

Life Sciences

Social Sciences

Technology

C H A P T E R

13

ANALYSIS OF VARIANCE

COMPARING TWO OR MORE POPULATION MEANS

Earlier we learned how to conduct mean comparisons between two groups, and we saw several illustrations of their practical application. In this chapter, we will learn how the methodology for two comparisons can be extended to any number. For instance, we will compare three different experimental processes for the treatment of water hardness caused by the presence of calcium and magnesium salts in the soil. Next, a comparison of four brands of spaghetti sauce will be made to determine if they differ in their amounts of beef. In the exercises, you will be asked to perform a variety of mean comparisons such as the

- *weight gains of puppies fed three brands of dog food,*
- *quality ratings of bond paper made with three sources of pulp,*
- *performance ratings for four computer word processing programs,*
- *amounts of DDT in lake trout from four lakes,*
- *beginning salaries of education majors in three states,*
- *drying times of four brands of furniture stain,*
- *radon levels in four school districts,*
- *prices of a camcorder at three types of retail outlets,*
- *costs of milk in three geographical regions of a state,*
- *life lengths of three brands of flashlight batteries,*

© Tim Davis/Photo Researchers, Inc.

- *hourly rates of automobile mechanics in three states,*
- *ascorbic acid (vitamin C) levels in Cortland apples grown in three geographical regions,*
- *benchmark test results for three Pentium-powered notebook computers,*
- *durability of four types of piston rings,*
- *prescription costs at four mail-order companies,*
- *salaries and bonuses received by CEOs,*
- *prices of store-brand products at four supermarket chains.*

LOOKING
▸
A H E A D

Photo courtesy UPI/Bettmann

The field of analysis of variance was developed by R. A. Fisher (1890–1962). Fisher brought the methodology to India, where it was used to analyze experimental designs applied to agricultural studies.

The title of this chapter, Analysis of Variance, might suggest that our primary intent will be analyzing variability in data. While we are concerned with this aspect, it is only a means to an end. The focus of the chapter is the **testing of hypotheses concerning several population means.** As we begin the chapter we will see that analysis of variance is simply a technique that allows one to generalize the pooled two-sample t-test for two means that was considered in Chapter 11. With this generalization, we will be able to test the null hypothesis that k population means are equal, where k may be two or more. The required assumptions for this procedure are the same as those for the pooled t-test, namely, that **independent random samples** have been selected from **normal populations** that have a **common variance σ^2.**

Analysis of variance is actually a general term that applies to several different types of statistical analyses, and the technique will be used in the concluding section to generalize the paired-samples t-test of Section 11.4. Until then, however, the chapter will focus on the method called a **one-way analysis of variance.** To understand the meaning of "one-way" recall from our study of regression that the **response variable** (dependent variable) is the variable of primary interest in the experiment. For instance, it might be the length of life of a dishwasher. We are usually concerned with how the response variable is affected by one or more independent variables. In analysis of variance, an independent variable is called a **factor,** and the different values of the factor are called its **levels.** For the dishwasher example, our primary concern might be to determine if there is a difference in the mean life of three dishwasher brands, such as A, B, and C. In this experiment, we would be interested in whether the means (μ_1, μ_2, μ_3) of the response variable (length of life) differ for the three levels (A, B, C) of the factor (dishwasher brand). The term **"one-way"** in a one-way analysis of variance indicates that the experiment is concerned with the effects of only **one factor** on the response variable.

In a one-way analysis of variance, the different levels (values) of the single factor are called the **treatments.** This term has been adopted because analysis of variance had its roots in agricultural experiments for which the factor levels were often different combinations of soil treatments.

Terminology used in a one-way analysis of variance is summarized on the following page.

KEY CONCEPTS

The **response variable** is the dependent variable and is the variable of primary interest in the experiment.

A **factor** is an independent variable.

The **levels** of a factor are its different values, and the levels are called the **treatments**.

The term **one-way** denotes that the experiment is concerned with the effects of only one factor on the response variable.

13.1 THE ANALYSIS OF VARIANCE TECHNIQUE

In Section 11.2, the pooled two-sample *t*-statistic was used to test the null hypothesis that two population means are equal. By using a technique referred to as **analysis of variance,** this type of problem can be generalized to testing hypotheses concerning the equality of two or more population means. The procedure is based on the same assumptions as were required for the pooled two-sample *t*-test. For the one-way analysis of variance we will assume the following (Figure 13.1).

NOTE

Assumptions for the One-Way Analysis of Variance:

1. The samples are random and independent.

2. Each population has a normal distribution.

3. The populations have the same variance σ^2.

Because the experiment is designed on the premise that independent random samples are used, the experiment is known as a **completely randomized design.**

We will use the following example to introduce the basic concept that underlies the analysis of variance methodology. Water hardness is principally caused by the presence of calcium and magnesium salts. The research and development division of a water conditioning company wants to evaluate the effectiveness of three differ-

© Milton Rand/Tom Stack & Associates

Each day, the United States uses about 40 billion gallons of water just for normal household purposes.

Figure 13.1
Assumptions for testing
$H_0: \mu_1 = \mu_2 = \mu_3 = \ldots \mu_k$.

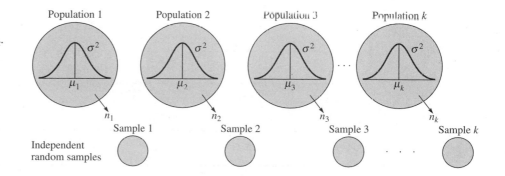

Population 1   Population 2   Population 3   Population k

σ^2 σ^2 σ^2 σ^2

μ_1 μ_2 μ_3 μ_k

n_1 n_2 n_3 n_k

Sample 1   Sample 2   Sample 3   Sample k

Independent random samples

TABLE 13.1

AMOUNTS (PPM) OF SALTS REMOVED BY THE THREE TREATMENTS

Process 1	Process 2	Process 3	
14	16	18	
12	14	16	
13	15	16	
15		17	
		19	
$n_1 = 4$	$n_2 = 3$	$n_3 = 5$	$n = 12$
$\bar{x}_1 = 13.5$	$\bar{x}_2 = 15$	$\bar{x}_3 = 17.2$	$\bar{x} = 15.417$
$s_1^2 = 1.667$	$s_2^2 = 1$	$s_3^2 = 1.7$	

ent experimental processes for the treatment of water hardness. For each treatment process, a random sample of treated water was selected and analyzed for calcium and magnesium salts. Table 13.1 gives the amounts (in parts per million, ppm) of these salts that were removed by the three treatment methods. The company wants to test the null hypothesis that the mean amount of salts removed is the same for each process.

$$H_0: \mu_1 = \mu_2 = \mu_3$$

$$H_a: \text{Not all the means are equal.}$$

In Table 13.1, we have denoted the total number of observations $(n_1 + n_2 + n_3)$ by n, and the mean of the entire group has been denoted by $\bar{x}$ and is called the **grand mean.** To illustrate graphically the three samples, we had MINITAB construct the dotplots that appear in Exhibit 13.1.

▬▬ **Exhibit 13.1**

Menu Commands:
Graph ➤
Character Graphs ➤
Dotplot
In **Variables** *box enter*
C1-C3
Check **Same scale for**
all variables *option*
Select OK

Dotplots for the Three Samples of Measurements

```
MTB > NAME C1 'PROCESS1' C2 'PROCESS2' C3 'PROCESS3'
MTB > DOTPLOT C1-C3;
SUBC> SAME.


         .        .        .        .
  -----+---------+---------+---------+---------+---------+-PROCESS1
                          .        .        .
  -----+---------+---------+---------+---------+---------+-PROCESS2
                                   .
                          .        .        .        .
  -----+---------+---------+---------+---------+---------+-PROCESS3
     12.0      13.5      15.0      16.5      18.0      19.5
```

Figure 13.2
Plot of the three samples combined (the sample number of each value is shown).

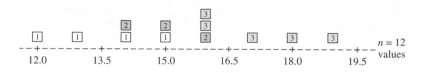

$n = 12$ values

We have seen that for two populations, the equality of their means is tested by measuring the difference between their sample means, $(\bar{x}_1 - \bar{x}_2)$. This procedure must be modified when dealing with more than two populations. To determine if the three population means in our example differ, we will analyze the variability in the samples (analysis of variance). First, think of the three samples as being combined into a single group of $n = 12$ values as illustrated in Figure 13.2, where each value is plotted by using its sample number.

A frequently used measure of the variability in a set of values is $SS(x)$, the sum of the squared deviations of the values from their mean. This quantity is called the **total sum of squares** and is denoted by SST. For the three samples, the total sum of squares is

$$SST = SS(x) = \Sigma x^2 - \frac{(\Sigma x)^2}{n}$$

$$= (14^2 + 12^2 + \cdots + 17^2 + 19^2) - \frac{(14 + 12 + \cdots + 17 + 19)^2}{12}$$

$$= 2{,}897 - \frac{(185)^2}{12} = 44.92.$$

In the analysis of variance procedure, the total sum of squares is partitioned into two parts that are called the **treatment sum of squares (SSTr)** and the **error sum of squares (SSE)**. The treatment sum of squares is a measure of the variability in the $n = 12$ measurements that can be attributed to differences among the three sample means. We will see that $SSTr$ measures the variability between the sample means and equals 31.12.

The error sum of squares (also called the **residual sum of squares**) is the difference between the total sum of squares (SST) and the treatment sum of squares $(SSTr)$. It can be thought of as that portion of SST that is left over (residual) after accounting for the differences between the treatment means.

The partitioning of the total sum of squares into the treatment and the error sum of squares is illustrated in Figure 13.3 on the following page.

To determine if the three population means differ, a test statistic can be derived by obtaining two independent estimates of σ^2, the assumed common variance of the three populations. One such estimate is based on SSE, and is just the extension to three samples of the pooled sample variance, s_p^2, that was used with the pooled two-sample t-test. Recall that s_p^2 is a weighted average of the sample variances, where the weights are the degrees of freedom associated with each variance. For three samples, the pooled sample variance is

$$s_p^2 = \frac{(n_1 - 1)s_1^2 + (n_2 - 1)s_2^2 + (n_3 - 1)s_3^2}{(n_1 - 1) + (n_2 - 1) + (n_3 - 1)}.$$

Figure 13.3
Partitioning of the total sum of squares.

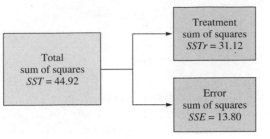

For the given samples, s_p^2 is obtained by multiplying the sample variances ($s_1^2 = 1.667$, $s_2^2 = 1$, and $s_3^2 = 1.7$) by their weights [$(4 - 1)$, $(3 - 1)$, and $(5 - 1)$], and dividing the result by the sum of the weights.

$$s_p^2 = \frac{(4 - 1)(1.667) + (3 - 1)(1) + (5 - 1)(1.7)}{(4 - 1) + (3 - 1) + (5 - 1)} = \frac{13.80}{9} = 1.533$$

Notice that the numerator of s_p^2 is 13.80, the value of the error sum of squares shown in Figure 13.3.

Since the pooled sample variance is a weighting of the three sample variances, s_p^2 (and its numerator SSE) can be thought of as a *measure of the variability within the samples*. An examination of the three dotplots in Exhibit 13.1 reveals that the observations within each sample are quite close to each sample mean, and this is why s_p^2 is small.

A second estimate of the three populations' common variance σ^2 can be obtained by dividing $SSTr$, the treatment sum of squares, by $k - 1$, where k is the number of sampled populations. This ratio is called the **mean square for treatments,** or more simply, the **treatment mean square.** It will be denoted by $MSTr$ and can be calculated as follows:

$$MSTr = \frac{SSTr}{k - 1} = \frac{n_1(\bar{x}_1 - \bar{x})^2 + n_2(\bar{x}_2 - \bar{x})^2 + n_3(\bar{x}_3 - \bar{x})^2}{k - 1}$$

where $\bar{x}_1$, $\bar{x}_2$, and $\bar{x}_3$ denote the three sample means, and $\bar{x}$ is the grand mean of the entire group of $n = n_1 + n_2 + n_3$ sample values. For the three samples, $\bar{x}_1 = 13.5$, $\bar{x}_2 = 15$, $\bar{x}_3 = 17.2$, and $\bar{x} = 15.417$ (see Table 13.1). Thus, the treatment mean square is

$$MSTr = \frac{4(13.5 - 15.417)^2 + 3(15 - 15.417)^2 + 5(17.2 - 15.417)^2}{3 - 1}$$

$$= \frac{31.12}{2} = 15.56.$$

The treatment mean square *measures the variability between the samples* by calculating the square of the deviation of each sample mean from the grand mean $\bar{x}$. If the three population means are equal, then we would expect that the three sample means will not differ much in value and will be close to the grand mean. Thus, a small value of $MSTr$ is consistent with the null hypothesis. On the other hand, a large value of $MSTr$ indicates considerable variation in the sample means and, thus, evidence against the null hypothesis of equal population means. For our example, the *MSTr* value of 15.56 is large compared to the pooled sample variance $s_p^2 = 1.533$, and the large value of *MSTr* reflects the presence of substantial variability between the three sample means. This is also evident by examining Figure 13.2. Although the *variation within* each sample is small, there is considerable *variability between* the three samples. The analysis of variance technique is based on a comparison of these two sources of variation. In particular, the test statistic is the following ratio, which can be shown to have an *F*-distribution.

$$F = \frac{MSTr}{s_p^2} = \frac{\text{variation between the samples}}{\text{variation within the samples}}$$

If the null hypothesis is true (and the stated assumptions hold), then the numerator and the denominator of the *F*-ratio are each estimates of σ^2 and, consequently, the *F*-ratio is expected to be near 1. When the null hypothesis is false, s_p^2 still estimates σ^2, but the numerator has an expected value larger than σ^2 and, thus, the *F*-ratio is expected to be greater than 1. The weight of evidence against H_0 is directly related to the size of *F* and, therefore, the rejection region is always upper tailed. It consists of values for which $F > F_\alpha$, where α is the specified significance level. For our example, the *F*-ratio is

$$F = \frac{MSTr}{s_p^2} = \frac{15.56}{1.533} = 10.15.$$

In the next section, we will see that this *F*-value is sufficiently large to warrant the rejection of the null hypothesis that the three population means are equal. We will thus conclude that the mean amount of salts removed is not the same for all three water treatment processes.

13.2 THE ANALYSIS OF VARIANCE TABLE AND COMPUTING FORMULAS

In performing an analysis of variance, various components involved in the calculation of the *F*-statistic are usually displayed in tabular form, referred to as the **ANOVA** (**AN**alysis **O**f **VA**riance) **table.** For the water treatment example in the previous section, the ANOVA table appears in Table 13.2.

The first column of the table lists the sources of variation in the data. The *SS*

TABLE 13.2

ANOVA TABLE FOR TESTING $H_0: \mu_1 = \mu_2 = \mu_3$

Source of Variation	df	SS	MS	F
Treatments	2	31.12	15.56	10.15
Error	9	13.80	1.533	
Total	11	44.92		

TABLE 13.3				
ANOVA TABLE FOR COMPARING *k* POPULATION MEANS				
Source of Variation	df	SS	MS	F
Treatments	$k - 1$	SSTr	$MSTr = \dfrac{SSTr}{k - 1}$	$\dfrac{MSTr}{MSE}$
Error	$n - k$	SSE	$MSE = \dfrac{SSE}{n - k}$	
Total	$n - 1$	SST		

column contains the sum of squares for the three sources, and the second column gives their associated degrees of freedom. The *MS* column contains the **mean square for treatments** *(MSTr)* and the **mean square for error** *(MSE)*. A mean square is obtained by dividing a sum of squares by its degrees of freedom. The last column gives the *F*-ratio that is obtained by dividing the mean square for treatments by the mean square for error. The general form of the ANOVA table based on *k* samples appears in Table 13.3.

To begin the construction of the ANOVA table, the sums of squares for the different sources of variation must be obtained. *SSTr* and *SSE* are seldom found by using the formulas of the previous section. Preferably, the sums of squares are obtained as part of the computer output from a statistical software package such as MINITAB. When this is not feasible, they can be obtained by employing shortcut formulas that we will now illustrate. We will construct the ANOVA table for the water treatment data considered earlier and reproduced below.

Amounts (ppm) of Salts Removed by the Three Treatments		
Process 1	Process 2	Process 3
14	16	18
12	14	16
13	15	16
15		17
		19

$$T_1 = 54 \qquad T_2 = 45 \qquad T_3 = 86 \qquad T = 185$$

$$n_1 = 4 \qquad n_2 = 3 \qquad n_3 = 5 \qquad n = 12$$

In the above, the totals for the samples have been calculated and are denoted by $T_1, T_2,$ and T_3. The combined total of the three samples is also needed and is denoted by T, where

$$T = \Sigma x = T_1 + T_2 + T_3 = 54 + 45 + 86 = 185.$$

To construct the ANOVA table, follow these steps.

1. **Calculate the total sum of squares.**

$$SST = \Sigma x^2 - \frac{T^2}{n}$$

$$= (14^2 + 12^2 + \cdots + 17^2 + 19^2) - \frac{185^2}{12}$$

$$= 2{,}897 - 2{,}852.08 = 44.92$$

2. **Calculate the treatment sum of squares.**

$$SSTr = \left(\frac{T_1^2}{n_1} + \frac{T_2^2}{n_2} + \frac{T_3^2}{n_3}\right) - \frac{T^2}{n}$$

$$= \left(\frac{54^2}{4} + \frac{45^2}{3} + \frac{86^2}{5}\right) - \frac{185^2}{12}$$

$$= 2{,}883.2 - 2{,}852.08$$

$$= 31.12$$

3. **Calculate the error sum of squares.**

$$SSE = SST - SSTr$$

$$= 44.92 - 31.12$$

$$= 13.80$$

The computational formulas are given below for obtaining the sums of squares for the general situation involving k samples.

PROCEDURE

Computing Formulas for the Sums of Squares in the ANOVA Table:

1. Calculate the total sum of squares.

(13.1) $$SST = \Sigma x^2 - \frac{T^2}{n}$$

2. Calculate the treatment sum of squares.

(13.2) $$SSTr = \left(\frac{T_1^2}{n_1} + \frac{T_2^2}{n_2} + \cdots + \frac{T_k^2}{n_k}\right) - \frac{T^2}{n}$$

3. Calculate the error sum of squares.

(13.3) $$SSE = SST - SSTr$$

In the above formulas,

k is the number of samples;

$n_1, n_2, \ldots, n_k$ are the sample sizes, and $n = \Sigma n_i$;

$T_1, T_2, \ldots, T_k$ are the sample totals, and $T = \Sigma T_j$.

After obtaining the treatment sum of squares *SSTr* and the error sum of squares *SSE,* the ANOVA table is completed by computing the associated mean squares and their *F*-ratio.

$$MSTr = \frac{SSTr}{df} = \frac{31.12}{2} = 15.56$$

$$MSE = \frac{SSE}{df} = \frac{13.80}{9} = 1.533$$

$$F = \frac{MSTr}{MSE} = \frac{15.56}{1.533} = 10.15$$

As discussed in the previous section, the *F*-ratio is used as the test statistic for deciding whether or not the null hypothesis H_0 is rejected. For instance, if we are testing at the 0.05 significance level, then H_0 would be rejected for values of $F > F_{.05} = 4.26$. This value is obtained from part b of Table 6 of Appendix A using $ndf = 2$ (*df* for *SSTr*) and $ddf = 9$ (*df* for *SSE*). In this instance H_0 would be rejected, since the *F*-value in the ANOVA table is $F = 10.15$, and this exceeds the table value 4.26. Consequently, we would conclude that there is a difference in the mean amounts of salts removed by the 3 water treatment processes.

SECTIONS 13.1 AND 13.2 EXERCISES

The Analysis of Variance Technique; The Analysis of Variance Table and Computing Formulas

In Exercises 13.1 through 13.4, independent random samples were selected from populations having a common variance σ^2. Use the given sample variances to calculate the pooled variance s_p^2.

13.1

Sample 1	Sample 2
$n_1 = 21$	$n_2 = 16$
$s_1^2 = 46$	$s_2^2 = 38$

13.2

Sample 1	Sample 2	Sample 3
$n_1 = 16$	$n_2 = 6$	$n_3 = 11$
$s_1^2 = 80$	$s_2^2 = 72$	$s_3^2 = 75$

13.3

Sample 1	Sample 2	Sample 3	Sample 4
$n_1 = 8$	$n_2 = 5$	$n_3 = 10$	$n_4 = 6$
$s_1^2 = 36$	$s_2^2 = 40$	$s_3^2 = 35$	$s_4^2 = 42$

13.4

Sample 1	Sample 2	Sample 3	Sample 4	Sample 5
$n_1 = 13$	$n_2 = 11$	$n_3 = 7$	$n_4 = 16$	$n_5 = 11$
$s_1^2 = 8$	$s_2^2 = 10$	$s_3^2 = 9$	$s_4^2 = 12$	$s_5^2 = 8$

13.5   Independent random samples were selected from three populations, each with variance σ^2. The results were as follows.

Sample 1	Sample 2	Sample 3
$\bar{x}_1 = 10$	$\bar{x}_2 = 22$	$\bar{x}_3 = 5$
$s_1^2 = 20$	$s_2^2 = 16$	$s_3^2 = 15$
$n_1 = 6$	$n_2 = 6$	$n_3 = 8$

a. Obtain the grand mean $\bar{x}$ by computing a weighted average of the three sample means.
b. Use the grand mean and the three sample means to calculate $SSTr$, the treatment sum of squares.
c. Calculate the pooled sample variance s_p^2.
d. Using the results from parts b and c, construct the ANOVA table.

13.6   Independent random samples were selected from four populations, each with variance σ^2. The results were as follows.

Sample 1	Sample 2	Sample 3	Sample 4
$n_1 = 10$	$n_2 = 6$	$n_3 = 5$	$n_4 = 9$
$\bar{x}_1 = 50$	$\bar{x}_2 = 20$	$x_3 = 80$	$\bar{x}_4 = 10$
$s_1^2 = 28$	$s_2^2 = 42$	$s_3^2 = 25$	$s_4^2 = 30$

a. Obtain the grand mean $\bar{x}$ by computing a weighted average of the four sample means.
b. Use the grand mean and the four sample means to calculate $SSTr$, the treatment sum of squares.
c. Calculate the pooled sample variance s_p^2.
d. Use the results from parts b and c to construct the ANOVA table.

13.7   A completely randomized design produced the following sample results.

Sample 1	Sample 2	Sample 3
19	10	11
23	11	6
20	15	9
18		6
20		

$$n_1 = 5 \qquad n_2 = 3 \qquad n_3 = 4 \qquad n = 12$$

$$\bar{x}_1 = 20 \qquad \bar{x}_2 = 12 \qquad \bar{x}_3 = 8 \qquad \bar{x} = 14$$

$$s_1^2 = 3.5 \qquad s_2^2 = 7 \qquad s_3^2 = 6$$

a. Use the grand mean $\bar{x}$ and the three sample means to calculate $SSTr$, the treatment sum of squares.

b. Use the shortcut Formula 13.2 to calculate $SSTr$.

c. Calculate the pooled sample variance s_p^2, and use it to obtain SSE, the error sum of squares.

d. Use the shortcut Formula 13.3 to calculate SSE.

e. Construct the analysis of variance table.

13.8   A completely randomized design produced the following sample results.

Sample 1	Sample 2	Sample 3	Sample 4
50	34	70	18
49	38	67	15
54	32	68	21
	31	67	16
	35		15

$$n_1 = 3 \qquad n_2 = 5 \qquad n_3 = 4 \qquad n_4 = 5 \qquad n = 17$$

$$\bar{x}_1 = 51 \qquad \bar{x}_2 = 34 \qquad \bar{x}_3 = 68 \qquad \bar{x}_4 = 17 \qquad \bar{x} = 40$$

$$s_1^2 = 7 \qquad s_2^2 = 7.5 \qquad s_3^2 = 2 \qquad s_4^2 = 6.5$$

a. Use the grand mean $\bar{x}$ and the four sample means to calculate $SSTr$, the treatment sum of squares.

b. Use the shortcut Formula 13.2 to calculate $SSTr$.

c. Calculate the pooled sample variance s_p^2, and use it to obtain SSE, the error sum of squares.

d. Use the shortcut Formula 13.3 to calculate SSE.

e. Construct the analysis of variance table.

13.9   Use the computational Formulas 13.1 through 13.3 to construct the ANOVA table for the following samples.

Sample 1	Sample 2	Sample 3
8	13	11
9	16	12
6	16	19
11	19	18
8		

13.10   Use the computational Formulas 13.1 through 13.3 to construct the ANOVA table for the following samples.

Sample 1	Sample 2	Sample 3	Sample 4
4.7	6.3	2.1	5.8
3.2	7.8	1.1	5.9
6.9	3.9	5.2	2.5
1.7	6.7	1.3	8.3
	3.2		5.3

13.11   Use the computational Formulas 13.1 through 13.3 to construct the ANOVA table for the following samples.

Sample 1	Sample 2	Sample 3	Sample 4	Sample 5
33	28	39	21	19
30	21	35	20	18
25	22	38	24	33
28	20	39	22	36
30		35	21	27

13.12   Consider the following partially completed ANOVA table for a completely randomized design.

Source of Variation	df	SS	MS	F
Treatments	5	—	200.2	—
Error	—	—	—	
Total	24	1,555		

a. How many treatments are involved in the experiment?
b. What is the total number of sample observations?
c. Complete the ANOVA table by filling in the missing blanks.

13.13   Consider the following partially completed ANOVA table for a completely randomized design.

Source of Variation	df	SS	MS	F
Treatments	—	21.7	3.1	—
Error	28	—	1.7	
Total	—	—		

a. Complete the ANOVA table by filling in the missing blanks.
b. How many treatments are involved in the experiment?
c. What is the total number of sample observations?

13.14   In Exercise 11.30, a pooled two-sample t-test was performed on the following samples.

Sample One	25.8	26.9	26.2	25.3	26.7	26.1	26.9
Sample Two	16.9	17.4	16.8	16.2	17.3	16.8	

Construct the ANOVA table.

13.3 APPLICATIONS OF THE ONE-WAY ANALYSIS OF VARIANCE

We will now consider some applications of the one-way analysis of variance technique for testing the null hypothesis that k population means are equal. The procedure discussed earlier is summarized below.

PROCEDURE

The One-Way Analysis of Variance for a Completely Randomized Design: To test the null hypothesis

$$H_0: \mu_1 = \mu_2 = \cdots = \mu_k,$$

obtain the ANOVA table from computer output or by using the following formulas to calculate the sums of squares.

$$SST = \Sigma x^2 - \frac{T^2}{n}$$

$$SSTr = \left(\frac{T_1^2}{n_1} + \frac{T_2^2}{n_2} + \cdots + \frac{T_k^2}{n_k}\right) - \frac{T^2}{n}$$

$$SSE = SST - SSTr$$

The test statistic is

$$F = \frac{MSTr}{MSE} = \frac{\dfrac{SSTr}{k-1}}{\dfrac{SSE}{n-k}}$$

The null hypothesis is rejected when $F > F_\alpha$, where α is the significance level of the test.

Assumptions:

1. The samples are random and independent.

2. Each population has a normal distribution.

3. The population variances are equal.

Note:

1. The F-distribution has $ndf = k - 1$, $ddf = n - k$.

2. In the above formulas,
   k is the number of sampled populations, n_i and T_i are the ith sample size and total, $n = \Sigma n_i$, and $T = \Sigma T_i$.

EXAMPLE 13.1   To determine if there is a difference in the average amount of beef contained in four brands of spaghetti sauce, jars of each brand were randomly selected, and the amount of beef contained in each jar was determined. The results appear below, where the units are grams per jar.

Brand 1	Brand 2	Brand 3	Brand 4
27	29	33	23
24	27	29	24
27	31	34	23
25	32	32	25
27	30	31	24
	31		

Is there sufficient evidence at the 0.05 significance level to conclude that there is a difference in the mean beef content for the four brands? Assume that the sampled populations are normally distributed with a common variance σ^2.

Solution

We will let μ_i denote the true mean amount of beef for the population of all jars of brand i, where $i = 1, 2, 3, 4$.

Step 1: Hypotheses.

$$H_0: \mu_1 = \mu_2 = \mu_3 = \mu_4$$

H_a: Not all the means are equal.

Step 2: Significance level.

$$\alpha = 0.05$$

Step 3: Calculations.

We need the total for each of the four samples, as well as the grand total of all the measurements.

$$T_1 = 130 \qquad T_2 = 180 \qquad T_3 = 159 \qquad T_4 = 119 \qquad T = 588$$

$$n_1 = 5 \qquad n_2 = 6 \qquad n_3 = 5 \qquad n_4 = 5 \qquad n = 21$$

Next, the sum of squares SST, $SSTr$, and SSE are obtained.

$$SST = \Sigma x^2 - \frac{T^2}{n}$$

$$= (27^2 + 24^2 + \cdots + 25^2 + 24^2) - \frac{588^2}{21}$$

$$= 16{,}710 - 16{,}464 = 246$$

$$SSTr = \left(\frac{T_1^2}{n_1} + \frac{T_2^2}{n_2} + \frac{T_3^2}{n_3} + \frac{T_4^2}{n_4} \right) - \frac{T^2}{n}$$

$$= \left(\frac{130^2}{5} + \frac{180^2}{6} + \frac{159^2}{5} + \frac{119^2}{5} \right) - \frac{588^2}{21}$$

$$= 16{,}668.4 - 16{,}464 = 204.4$$

$$SSE = SST - SSTr$$
$$= 246 - 204.4 = 41.6$$

With the sums of squares, the ANOVA table can now be constructed.

Source	df	SS	MS	F
Treatments	3	204.4	68.13	27.84
Error	17	41.6	2.447	
Total	20	246.0		

From the ANOVA table, the value of the test statistic is $F = 27.84$.

Step 4: Rejection region.

The rejection region is right-tailed and consists of values for which $F > F_\alpha = F_{.05} = 3.20$. The value is obtained from part b of Table 6 of Appendix A using $ndf = k - 1 = 3$ and $ddf = n - k = 17$. Note that these are the degrees of freedom for the numerator ($MSTr$) and for the denominator (MSE) of the test statistic F. The rejection region is shown in Figure 13.4.

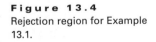
Figure 13.4
Rejection region for Example 13.1.

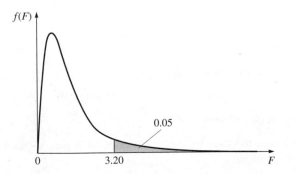

Step 5: Conclusion.

Since the value of the test statistic in the ANOVA table is $F = 27.84$, and this exceeds the table value 3.20, H_0 is rejected. Thus, at the 5 percent significance level we conclude that at least 2 of the brands differ in the mean amount of beef content.

MINITAB EXAMPLE 13.2   Use MINITAB to obtain the analysis of variance table for the four samples in Example 13.1.

Solution

The measurements for the four samples are placed in columns C1 through C4.

```
SET C1
27 24 27 25 27
END
SET C2
29 27 31 32 30 31
END
SET C3
33 29 34 32 31
END
SET C4
23 24 23 25 24
END
```

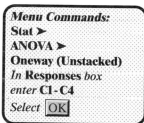

Menu Commands:
Stat ➤
ANOVA ➤
Oneway (Unstacked)
In Responses box
enter **C1-C4**
Select OK

To have MINITAB produce the analysis of variance table, use the command **AOVONEWAY.**

AOVONEWAY C1-C4

MINITAB will respond by producing the output in Exhibit 13.2. In the ANOVA table, MINITAB gives the *p*-value for the value of the test statistic *F*. Since the *p*-value is smaller than the specified α value of 0.05, the null hypothesis would be rejected.

In addition to the ANOVA table, MINITAB gives the sample means and standard deviations for the four treatments (the levels of the factor—brand of sauce). It also visually displays 95 percent confidence intervals for the treatment means. Consider, for example, the confidence interval for the mean amount of beef in brand 3. Each tick mark on the horizontal axis represents 0.3 (10 ticks cover a spread of 3 units). The interval for μ_3 appears to extend from about 30.3 to 33.3. In the next section, we will show how this interval is constructed and that the actual interval is from 30.32 to 33.28 grams.

The one-way analysis of variance can be used instead of the pooled two-sample *t*-test for determining if a difference exists between two population means. This is illustrated in the following example.

Exhibit 13.2

```
MTB > AOVONEWAY C1-C4

ANALYSIS OF VARIANCE
SOURCE      DF         SS        MS        F         p
FACTOR       3     204.40     68.13     27.84     0.000
ERROR       17      41.60      2.45
TOTAL       20     246.00
                                   INDIVIDUAL 95 PCT CI'S FOR MEAN
                                   BASED ON POOLED STDEV
LEVEL        N       MEAN     STDEV  ------+---------+---------+---------+-
C1           5     26.000     1.414          (----*----)
C2           6     30.000     1.789                         (---*---)
C3           5     31.800     1.924                           (----*----)
C4           5     23.800     0.837  (----*----)
                                     ------+---------+---------+---------+-
POOLED STDEV =      1.564           24.0      27.0      30.0      33.0
```

EXAMPLE 13.3   In Example 13.1, suppose one were only interested in comparing the amount of beef for brands 1 and 2. The samples for these 2 brands are copied below. At the 0.05 significance level, test to determine if a difference exists in the mean beef content for these 2 brands by using

I. a pooled two-sample t-test,

II. a one-way analysis of variance.

Brand 1	Brand 2
27	29
24	27
27	31
25	32
27	30
	31

Solution

For each test we need to assume independent random samples from normal populations having a common variance σ^2.

I. Pooled two-sample t-test.

Step 1: Hypotheses.

$$H_0: \mu_1 = \mu_2$$

$$H_a: \mu_1 \neq \mu_2$$

Step 2: Significance level.

$$\alpha = 0.05$$

Step 3: Calculations.

$$\bar{x}_1 = \frac{T_1}{n_1} = \frac{130}{5} = 26, \qquad \bar{x}_2 = \frac{T_2}{n_2} = \frac{180}{6} = 30$$

$$s_1^2 = \frac{(27^2 + 24^2 + 27^2 + 25^2 + 27^2) - \dfrac{130^2}{5}}{5 - 1} = 2$$

$$s_2^2 = \frac{(29^2 + 27^2 + 31^2 + 32^2 + 30^2 + 31^2) - \dfrac{180^2}{6}}{6 - 1} = 3.2$$

The assumed common variance of the 2 populations is estimated by the pooled sample variance s_p^2.

$$s_p^2 = \frac{(n_1 - 1)s_1^2 + (n_2 - 1)s_2^2}{n_1 + n_2 - 2} = \frac{4(2) + 5(3.2)}{9} = 2.667$$

The value of the test statistic is

$$t = \frac{(\bar{x}_1 - \bar{x}_2) - (\mu_1 - \mu_2)}{\sqrt{s_p^2 \left(\dfrac{1}{n_1} + \dfrac{1}{n_2}\right)}} = \frac{(26 - 30) - 0}{\sqrt{2.667 \left(\dfrac{1}{5} + \dfrac{1}{6}\right)}} = -4.04.$$

Step 4: Rejection region.

Since the alternative hypothesis is two sided, the rejection region is two tailed. The degrees of freedom are $df = n_1 + n_2 - 2 = 9$, and H_0 is rejected for $t < -t_{\alpha/2} = -t_{.025} = -2.262$ and $t > t_{.025} = 2.262$. The rejection region appears in Figure 13.5.

Figure 13.5
Rejection region for Example 13.3 (I).

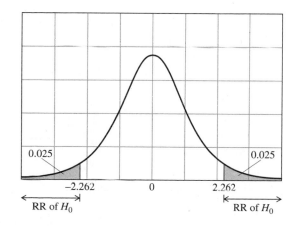

0.025   0.025

−2.262   0   2.262

 RR of H_0 RR of H_0

Step 5: Conclusion.

Since $t = -4.04$ is in the rejection region, H_0 is rejected, and we conclude that the mean beef contents for brands 1 and 2 do differ.

II. One-way analysis of variance.

Step 1: Hypotheses.

$$H_0: \mu_1 = \mu_2$$

$$H_a: \mu_1 \neq \mu_2$$

Step 2: Significance level.

$$\alpha = 0.05$$

Step 3: Calculations.

$$T = T_1 + T_2 = 130 + 180 = 310, \qquad n = n_1 + n_2 = 11$$

$$SST = \Sigma x^2 - \frac{T^2}{n}$$

$$= (27^2 + 24^2 + \cdots + 30^2 + 31^2) - \frac{310^2}{11}$$

$$= 8,804 - 8,736.36 = 67.64$$

$$SSTr = \left(\frac{T_1^2}{n_1} + \frac{T_2^2}{n_2}\right) - \frac{T^2}{n}$$

$$= \left(\frac{130^2}{5} + \frac{180^2}{6}\right) - \frac{310^2}{11}$$

$$= 8,780 - 8,736.36 = 43.64$$

$$SSE = SST - SSTr$$
$$= 67.64 - 43.64 = 24$$

The sums of squares are now used to form the ANOVA table.

Source	df	SS	MS	F
Treatments	1	43.64	43.64	16.36
Error	9	24	2.667	
Total	10	67.64		

Step 4: Rejection region.

The rejection region is based on the F-distribution with $ndf = 1$ and $ddf = 9$. From part b of Table 6, H_0 is rejected for $F > F_\alpha = F_{.05} = 5.12$ (see Figure 13.6).

Figure 13.6
Rejection region for Example
13.3 (II).

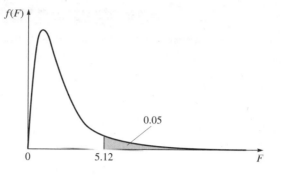

Figure 13.6
Rejection region for Example
13.3 (II).

Step 5: Conclusion.

From the ANOVA table, the value of the test statistic is $F = 16.36$. Since this exceeds the table value of 5.12, H_0 is rejected and we conclude that the mean beef contents for the 2 brands are different.

The pooled two-sample t-test and the one-way analysis of variance used in Example 13.3 are equivalent for determining if two population means differ. Each is based on the same assumptions of independent random samples from normal populations with the same variance. Also, except for round off error, the square of the value of the t-statistic equals the value of the F-statistic ($4.04^2 \approx 16.36$). The same relation between t and F applies to the table values used to determine the rejection regions ($2.262^2 = 5.12$). Furthermore, the *MSE* value in the ANOVA table equals the pooled sample variance used in the t-statistic. Of course, the analysis of variance is more versatile in that it can be used to compare more than two population means, while the pooled t-test is restricted to two means.

13.4 ESTIMATION OF MEANS

In addition to conducting an analysis of variance to determine if the population means differ, one may want to estimate either a single mean or the difference between two means. The form of each confidence interval is very similar to those used in earlier chapters. Because the mean square error *(MSE)* in the ANOVA table is a pooled sample variance (s_p^2), it is used to estimate the assumed common variance of the populations.

PROCEDURE

$(1 - \alpha)$ Confidence Interval for One Population Mean μ_i:

$$\textbf{(13.4)} \qquad \bar{x}_i \pm t_{\alpha/2} \frac{s_p}{\sqrt{n_i}}$$

Assumptions:

1. The samples are random and independent.

2. Each population has a normal distribution.

3. The population variances are equal.

Note:

1. $\bar{x}_i = T_i/n_i$ and $s_p = \sqrt{MSE}$

2. The t-distribution is based on $df = n - k$, where k is the number of samples and $n = n_1 + n_2 + \cdots + n_k$.

PROCEDURE

$(1 - \alpha)$ Confidence Interval for $(\mu_i - \mu_j)$:

$$\textbf{(13.5)} \qquad (\bar{x}_i - \bar{x}_j) \pm t_{\alpha/2} \sqrt{s_p^2 \left(\frac{1}{n_i} + \frac{1}{n_j} \right)}$$

Assumptions:

1. The samples are random and independent.

2. Each population has a normal distribution.

3. The population variances are equal.

Note:

1. $\bar{x}_i = T_i/n_i$, $\bar{x}_j = T_j/n_j$, and $s_p^2 = MSE$.

2. The t-distribution is based on $df = n - k$, where k is the number of samples and $n = n_1 + n_2 + \cdots + n_k$.

EXAMPLE 13.4   Use the results of Example 13.1 to obtain a 95 percent confidence interval for

1. μ_3, the mean beef content of brand 3,

2. $(\mu_3 - \mu_2)$, the difference in the mean beef content for brands 3 and 2.

Solution

From Example 13.1, we have

$$T_2 = 180, \, T_3 = 159, \, n_2 = 6, \, n_3 = 5, \text{ and } MSE = 2.447,$$

$$\bar{x}_2 = \frac{T_2}{n_2} = \frac{180}{6} = 30,$$

$$\bar{x}_3 = \frac{T_3}{n_3} = \frac{159}{5} = 31.80,$$

$$s_p = \sqrt{MSE} = \sqrt{2.447} = 1.564,$$

and

$$t_{\alpha/2} = t_{.025} = 2.110$$

from Table 4 using $df = n - k = 21 - 4 = 17$.

1. A 95 percent confidence interval for μ_3 is

$$\bar{x}_3 \pm t_{\alpha/2} \frac{s_p}{\sqrt{n_3}}$$

$$31.80 \pm 2.11 \frac{1.564}{\sqrt{5}}$$

$$31.80 \pm 1.48.$$

Thus, with 95 percent confidence, we estimate that the mean amount of beef for brand 3 is between 30.32 and 33.28 grams.

2. A 95 percent confidence interval for $(\mu_3 - \mu_2)$ is

$$(\bar{x}_3 - \bar{x}_2) \pm t_{\alpha/2} \sqrt{s_p^2 \left(\frac{1}{n_3} + \frac{1}{n_2} \right)}$$

$$(31.80 - 30) \pm 2.11 \sqrt{(2.447) \left(\frac{1}{5} + \frac{1}{6} \right)}$$

$$1.80 \pm 2.00.$$

Therefore, we are 95 percent confident that $(\mu_3 - \mu_2)$ lies in the interval from -0.20 to 3.80. Since this interval contains both negative and positive values, we cannot conclude with 95 percent confidence that μ_3 is larger than μ_2.

In conducting an analysis of variance, we are testing the null hypothesis that k population means are equal. When the test results indicate that the null hypothesis should be rejected, then we conclude that at least two of the population means are not the same. At this point we may want to pursue the problem further and determine which means are different. For instance, in Example 13.1 we performed an analysis of variance to determine if there is a difference in the mean amount of beef contained in four brands of spaghetti sauce. At the completion of the test we concluded that at least two brands differ in their average beef content. We could begin our exploration of which means differ by visually inspecting the confidence intervals for the

Figure 13.7
The 95 percent confidence intervals for the four means.

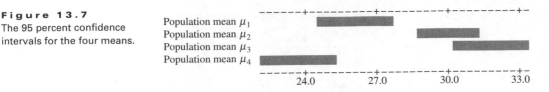

four population means that were produced by MINITAB in Exhibit 13.2. The visual display of the intervals from Exhibit 13.2 is reproduced in Figure 13.7.

An examination of Figure 13.7 indicates that the confidence intervals for μ_1 and μ_4 overlap, and there is also an overlap between the intervals for μ_2 and μ_3. This separation of the confidence intervals into the two groups

$$\mu_1, \mu_4 \quad \text{and} \quad \mu_2, \mu_3$$

suggests that means one and four are similar, as are means two and three. That there is no interval overlapping between the two groups further suggests that μ_1 and μ_4 may be different from μ_2 and μ_3.

We emphasize that the above analysis is only exploratory and does not conclusively indicate which population means are similar and which are different. To support our initial and tentative assessments, a formal evaluation needs to be conducted such as the performance of multiple comparisons between all possible pairs of population means. Several multiple comparison procedures are available and can readily be found in books on analysis of variance and experimental design.

In concluding our discussion of the one-way analysis of variance procedure, it should be emphasized that its validity is dependent on its underlying assumptions. The assumption concerning independent random samples is critical and must not be violated. As with the pooled two-sample t-test, it is also important that the variances of the populations are approximately equal, particularly when the sample sizes are different. The analysis of variance procedure, however, is quite robust with regard to the normality assumption.

When one has doubts concerning the reasonableness of the normality or variance assumptions, an alternative method known as the Kruskal-Wallis H-test can be used. This is a nonparametric test discussed in Chapter 15.

SECTIONS 13.3 and 13.4 EXERCISES

Applications of the One-Way Analysis of Variance; Estimation of Means

In the following exercises, assume that independent random samples were selected from populations that have approximately normal distributions with the same variance.

13.15   Consider the following three samples.

Sample 1	Sample 2	Sample 3
17	16	12
21	12	7
18	11	10
16		11
18		

    a. Construct a dotplot of each sample and form an opinion as to whether or not the sample means differ significantly.

    b. Construct the analysis of variance table and test the null hypothesis $H_0: \mu_1 = \mu_2 = \mu_3$. Use a significance level of 5 percent.

13.16   Consider the following four samples.

Sample 1	Sample 2	Sample 3	Sample 4
9.9	4.3	3.1	5.3
6.2	5.8	2.1	5.9
7.7	1.9	6.2	2.8
4.7	4.7	2.3	8.5
	1.2		5.3

    a. Construct a dotplot of each sample and form an opinion as to whether or not the sample means differ significantly.

    b. Construct the analysis of variance table and test the null hypothesis $H_0: \mu_1 - \mu_2 - \mu_3 = \mu_4$. Use $\alpha = 0.05$.

13.17   Refer to Exercise 13.15 and obtain a 95 percent confidence interval for the following.

    a. μ_1 b. $(\mu_1 - \mu_2)$

13.18   Refer to Exercise 13.16 and obtain a 95 percent confidence interval for the following.

    a. μ_3 b. $(\mu_4 - \mu_3)$

13.19   Determine the approximate p-value for the analysis of variance test in Exercise 13.15.

13.20   Determine the approximate p-value for the analysis of variance test in Exercise 13.16.

13.21   Exercise 11.39 considered an economist's investigation of the hourly labor rates of automobile mechanics in two states. A pooled t-test was used on the following samples to determine if a difference exists in the mean hourly rates for the two states. Use the analysis of variance procedure to perform this test at the 0.05 significance level.

First State	Second State
$40.00	$35.00
$38.00	$37.00
$38.00	$31.00
$37.00	$39.00
$36.00	$31.50
$39.00	$35.00
$41.50	$32.50
$38.00	$34.00
$39.50	$39.00
$37.50	$36.00
$35.00	
$40.00	

13.22 Use the *F*-table to obtain the approximate *p*-value for the analysis of variance test in Exercise 13.21.

13.23 Use the results of Exercise 13.21 to construct a 99 percent confidence interval for the difference in the mean hourly rates of automobile mechanics in the two states.

13.24 A kennel owner studied the effects of 3 different brands of dog food on the weight gain of puppies. From a litter of 12 puppies, 4 were randomly selected and fed brand A, 4 were randomly selected from the remaining 8 and fed brand B, and the remaining 4 were fed brand C. Their weight gains after 6 weeks are given below.

Brand of Dog Food		
A	B	C
1.3	3.1	3.9
1.9	2.8	3.4
1.8	3.0	4.3
1.4	2.9	4.7

Test at the 0.01 significance level if there is a difference in the mean weight gains for the 3 brands of dog food.

13.25 Determine the approximate *p*-value for the analysis of variance test in Exercise 13.24.

13.26 For Exercise 13.24, estimate with 99 percent confidence the mean weight gain for brand C.

13.27 In Exercise 13.24, construct a 99 percent confidence interval to estimate the difference in the mean weight gains for brands C and A.

13.28 Radon, the second largest cause of lung cancer after smoking, is a radioactive gas produced by the natural decay of radium in the ground. Radon can seep into a building through openings in the foundation, and the EPA recommends that corrective measures be taken if levels reach 4 or more picocuries per liter (pc/l). To investigate radon levels in 4 public schools, an official took a sample of 6 readings at each school. The results, in pc/l, appear below.

School 1	School 2	School 3	School 4
1.7	5.3	5.1	1.2
1.3	4.2	4.1	2.9
0.8	3.9	5.2	1.5
1.5	5.7	3.3	2.3
0.9	3.8	2.6	2.1
1.1	5.2	4.9	1.8

Do the data suggest that there is a difference in the mean radon levels at the 4 schools? Test with $\alpha = 5$ percent.

13.29   Estimate with 95 percent confidence the mean radon level for school 3 in Exercise 13.28.

13.30   In Exercise 13.28, construct a 95 percent confidence interval to estimate the difference in the mean radon levels for schools 2 and 4.

13.31   An agronomist wanted to determine if the amount of ascorbic acid (vitamin C) in Cortland apples depends on the soil. The amounts of ascorbic acid were measured for samples of apples from 3 geographical regions, and the results are given below. The figures are in milligrams of ascorbic acid per 100 grams.

Region 1	Region 2	Region 3
9	13	10
8	10	12
11	9	13
10	10	10
11	11	11
9	8	11
10	14	13
10	13	

Is there sufficient evidence to indicate a difference among mean ascorbic acid levels for Cortland apples grown in the 3 regions? Test at the 0.05 significance level.

13.32   Determine the approximate p-value for the analysis of variance test in Exercise 13.31.

13.33 For Exercise 13.31, estimate with 90 percent confidence the difference in mean ascorbic acid levels for Cortland apples grown in regions 1 and 3.

13.34 Dangerous chemicals from industrial wastes linked to cancer and other diseases can enter the food chain through their presence in lake sediments. The amounts of DDT were determined for samples of trout from 4 lakes and are given below.

Lake	Levels of DDT in Parts per Million					
One	1.7	1.4	1.9	1.1	2.1	1.8
Two	0.3	0.7	0.5	0.1	1.1	0.9
Three	2.7	1.9	2.0	1.5	2.6	
Four	1.2	3.1	1.9	3.7	2.8	3.5

Do the data provide sufficient evidence at the 0.05 level to conclude that there is a difference in mean DDT levels for the 4 lakes?

13.35 Determine the approximate p-value for the analysis of variance test in Exercise 13.34.

13.36 For Exercise 13.34, compare the mean DDT levels in lakes 2 and 4 by constructing a 90 percent confidence interval.

13.37 Suppose in Exercise 13.34 the experimenter hypothesized before collecting the data that the mean DDT levels were not all the same for lakes 1, 3, and 4. Is there sufficient evidence with $\alpha = 0.05$ to support the researcher's belief?

13.38 A university's career development director conducted a survey of starting salaries offered to education majors in three states. Ten offers were recorded for each state, and the analysis of variance output produced by MINITAB appears below.

```
ANALYSIS OF VARIANCE
SOURCE    DF       SS        MS        F        P
FACTOR     2   17470016   8735008    9.99    0.001
ERROR     27   23611000    874482
TOTAL     29   41081016
                           INDIVIDUAL 95 PCT CI'S FOR MEAN
                           BASED ON POOLED STDEV
  LEVEL    N    MEAN  STDEV   ----------+---------+---------+------
STATE 1   10   23958    968   (------*-------)
STATE 2   10   25512    682                      (-------*------)
STATE 3   10   23836   1105  (-------*-------)
                           ----------+---------+---------+------
POOLED STDEV = 935             24000     24800     25600
```

a. Do the data provide sufficient evidence to indicate that the mean starting salary is different for at least 2 states? Test at the 0.01 significance level.

b. Based on the visual display of the confidence intervals in the above output, form a tentative opinion as to which population means may be equal and which may differ.

13.39　Use the MINITAB output in Exercise 13.38 to estimate with 95 percent confidence the difference in mean starting salaries in states 1 and 2.

MINITAB Assignments

M➤ 13.40　A furniture manufacturer wanted to compare the mean drying times for 4 brands of stains. Each stain was applied to 10 chairs, and the drying times in minutes appear below.

Brand 1	Brand 2	Brand 3	Brand 4
80.6	91.6	90.5	86.7
81.3	83.5	98.5	75.4
82.8	83.4	97.5	79.7
81.5	88.6	99.9	76.5
80.4	96.7	96.9	75.7
79.7	84.8	90.5	84.7
82.3	88.4	96.7	74.5
81.7	89.5	93.8	83.3
80.6	84.4	97.8	84.2
81.5	85.1	96.8	75.3

a. Use MINITAB to test at the 0.05 level if there is sufficient evidence to conclude that a difference exists in the mean drying times for the 4 brands.
b. Based on the visual display of the confidence intervals in the MINITAB output, form a tentative opinion as to which population means may be equal and which may differ.

M➤ 13.41　Use the MINITAB output produced in Exercise 13.40 to estimate with 95 percent confidence the difference in mean drying times for brands 2 and 4.

M➤ 13.42　Use MINITAB to perform an analysis of variance to determine if there is a difference in the mean drying times for brands 3 and 4 in Exercise 13.40. Use $\alpha = 0.05$.

M➤ 13.43　Use MINITAB to obtain the ANOVA table for the data in Exercise 13.28.

M➤ 13.44　Use MINITAB to obtain the analysis of variance table for the data in Exercise 13.31.

M➤ 13.45　Use MINITAB to obtain the ANOVA table for the samples in Exercise 13.34.

13.5 RANDOMIZED BLOCK DESIGNS AND THE ANALYSIS OF VARIANCE

In this chapter, we have seen that the analysis of variance technique for a completely randomized design can be thought of as a generalization of the pooled two-sample t-test discussed in Section 11.2. While the latter is limited to comparisons of two means, the analysis of variance can be used to compare any number of population means. In this section, we will illustrate an analogous generalization for the paired-samples t-test of Section 11.4.

Recall that in a paired-samples experiment each value from the first sample is related by some characteristic to a value from the second sample. The pairing of similar experimental units allows for the elimination of unwanted variation in the response variable, thus resulting in dependent samples that can increase the inferential ability to distinguish between two population means. As an illustration we considered an experiment in which ten infants were involved in a study to compare the effectiveness of two medications for the treatment of diaper rash. For each baby, two areas of approximately the same size and rash severity were selected, and one randomly selected area was treated with medication A and the other was treated with B. The number of hours for the rash to disappear was recorded for each medication and each infant, and the values obtained earlier are repeated in Table 13.4.

We could have designed the experiment so that medication A was applied to 10 randomly selected infants and medication B was used by 10 other randomly selected babies (an *independent samples* design). This, however, would be inferior to our *matched comparison* of treatments A and B for each infant, because the latter allows us to exclude from each pair any extraneous differences caused by sensitivity variations among different babies. In the parlance of experimental design, the ten infants and their paired values are called **blocks,** and the entire arrangement is referred to as a **randomized block design.** In this particular illustration, there are $b = 10$ blocks

TABLE 13.4

NUMBER OF HOURS FOR THE RASH TO DISAPPEAR WITH MEDICATIONS A AND B

	Medication	
Infant	A	B
1	46	43
2	50	49
3	46	48
4	51	47
5	43	40
6	45	40
7	47	47
8	48	44
9	46	41
10	48	45

and $k = 2$ treatments, but the number of treatments can be more than two. For instance, in Example 13.5 we will compare $k = 3$ medications by selecting 3 areas on each infant and randomly assigning them to the 3 treatments.

KEY CONCEPT

A Randomized Block Design with k Treatments and b Blocks:

The experimental design consists of b blocks, each of which contains k matched experimental units. The k units within each block are randomly assigned to the k treatments.

Note:

The blocking (matching) should be done so that the experimental units within each block are homogeneous as to their influence on the response variable.

In analyzing a randomized block design, we can distinguish between the variation caused by the treatments (the medications) and the variation due to differences between the blocks (the infants). The analysis of variance procedure for a randomized block design is very similar to that considered in the preceding sections for a completely randomized design. For that design the **total sum of squares (SST)** was partitioned into two parts: the **treatment sum of squares ($SSTr$)** and the **error sum of squares (SSE).** In a randomized block design, a third component corresponding to the blocks is separated from SST (see Figure 13.8). It is called the **block sum of squares (SSB)** and is a measure of variation in the data that is attributed to differences between the blocks. More specifically, just as $SSTr$ measures the variation among treatment means, SSB measures the variation among block means. SSB and the three other sums of squares in a randomized block design are related by the equation

$$SST = SSTr + SSB + SSE$$

The interpretation and calculation of SST are identical to those discussed earlier for the completely randomized design. SSE, a measure of how the data fluctuate *within the blocks*, is the unexplained variation and is found indirectly by subtracting $SSTr$ and SSB from SST.

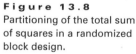

Figure 13.8
Partitioning of the total sum of squares in a randomized block design.

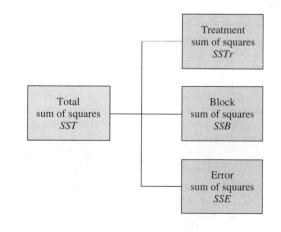

As was done earlier to test if the treatment means differ significantly, we compare the **mean square for treatments (MSTr)** to the **mean square for error (MSE)** by calculating the F-ratio

$$F = \frac{MSTr}{MSE}$$

Computational formulas and the analysis of variance procedure for the randomized block design are given below and illustrated in Example 13.5.

PROCEDURE

The Analysis of Variance for a Randomized Block Design:
To test the null hypothesis for k treatment means,

$$H_0: \mu_1 = \mu_2 = \cdots = \mu_k,$$

obtain the ANOVA table from computer output or by using the following formulas to calculate the sums of squares.

$$SST = \Sigma x^2 - \frac{T^2}{n}$$

$$SSTr = \frac{T_1^2 + T_2^2 + \cdots + T_k^2}{b} - \frac{T^2}{n}$$

$$SSB = \frac{B_1^2 + B_2^2 + \cdots + B_b^2}{k} - \frac{T^2}{n}$$

$$SSE = SST - SSTr - SSB$$

The ANOVA table is

Source of Variation	df	SS	MS	F
Treatments	$k - 1$	$SSTr$	$MSTr = SSTr/(k - 1)$	$F = MSTr/MSE$
Blocks	$b - 1$	SSB	$MSB = SSB/(b - 1)$	
Error	$(k - 1)(b - 1)$	SSE	$MSE = SSE/(k - 1)(b - 1)$	
Total	$n - 1$	SST		

The null hypothesis is rejected when $F > F_\alpha$, where α is the level of significance, and the F-distribution has $ndf = k - 1$, $ddf = (k - 1)(b - 1)$.

Note:

In the above formulas, k is the number of treatments, b is the number of blocks, T_i is the total of the ith treatment, B_j is the total of the jth block, $T = \Sigma T_i$ and $n = kb$.

Assumptions:

For each possible combination of a treatment level with a block level, the sample observations are independent and their probability distribution is normal with common variance σ^2.

EXAMPLE 13.5   Ten infants participated in a study to compare the effectiveness of three medications for the treatment of diaper rash. For each baby, three areas of approximately the same size and rash severity were selected, and one area was treated with medication A, one with B, and one with C. The treatments were randomly assigned for each infant. The number of hours for the rash to disappear was recorded for each medication and each infant, and the values obtained appear below (also shown are treatment totals T_i and block totals B_j).

Infant	Medication			
	A	B	C	
1	46	43	42	$B_1 = 131$
2	50	49	47	$B_2 = 146$
3	46	48	41	$B_3 = 135$
4	51	47	46	$B_4 = 144$
5	43	40	40	$B_5 = 123$
6	45	40	41	$B_6 = 126$
7	47	47	42	$B_7 = 136$
8	48	44	41	$B_8 = 133$
9	46	41	42	$B_9 = 129$
10	48	45	43	$B_{10} = 136$
	$T_1 = 470$	$T_2 = 444$	$T_3 = 425$	$T = 1{,}339$

Is there sufficient evidence at the 1% level to conclude that a difference exists in the mean healing times for the 3 medications?

Solution

Let μ_A, μ_B, and μ_C denote the mean number of hours for the rash to disappear for medications A, B, and C, respectively.

Step 1. Hypotheses.

$$H_0: \mu_A = \mu_B = \mu_C$$

H_a: Not all the means are equal.

Step 2: Significance level.

$$\alpha = 0.01$$

Step 3: Calculations.

To obtain the sums of squares for the ANOVA table, we need the following totals for the $k = 3$ treatments (medications) and the $b = 10$ blocks (infants).

$$B_1 = 131 \quad B_2 = 146 \quad B_3 = 135 \quad B_4 = 144 \quad B_5 = 123$$
$$B_6 = 126 \quad B_7 = 136 \quad B_8 = 133 \quad B_9 = 129 \quad B_{10} = 136$$
$$T_1 = 470 \quad T_2 = 444 \quad T_3 = 425 \quad T = 1{,}339$$

Using these totals, we next calculate the total sum of squares *SST*, the treatment sum of squares *SSTr*, the block sum of squares *SSB*, and the error sum of squares *SSE*.

$$SST = \Sigma x^2 - \frac{T^2}{n}$$

$$= (46^2 + 50^2 + 46^2 + \cdots + 43^2) - \frac{1,339^2}{30}$$

$$= 60,063 - 59,764.033 = 298.967$$

$$SSTr = \frac{T_1^2 + T_2^2 + T_3^2}{b} - \frac{T^2}{n}$$

$$= \frac{470^2 + 444^2 + 425^2}{10} - \frac{1,339^2}{30}$$

$$= 59,866.1 - 59,764.033 = 102.067$$

$$SSB = \frac{B_1^2 + B_2^2 + \cdots + B_{10}^2}{k} - \frac{T^2}{n}$$

$$= \frac{131^2 + 146^2 + 135^2 + 144^2 + 123^2 + 126^2 + 136^2 + 133^2 + 129^2 + 136^2}{3} - \frac{1,339^2}{30}$$

$$= 59,921.667 - 59,764.033 = 157.634$$

$$SSE = SST - SSTr - SSB$$
$$= 298.967 - 102.067 - 157.634 = 39.266$$

Using the above sums of squares, the following ANOVA table is constructed.

Source	df	SS	MS	F
Treatments	2	102.067	51.033	23.40
Blocks	9	157.634	17.515	
Error	18	39.266	2.181	
Total	29	298.967		

Step 4: Rejection Region.

The rejection region consists of F-values for which $F > F_\alpha = F_{.01} = 6.01$. This value is obtained from part d of Table 6 in Appendix A using $ndf = k - 1 = 2$ and $ddf = (k - 1)(b - 1) = 18$. Note that these are the degrees of freedom in the ANOVA table that are associated with the numerator ($MSTr$) and the denominator (MSE) used to calculate the test statistic F.

Step 5: Conclusion.

The test statistic value $F = 23.40$ exceeds the table value 6.01. Consequently, we reject H_0 and accept the alternative hypothesis. Thus, sufficient evidence exists at the 1 percent level to conclude that at least 2 of the treatments differ in the mean time required to eliminate the rash.

The previous example illustrates how the sums of squares are calculated for the analysis of variance table in a randomized block design. In practice, they are usually obtained with computer software packages such as MINITAB. This procedure is illustrated in the following example.

MINITAB EXAMPLE 13.6   Use MINITAB to obtain the analysis of variance table for the medication data in Example 13.5.

Solution

MINITAB offers a choice of methods for obtaining the analysis of variance table. We will illustrate the use of the command **ANOVA.** With this command all values of the response variable (healing time) are stacked in a single column C1, and two auxiliary columns C2 and C3 are used to indicate for each response value the associated treatment and block numbers.

```
NAME C1 'HealTime' C2 'Treatmnt' C3 'Infant'
READ C1 C2 C3
46 1  1
50 1  2
46 1  3
51 1  4
43 1  5
45 1  6
 .  .  .

 .  .  .
41 3  8
42 3  9
43 3 10
END
```

To clarify the contents of columns C1, C2, and C3, consider the fourth row of data (51 1 4). The response value *51* hours is associated with the *1st* treatment (medication A) and the *4th* block (infant 4). Similarly, for the last data row (43 3 10) the *43* hours corresponds to the *3rd* treatment and the *10th* block. The following command instructs MINITAB to produce the analysis of variance table.

Menu Commands:
Stat ➤
ANOVA ➤
Balanced ANOVA
In **Responses** *box*
enter **C1**
In **Model** *box enter*
C2 C3
Select OK

```
ANOVA C1 = C2 C3   # Data in C1, Trt & Blk indices in C2,C3
```

The output produced by MINITAB appears in Exhibit 13.3. Note that the p-value 0.000 for the test statistic value $F = 23.39$ is given in the analysis of variance table. Since it is less than the significance level $\alpha = 0.01$, the null hypothesis would be rejected.

■■■ **Exhibit 13.3**

```
MTB > ANOVA C1 = C2 C3  # Data in C1, Trt & Blk indices in C2,C3

Analysis of Variance for HealTime

Source      DF        SS          MS        F       P
Treatmnt     2    102.067      51.033    23.39   0.000
Infant       9    157.633      17.515     8.03   0.000
Error       18     39.267       2.181
Total       29    298.967
```

Testing the Effectiveness of Blocking

By blocking the experimental units, instead of using independent random samples, the experimenter hopes to appreciably reduce the data's unexplained variation SSE, and thus enhance the power of the test. To affirm the success of this objective, the analysis of variance table can be used to also test the effectiveness of blocking. We accomplish this by testing if the block means are significantly different. This is done by comparing the mean square for blocks MSB to the mean square error MSE by calculating an F-ratio similar to that used to compare the treatment means. The test procedure is summarized below.

PROCEDURE

Test to Compare Block Means in a Randomized Block Design:
To test the null hypothesis for b block means,

$$H_0: \mu_1 = \mu_2 = \cdots = \mu_b,$$

use the ANOVA table to determine the F-ratio

$$F = \frac{MSB}{MSE}$$

The null hypothesis is rejected when $F > F_\alpha$, where α is the level of significance, and the F-distribution has $ndf = b - 1$, $ddf = (k - 1)(b - 1)$.

EXAMPLE 13.7   With reference to Example 13.5, test if blocking was effective. That is, does sufficient evidence exist at the 5 percent level to conclude that there is a difference in the mean healing times for the 10 infants (blocks)?

Solution

Let μ_1 through μ_{10} denote the mean number of hours for a rash to disappear for infants 1 through 10.

Step 1: Hypotheses.

$$H_0: \mu_1 = \mu_2 = \mu_3 = \cdots = \mu_{10}$$

H_a: Not all the means are equal.

Step 2: Significance level.

$$\alpha = 0.05$$

Step 3: Calculations.

From the ANOVA table in Example 13.5, we have $MSB = 17.515$ and $MSE = 2.181$. The value of the test statistic is

$$F = \frac{MSB}{MSE} = \frac{17.515}{2.181} = 8.03$$

Note that in the analysis of variance table in Exhibit 13.3, MINITAB gives this F-ratio in addition to the F-value for testing treatment means.

Step 4: Rejection Region.

The rejection region consists of F-values for which $F > F_\alpha = F_{.05} = 2.46$. This value is obtained from part b of Table 6 in Appendix A using $ndf = b - 1 = 9$ and $ddf = (k - 1)(b - 1) = 18$. These are also the degrees of freedom given in the ANOVA table that are associated with the numerator MSB and the denominator MSE in the F-ratio for blocks.

Step 5: Conclusion.

Since the calculated value $F = 8.03$ is greater than the table value 2.46, we reject H_0 and accept the alternative hypothesis. There is sufficient evidence at the 5 percent level to conclude that the block means are not all the same and that blocking was effective.

In the last example, our test to compare block means supports the researcher's use of a randomized block design. We emphasize that when the block test fails to reject the null hypothesis, we should not conclude that blocking was inappropriate. In designing the diaper rash experiment, the researcher believed that skin patches would be more homogeneous *within blocks* (for the same infant) than *between blocks* (for different infants). If this is the expectation, then a randomized block design should be used. The test for block means is conducted primarily to lend support for the choice of the original design and to provide guidance for its use in followup experiments. When the test results seem to suggest there is little difference in the homogeneity of experimental units within blocks and between blocks, then the researcher can re-assess the randomized block design's appropriateness in similar future experiments.

In concluding, we note that there are occasions when it can be a costly mistake to prefer the randomized block over the completely randomized design, because it is possible for the latter to yield more information. Specifically, the mean square error MSE might actually be larger for the randomized block design, even though its SSE is smaller. This possible inflation in MSE can occur because its degrees of freedom are reduced in changing from a completely randomized design to a comparable randomized block design.

Estimating the Difference Between Two Means

In planning a randomized block design, the experimenter wants to determine if a difference exists among the k treatment means. Additionally, he or she may focus interest on a particular pair of means, say μ_i and μ_j, and want to ultimately estimate by how much they differ. With the aid of the ANOVA table, this can be accomplished by constructing a confidence interval for $(\mu_i - \mu_j)$, the difference between the specified means. The confidence interval formula is the same as that given in Section 13.4 for the completely randomized design. By using in Equation 13.5 the fact that the sample sizes n_i and n_j are equal to b, we have the following confidence interval formula for $(\mu_i - \mu_j)$.

PROCEDURE

$(1 - \alpha)$ Confidence Interval for the Difference Between Two Treatment Means $(\mu_i - \mu_j)$:

(13.6)
$$(\bar{x}_i - \bar{x}_j) \pm t_{\alpha/2} \sqrt{MSE \left(\frac{2}{b} \right)}$$

Note:

1. $\bar{x}_i = T_i/b$, $\bar{x}_j = T_j/b$, where T_i and T_j are the ith and jth treatment totals, respectively, and b is the number of blocks.

2. The t-distribution is based on the degrees of freedom for the mean square error MSE. These are $df = (k - 1)(b - 1)$, where k is the number of treatments.

EXAMPLE 13.8   For the medication data in Example 13.5, construct a 95 percent confidence interval to estimate the difference in the mean healing times for medications A and B.

Solution
From Example 13.5 the following values are obtained: $T_1 = 470$, $T_2 = 444$, $b = 10$, $MSE = 2.181$ with $df = 18$.

$$\bar{x}_1 = \frac{T_1}{b} = \frac{470}{10} = 47 \quad \text{and} \quad \bar{x}_2 = \frac{T_2}{b} = \frac{444}{10} = 44.4$$

For a confidence level of 95 percent and $df = 18$, the t-value is $t_{\alpha/2} = t_{.025} = 2.101$. A 95 percent confidence interval for $(\mu_A - \mu_B)$ is

$$(\bar{x}_1 - \bar{x}_2) \pm t_{\alpha/2} \sqrt{MSE \left(\frac{2}{b} \right)}$$

$$(47 - 44.4) \pm 2.101 \sqrt{2.181 \left(\frac{2}{10} \right)}$$

$$2.6 \pm 1.39$$

Thus, with 95 percent confidence, we estimate that medication A requires an average of 1.21 to 3.99 more hours than medication B for the elimination of a diaper rash.

Exhibit 13.4

```
MTB > NAME C1 'HealTime' C2 'Treatmnt' C3 'Infant'
MTB > %MAIN C2 C3;
SUBC> RESPONSE C1.
```

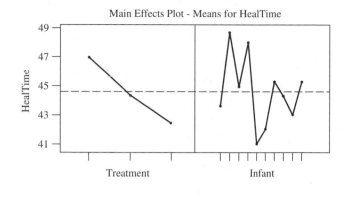

For the completely randomized design, we gave a confidence interval formula for $(\mu_i - \mu_j)$ and also a formula for estimating a single treatment mean. For the randomized block design, a confidence interval for one mean is usually not appropriate, because it would require that the blocks used in the experiment had been randomly selected from a theoretical population of blocks. In practice this is often not the case since the choice of blocks is usually the prerogative of the experimenter. However, MINITAB does provide a convenient way to visually display in a single graph all the treatment means, as well as the block means. To illustrate, recall that in Example 13.6 we stored the values of the response variable (healing time) in column C1, and columns C2 and C3 were used to identify the associated treatment (medication) and block (infant) numbers. To produce the graph shown in Exhibit 13.4, type the following commands.

```
%MAIN C2 C3;
RESPONSE C1.
```

The main command **%MAIN** in the above includes the columns C2 and C3 that contain the treatment and the block numbers. The subcommand **RESPONSE** is used to specify where the values of the response variable are stored. Displayed in Exhibit 13.4 is the mean number of hours for a rash to disappear for each of the three medications and for each of the ten infants. It is readily apparent that the largest difference in healing times occurred for the first (medication A) and third (medication C) treatments. We also note from the display of the block means that considerable variation exists in mean healing times between the infants, thus lending support for the use of a randomized block design for this experiment.

As with the one-way analysis of variance, for the randomized block design there are multiple comparison procedures that can be used to formally evaluate all

Menu Commands:
Stat ➤
ANOVA ➤
Main Effects Plot
In **Factors** *box enter*
C2 C3
In **Raw response**
data in *box enter* **C1**
Select OK

possible pairs of treatment means. Several multiple comparison techniques are discussed in detail in the references on experimental design that appear in Appendix C.

In Chapter 15, we will consider an alternate method for analyzing a randomized block design. The Friedman F-test, a nonparametric procedure, can be used when one has reservations about the validity of the underlying assumptions that are necessary for analyzing a randomized block design.

SECTION 13.5 EXERCISES

Randomized Block Designs and the Analysis of Variance

In the following exercises, assume that the assumptions that are necessary for analyzing a randomized block design are satisfied.

13.46 A consumer research organization wanted to compare mileage achievement for five national brands of gasoline. Explain how the experiment could be planned so that its analysis would use a
a. completely randomized design,
b. randomized block design.

13.47 With reference to Exercise 13.46, which design do you believe would be more appropriate? Explain your answer.

13.48 The following is a partially completed ANOVA table for a randomized block design.

Source	df	SS	MS	F
Treatments	—	300.00	—	15.00
Blocks	—	—	—	4.50
Error	20	—	5.00	
Total	—	512.50		

a. Complete the table by filling in the missing blanks.
b. How many treatments are involved in the experiment?
c. What is the number of blocks involved?

13.49 Consider the following partially completed ANOVA table for a randomized block design.

Source	df	SS	MS	F
Treatments	5	—	3.13	—
Blocks	8	23.12	—	—
Error	—	63.20	1.58	
Total	—	—		

a. What is the number of treatments in the experiment?
b. What is the number of blocks?
c. Complete the analysis of variance table by filling in the missing blanks.

13.50   Use the analysis of variance table in Exercise 13.48 to test if the treatment means differ significantly at the 5 percent level.

13.51   With reference to Exercise 13.49, is there sufficient evidence to conclude that a difference exists in the treatment means? Use a significance level of 5 percent.

13.52   Refer to Exercise 13.48 and test if blocking was effective in removing a significant amount of variation in the data. Use $\alpha = 0.10$.

13.53   With reference to Exercise 13.49, test if blocking was effective. Use a significance level of 5 percent.

13.54   The following data pertain to a randomized block design that was used to compare the means of treatments A, B, and C. Test if there is a significant difference at the 1 percent level.

	Treatment		
Block	A	B	C
1	16	12	11
2	20	19	15
3	13	12	11
4	19	14	10
5	15	13	13

13.55   A randomized block design was used to compare four treatment means. The data appear below. Does a significant difference exist between the treatment means? Test using $\alpha = 0.05$.

	Treatment			
Block	A	B	C	D
1	1.8	0.9	2.8	1.9
2	2.1	1.1	3.0	2.1
3	1.1	0.1	1.9	1.8
4	1.9	0.5	2.9	2.0
5	1.4	0.3	2.3	1.9
6	1.7	0.7	2.7	2.0
7	0.9	0.1	1.6	1.7
8	2.4	1.4	3.5	2.1

13.56   Do the data in Exercise 13.54 provide sufficient evidence at the 5 percent level to indicate that a randomized block design was useful?

13.57   For Exercise 13.55, test at the 10 percent level if the use of a randomized block design was worthwhile.

13.58 With reference to Exercise 13.54, construct a 95 percent confidence interval to estimate the difference between means for treatments B and C.

13.59 Refer to Exercise 13.55 and estimate the difference between the treatment means for C and D. Use a confidence level of 90 percent.

13.60 1995 was a watershed year for the use of Intel's Pentium processor in notebook computers. At the beginning of the year a Pentium-based notebook was a rarity, but by year's-end sales of such systems dominated the notebook market. *Computer Reseller News* subjected three high-end systems for technical evaluation (August 28, 1995). For each system a benchmark test was performed on four test platforms: Business Graphics, Data Base, Spreadsheet, and Word Processing. The test scores appear below.

Notebook Model	Test Platform			
	Business Graphics	Data Base	Spreadsheet	Word Processing
610CT	7.5	5.3	6.4	6.6
5000	6.4	5.1	5.6	5.9
250P	5.6	4.5	5.1	5.3

a. Identify the likely treatments and blocks for this experiment.
b. Test at the 5 percent level if a significant difference exists in the mean scores for the 3 notebook models.
c. Is there sufficient evidence to indicate differences in mean scores among the four test platforms? Test at the 5 percent significance level.

13.61 Find the approximate *p*-value for each test in Exercise 13.60.

13.62 A manufacturer of automotive piston rings experimented with the use of four types of composite materials that were code named EX-105, EX-106, EX-107, and EX-108. Five 4-cylinder cars were equipped with each of the 4 types of rings and driven 50,000 miles. The engines were disassembled and the percentage of wear was recorded for each of the 20 rings. For the results displayed below and $\alpha = 0.01$, test if there is a significant difference in the mean percentages of wear for the four types of composite materials.

Car	Type of Material			
	EX-105	EX-106	EX-107	EX-108
1	5.3	4.5	2.2	3.6
2	6.4	6.8	1.7	1.9
3	5.0	3.9	1.8	2.7
4	5.5	5.6	1.7	1.4
5	4.4	4.0	1.9	3.4

13.63   Find the approximate p-value for the test in Exercise 13.62.

13.64   In Exercise 13.62, was a significant amount of variation accounted for by blocking on the cars? Test at the 10 percent significance level.

13.65   Find the approximate p-value for the test in Exercise 13.64.

13.66   With reference to Exercise 13.62, construct a 95 percent confidence interval to estimate the difference in the mean percentages of wear for piston rings made with materials EX-106 and EX-107.

MINITAB Assignments

M▶ 13.67   In Exercise 13.40, a furniture manufacturer used a completely randomized design to compare drying times for four brands of stain. Later it was decided to utilize a randomized block design involving five types of wood. Each of the 4 brands was used to stain 5 chairs constructed from pine, spruce, oak, walnut, and cherry. The drying times in minutes are given below.

Wood	Brand of Stain			
	1	2	3	4
Pine	85.7	89.7	90.5	88.9
Spruce	84.8	88.9	89.6	90.1
Oak	92.8	92.7	93.1	93.9
Walnut	91.7	95.8	95.7	93.8
Cherry	91.9	96.5	97.5	94.6

Use MINITAB to test if a significant difference exists in the mean drying times for the four brands. Let $\alpha = 0.01$.

M▶ 13.68   Use the results from Exercise 13.67 to test if a significant amount of variability was accounted for by the use of blocks in the experiment. Use a significance level of 10 percent.

M▶ 13.69   With reference to Exercise 13.67, use MINITAB to produce a visual display of the mean drying times for the 4 brands of stain and the 5 types of wood.

M▶ 13.70   Total compensation for CEOs at the nation's 350 largest companies averaged $1,779,663 in 1994. In several instances the CEO's bonus exceeded his/her salary. *The Wall Street Journal/William M. Mercer CEO Compensation Survey* gave the following compensations for 29 utilities (*The Wall Street Journal,* April 12, 1995). The figures are in thousands of dollars.

Utilities Company	Compensation ($1,000's)	
	Salary	Bonus
American Electric Power	620.0	209.4
Ameritech	709.6	754.2
Bell Atlantic	831.2	778.7
BellSouth	588.5	765.0
Centerior Energy	360.0	74.5
Central & South West	599.8	162.7
Columbia Gas System	682.0	50.0
Con Edison	603.3	353.6
Detroit Edison	540.0	33.5
Dominion Resources	571.1	273.7
Duke Power	558.5	103.5
FLP	795.8	650.0
General Public Utilities	573.8	292.5
GTE	784.6	1219.5
MCI Communications	850.0	900.0
Niagra Mohawk	457.9	0.0
Nynex	800.0	885.0
Ohio Edison	461.7	178.8
PSEG	652.5	265.3
Pacific Enterprises	641.0	428.6
Pacific Gas & Electric	550.0	134.2
Pacific Telesis	541.5	391.7
PacifiCorp	504.1	302.1
Peco Energy	485.3	305.7
SCEcorp	664.0	0.0
Southwestern Bell	762.0	1190.0
Sprint	863.9	1085.6
U S West	700.0	560.0
Union Electric	400.0	120.0

Use MINITAB and an analysis of variance to test if the mean difference between salary and bonus is statistical significant at the 5 percent level.

M▶ 13.71   Solve Exercise 13.70 by using MINITAB to perform a paired t-test. What is the relationship between the value of the test statistic t and the F-ratio in the ANOVA table?

LOOKING
B A C K

Analysis of variance is a technique for testing the null hypothesis that the means of two or more populations are equal. With the **one-way analysis of variance,** the experimenter investigates the effects of only one factor on the response variable. A one-way analysis of variance is a generalization of the two-sample t-test that was discussed earlier in Chapter 11. Both methods are based on the **assumptions**

that **independent random samples** have been selected from **normal populations** that have the **same variance.**

In performing a one-way analysis of variance, the various components involved in the calculation of the test statistic F are displayed in tabular form, referred to as the **ANOVA table.**

ONE-WAY ANOVA TABLE FOR COMPARING k POPULATION MEANS

Source of Variation	df	SS	MS	F
Treatments	$k - 1$	$SSTr$	$MSTr = SSTr/(k - 1)$	$MSTr/MSE$
Error	$n - k$	SSE	$MSE = SSE/(n - k)$	
Total	$n - 1$	SST		

The first step in constructing the table is to calculate the three **sums of squares.** These are obtained by using the following shortcut formulas.

PROCEDURE

Total *SS:*

$$SST = \Sigma x^2 - \frac{T^2}{n}$$

Treatment *SS:*

$$SSTr = \left(\frac{T_1^2}{n_1} + \frac{T_2^2}{n_2} + \cdots + \frac{T_k^2}{n_k}\right) - \frac{T^2}{n}$$

Error *SS:*

$$SSE = SST - SSTr$$

In the above formulas, n_i and T_i denote the sample sizes and totals, respectively, $n = \Sigma n_i$, and $T = \Sigma T_i$.

Confidence intervals for one mean or the difference between two population means can be constructed by using the following formulas.

PROCEDURE

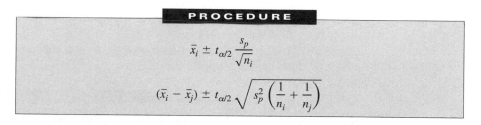

$$\bar{x}_i \pm t_{\alpha/2} \frac{s_p}{\sqrt{n_i}}$$

$$(\bar{x}_i - \bar{x}_j) \pm t_{\alpha/2} \sqrt{s_p^2 \left(\frac{1}{n_i} + \frac{1}{n_j}\right)}$$

The degrees of freedom for determining $t_{\alpha/2}$ are $df = n - k$, and $s_p^2 = MSE$ from the ANOVA table.

The analysis of variance technique can also be used to analyze a **randomized**

block design, which is a generalization of the paired-samples t-test from Chapter 11. A randomized block design consists of b **blocks,** each of which contains k matched experimental units. The k units within each block are randomly assigned to the k treatments. To test the null hypothesis that the k treatment means are equal, the following ANOVA table is constructed.

ANOVA TABLE FOR A RANDOMIZED BLOCK DESIGN

Source of Variation	df	SS	MS	F
Treatments	$k - 1$	$SSTr$	$MSTr = SSTr/(k - 1)$	$F = MSTr/MSE$
Blocks	$b - 1$	SSB	$MSB = SSB/(b - 1)$	
Error	$(k - 1)(b - 1)$	SSE	$MSE = SSE/(k - 1)(b - 1)$	
Total	$n - 1$	SST		

The four sums of squares in the ANOVA table are calculated using the following shortcut formulas.

PROCEDURE

Total *SS:*

$$SST = \Sigma x^2 - \frac{T^2}{n}$$

Treatment *SS:*

$$SSTr = \frac{T_1^2 + T_2^2 + \cdots + T_k^2}{b} - \frac{T^2}{n}$$

Block *SS:*

$$SSB = \frac{B_1^2 + B_2^2 + \cdots + B_b^2}{k} - \frac{T^2}{n}$$

Error *SS:*

$$SSE = SST - SSTr - SSB$$

In the above formulas, k is the number of treatments, b is the number of blocks, T_i is the total of the ith treatment, B_j is the total of the jth block, $T = \Sigma T_i$, and $n = kb$.

The difference between two treatment means can be estimated using the following confidence interval.

PROCEDURE

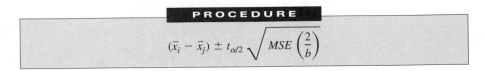

$$(\bar{x}_i - \bar{x}_j) \pm t_{\alpha/2} \sqrt{MSE \left(\frac{2}{b} \right)}$$

The degrees of freedom for determining $t_{\alpha/2}$ are $df = (k - 1)(b - 1)$.

Key Words

In reviewing this chapter, you should be able to define, explain, and illustrate each of the following.

analysis of variance *(page 587)*

one-way analysis of variance *(page 587)*

response variable *(page 587)*

factor *(page 587)*

levels *(page 587)*

treatments *(page 587)*

completely randomized design *(page 587)*

total sum of squares (*SST*) *(page 589)*

treatment sum of squares (*SSTr*) *(page 589)*

error sum of squares (*SSE*) *(page 589)*

mean square for treatments (*MSTr*) *(page 590)*

F-ratio *(page 591)*

mean square for error (*MSE*) *(page 592)*

ANOVA table *(page 591)*

blocks *(page 614)*

randomized block design *(page 614)*

block sum of squares (*SSB*) *(page 615)*

mean square for blocks (*MSB*) *(page 620)*

M▶ MINITAB Commands

NAME _ _ *(page 588)*

DOTPLOT _ _; *(page 588)*
SAME.

AOVONEWAY _ _ *(page 601)*

SET _ *(page 601)*

END *(page 601)*

READ _ *(page 619)*

ANOVA _ = _ _ *(page 619)*

%MAIN _ _; *(page 623)*
RESPONSE _.

REVIEW EXERCISES

13.72   A particular camcorder is sold by mail order companies, camera shops, and general merchandise outlets. The manufacturer is interested in determining if the mean selling price differs for these three sources. Seven stores of each type were randomly selected, and the following prices were obtained:

Type of Store	Selling Price in Dollars						
Mail order	899	929	900	979	925	950	959
Camera shop	995	935	950	979	979	995	929
General merch.	979	999	950	995	925	975	989

Formulate a suitable set of hypotheses and test at the 0.01 significance level.

13.73   Determine the approximate *p*-value for the test in Exercise 13.72.

13.74   With reference to Exercise 13.72, construct a 90 percent confidence interval to estimate the mean price charged by mail order companies.

13.75 With reference to Exercise 13.72, construct a 90 percent confidence interval to estimate the difference in mean prices charged by mail order and general merchandise companies.

13.76 Calculate the pooled sample variance for the three samples whose variances are given below.

Sample 1	Sample 2	Sample 3
$n_1 = 7$	$n_2 = 10$	$n_3 = 9$
$s_1^2 = 20$	$s_2^2 = 16$	$s_3^2 = 22$

13.77 *MSE*, the mean square error in a one-way analysis of variance table, is a weighted average of the sample variances $s_1^2, s_2^2, s_3^2, \ldots, s_k^2$, where the weights are $(n_1 - 1), (n_2 - 1), (n_3 - 1), \ldots, (n_k - 1)$, respectively. Show that when all the sample sizes are the same, *MSE* is just

$$\frac{(s_1^2 + s_2^2 + s_3^2 + \cdots + s_k^2)}{k}.$$

13.78 Independent random samples from four populations produced the following results:

Sample 1	Sample 2	Sample 3	Sample 4
$n_1 = 9$	$n_2 = 10$	$n_3 = 5$	$n_4 = 6$
$\bar{x}_1 = 42$	$\bar{x}_2 = 39$	$\bar{x}_3 = 54$	$\bar{x}_4 = 50$
$s_1^2 = 18$	$s_2^2 = 25$	$s_3^2 = 22$	$s_4^2 = 20$

a. Calculate the treatment sum of squares *SSTr*.
b. Calculate the pooled sample variance, s_p^2.
c. Construct the analysis of variance table.
d. With $\alpha = 0.05$, test the null hypothesis $H_0: \mu_1 = \mu_2 = \mu_3 = \mu_4$.

13.79 In Exercise 13.78, estimate μ_2 by obtaining a 90 percent confidence interval.

13.80 In Exercise 13.78, construct a 90 percent confidence interval for $(\mu_3 - \mu_1)$.

13.81 To determine if the means of four populations differ, the following independent random samples were selected:

Sample 1	Sample 2	Sample 3	Sample 4
6	9	12	13
7	10	11	11
9	9	9	10
8	13	10	13
5			15

a. Construct a dotplot of each sample and form an opinion as to whether the sample means differ significantly.

b. Construct the analysis of variance table and test the null hypothesis H_0: $\mu_1 = \mu_2 = \mu_3 = \mu_4$. Use $\alpha = 0.05$.

13.82   A partially completed ANOVA table is given below for a completely randomized design.

Source of Variation	df	SS	MS	F
Treatments	—	493	70.43	—
Error	22	—	—	
Total	—	721		

Fill in the missing values in the analysis of variance table, and test at the 0.01 level the null hypothesis that the sampled populations have the same mean.

13.83   In a state for which the price of milk is not regulated, 8 stores were randomly selected in the western, central, and eastern parts of the state. The prices charged by the 24 stores for a quart of milk appear below. Do the data provide sufficient evidence to conclude that there is a difference in the mean prices for a quart of milk in the 3 geographical regions? Test at the 0.05 significance level.

Region	Cost in Cents for a Quart of Milk							
Western	63	64	63	60	55	62	60	62
Central	67	62	69	68	65	65	65	66
Eastern	68	68	73	64	69	72	69	68

13.84   Determine the approximate p-value for the test in Exercise 13.83.

13.85   For Exercise 13.83, estimate with 95 percent confidence the mean price of a quart of milk in the central part of the state.

13.86   For Exercise 13.83, construct a 95 percent confidence interval to estimate the difference in the mean price of a quart of milk in the east and in the west.

13.87   A study was conducted to compare the four leading computer word processing programs. Each of four groups of students received training in one program for a four-week period. At the conclusion of the instruction, all participants were rated on their ability to perform a given task. Their ratings are given on the following page.

Program 1	Program 2	Program 3	Program 4
24	29	24	25
22	27	38	23
21	21	25	24
20	25	24	25
21	22	29	23
	28		

Is there sufficient evidence at the 0.05 level to conclude that the mean performance ratings for the four word processing programs are not all the same?

13.88 Obtain the approximate p-value for the analysis of variance test in Exercise 13.87.

13.89 For Exercise 13.87, estimate with 90 percent confidence the difference in mean performance ratings for programs 3 and 1.

13.90 A consumer products evaluation magazine tested three brands of flashlight batteries. Each brand was used in five flashlights, and the lights were left on until the batteries failed. The life lengths in hours are given below.

Brand	Length of Life in Hours				
A	7.3	6.9	5.8	7.9	8.2
B	6.7	7.1	6.0	6.5	5.9
C	7.5	8.3	7.9	8.4	8.3

Is there a difference in the mean length of life among the three brands of batteries? Test at the 0.05 level.

13.91 Many consumers have their medical prescriptions filled through mail order companies. *Smart Money* magazine (November, 1995) checked the prices charged by four such companies for the commonly used prescription drugs Seldane (allergies), Procardia XL (high blood pressure), Ortho-Novum (birth control), and Prozac (antidepressant). Use the following data to test at the 10 percent level if there is a difference in the mean prescription costs for the four mail order companies.

Prescription Drug	Mail Order Company			
	1	2	3	4
Seldane, 60 mg, 30 tab	$27.25	$26.99	$31.19	$36.99
Procardia XL, 60 mg, 100 tab	$178.40	$180.39	$187.20	$206.75
Ortho-Novum 1/35, 21-day pk	$21.25	$21.39	$20.83	$22.44
Prozac, 10 mg, 50 tab	$97.20	$93.69	$110.31	$104.09

13.92 Determine the approximate p-value for the test in Exercise 13.91.

13.93   With reference to Exercise 13.91, estimate the difference in mean prescription costs for companies 4 and 1 by constructing a 90 percent confidence interval.

13.94   Consider the following partially completed ANOVA table for a randomized block design.

Source	df	SS	MS	F
Treatments	3	—	4.57	—
Blocks	9	37.08	—	—
Error	—	49.14	1.82	
Total	—	—		

a. What is the number of treatments in the experiment?
b. What is the number of blocks?
c. Complete the analysis of variance table by filling in the missing blanks.

13.95   Use the analysis of variance table in Exercise 13.94 to test if the treatment means differ significantly at the 5 percent level.

13.96   Refer to Exercise 13.94 and test if blocking was effective in removing a significant amount of variation in the data. Use $\alpha = 0.10$.

13.97   The following data pertain to a randomized block design that was used to compare the means of four treatments A, B, C, and D. Test if there is a significant difference at the 5 percent level.

	Treatment			
Block	A	B	C	D
1	12	20	17	11
2	10	13	20	14
3	12	14	18	10
4	15	12	19	13
5	10	11	16	10

13.98   In Exercise 13.97, was a significant amount of variation accounted for by blocking? Test at the 10 percent significance level.

13.99   Three light bulb manufacturers, A, B, and C, claim that their brands have an expected life of 750 hours. An independent laboratory tested a 40-watt, 60-watt, 75-watt, and 100-watt bulb of each brand, and the following life-lengths in hours were obtained.

	Wattage of Bulb			
Brand	**40**	**60**	**75**	**100**
A	798	801	743	753
B	817	813	792	788
C	806	803	758	717

Is there a significant difference in the mean length of life for the three brands? Test with $\alpha = 0.05$.

13.100 Obtain the approximate p-value for the test in Exercise 13.99.

13.101 For Exercise 13.99, obtain a 90 percent confidence interval to estimate the difference between the mean life length for brand A and brand C.

MINITAB Assignments

M▶ 13.102 In Exercise 13.21, a test of hypothesis was conducted to determine if there is a difference in the mean hourly labor rates of automobile mechanics in two states. Suppose it was decided to include a third state in the comparison. For each of the three states, dealerships were randomly selected, and the following hourly charges in dollars were obtained.

First State	Second State	Third State
40.00	35.00	40.00
38.00	37.00	42.00
38.00	31.00	40.00
37.00	39.00	39.50
36.00	31.50	37.50
39.00	35.00	45.00
41.50	32.50	43.00
38.00	34.00	42.50
39.50	39.00	39.50
37.50	36.00	41.00
35.00		44.00
40.00		

a. Use MINITAB to determine if a difference exists in the mean hourly rates for these three states. Test at the 0.01 significance level.

b. Based on the visual display of the confidence intervals in the MINITAB output, form a tentative opinion as to which population means may be equal and which may differ.

M▶ 13.103 Use the output produced by MINITAB in Exercise 13.102 to estimate with 95 percent confidence the difference in the mean hourly rates for mechanics in the first and third states.

M▶ 13.104 A manufacturer of bond writing paper conducted a study to compare the quality of paper for 3 different sources of pulp used in the manufacturing process. For each type of pulp, 16 sheets of paper were randomly selected and assigned a quality rating. The results were as follows:

Pulp A		Pulp B		Pulp C	
78	74	73	62	79	91
69	89	65	80	87	90
75	71	78	76	84	87
89	80	76	72	98	92
76	88	73	56	87	85
78	84	65	69	92	78
65	82	69	61	91	86
78	91	74	71	87	92

Is there sufficient evidence to conclude that a difference exists in the mean quality ratings for the 3 sources of pulp? Use MINITAB to test at the 0.05 significance level.

M▶ 13.105 With reference to Exercise 13.104, use MINITAB to perform an analysis of variance to determine if there is a difference in the mean quality ratings for sources A and C. Test at the 5 percent level.

M▶ 13.106 A national supermarket chain conducted a survey of its three largest competitors primarily for the purpose of comparing prices of store-brand products. Prices were obtained for 20 popular items of the same size, and the costs appear below.

	Supermarket Chain			
Store-Brand Product	Investigating Chain	First Competitor	Second Competitor	Third Competitor
Paper towels	$0.79	$0.69	$0.85	$0.57
Trash bags	$2.98	$3.19	$2.99	$3.02
Skim milk	$2.05	$2.19	$2.11	$2.01
Lettuce	$0.79	$0.89	$0.79	$0.74
Jam	$1.59	$1.65	$1.69	$1.47
Coffee	$3.49	$3.59	$3.59	$3.29
Corn flakes	$2.19	$2.25	$2.29	$2.15
Frozen peas	$0.79	$0.85	$0.79	$0.65
Orange juice	$1.75	$1.99	$1.89	$1.71
Bananas	$0.39	$0.39	$0.35	$0.37
Dish detergent	$0.99	$1.15	$1.19	$0.95
Canned dog food	$0.57	$0.65	$0.59	$0.49
Vegetable soup	$0.67	$0.69	$0.55	$0.73
Butter	$1.39	$1.45	$1.49	$1.19
Potatoes	$2.19	$2.39	$2.09	$2.25
Bread	$0.99	$0.99	$0.99	$0.99
Ground beef	$2.25	$2.59	$2.45	$2.31
Onions	$0.49	$0.59	$0.53	$0.44
Tissues	$1.19	$1.35	$1.29	$1.05
Ice cream	$2.59	$2.79	$2.75	$2.47

Use MINITAB to test if a significance difference exists in the mean prices charged by the four supermarket chains. Let $\alpha = 0.05$.

M▶ 13.107 Refer to Exercise 13.106 and use MINITAB to test at the 5% level if a significance difference exists in the mean prices charged by the three competing chains.

M▶ 13.108 As part of its admissions evaluation process, an actuarial science department administers an examination to each applicant. To determine if there is a significant difference in the grading of the exams by 2 faculty members, 10 exams were graded by each professor, and the following grades were assigned.

Applicant	1	2	3	4	5	6	7	8	9	10
Professor A's Grade	75	87	89	63	93	54	83	71	88	71
Professor B's Grade	69	84	80	57	95	49	79	65	88	67

Have MINITAB perform an analysis of variance to determine if the data provide sufficient evidence of a difference in the mean grades assigned by the two professors. Test at the 0.01 significance level.

M▶ 13.109 With reference to Exercise 13.108, use MINITAB to produce a visual display of the mean grades for the 2 professors and the 10 applicants.

LINEAR
REGRESSION
ANALYSIS

NASA

FORECASTING HURRICANES WITH REGRESSION MODELS

William Gray, a meteorologist at Colorado State University, developed a **mathematical model** for predicting seasonal hurricane activity. The model consists of an equation involving variables such as the spring weather conditions in Caribbean waters, equatorial winds, West African rainfall, El Niño winds, precipitation levels in western Sahel and the Gulf of Guinea, and several other factors.

The model is used by Gray each November to forecast activity for the following year's hurricane season. Knowledge of the variables in the mathematical model has enabled Gray to establish a global reputation by accurately predicting hurricane activity in 9 of 11 recent years. (Source: Popular Science, September 1995.)

This chapter is concerned with investigating the relation between two variables and using a mathematical model to describe the hypothesized relationship.

LOOKING
A H E A D

Chapter 3 provided an introduction to the topic of modeling bivariate data. Many problems that are subjected to statistical analyses are concerned with investigating the relationship between two variables x and y.

- A social scientist might be concerned with how a city's crime rate (y) is related to its unemployment rate (x).

- A nutritionist might try to relate the quantity (y) of breakfast cereal consumed to the cereal's percentage of sugar (x).

- An economist might be interested in studying the relationship between consumer spending (y) and the prime interest rate (x).

- An investment analyst might investigate how the price (y) of a stock is related to its price/earnings ratio (x).

Seldom will an exact mathematical relationship exist between two variables x and y and, thus, the researcher must be content with obtaining a model (equation) that will describe it approximately. In such situations it is important to have some measure of how well the data support the hypothesized model.

In fitting a bivariate model to a set of data, one can proceed systematically by first making a scatter diagram to discern what type of relationship might exist between x and y. This chapter will focus on describing relationships that can be modeled by a straight line. These linear models are called **simple** because they involve only one predictor variable x. The process of fitting a simple linear model to a set of data is called a **simple linear regression analysis.**

If the scatter diagram indicates that the data points tend to lie near a straight line, then the **least squares line** will be obtained and considered as a possible model. Before adopting it, however, it is necessary to evaluate the model's potential usefulness. In Chapter 3, the correlation coefficient r was considered as a means of evaluating the utility of a model. We saw that this is a measure of the strength of the linear relationship between x and y. A value of r that is "near" -1 or $+1$ indicates the existence of a good fit between the data points and the least squares line. In Chapter 3, we had no way of determining if r was sufficiently "near" -1 or $+1$. Now that we have a background in hypothesis testing, this issue can be addressed. In fact, we will see that additional criteria can be used to assess the utility of a model.

Finally, once we have obtained supporting evidence of the usefulness of a model, we will then show how it can be utilized for estimation and prediction.

14.1 THE SIMPLE LINEAR MODEL AND RELATED ASSUMPTIONS

In Chapter 3, we considered an example in which the mathematics department at a liberal arts college used a placement test to assist in assigning appropriate math courses to incoming freshmen. A 25-point test is administered during freshman orientation to measure quantitative and analytical skills. The department believes that the test is a good predictor of a student's final numerical grade in its introductory statistics course. Table 14.1 contains the placement test scores and final course grades for a sample of 15 students who took the course. In Chapter 3, these 15 data points were used to build a model that uses a student's placement test score (x) to predict his or her final statistics grade (y). We will review and, in the process, elucidate the steps involved.

To determine what type of relationship might exist between x and y, we begin by constructing a **scatter diagram,** which is a graph of the data points. This appears in Figure 14.1. From the scatter diagram it appears that there might be approximately a straight line relationship between placement test score x and course grade y. Based on the appearance of the scatter diagram, we will tentatively assume as our model that the mean course grade μ_y is related to test score x by a straight line. Mathematically, we write this as

(14.1)
$$\mu_y = \beta_0 + \beta_1 x,$$

where β_0 and β_1 are unknown constants (parameters) that denote the y-intercept and the slope, respectively, of the line. This line of means is called the **regression line** and is illustrated in Figure 14.2.

TABLE 14.1

PLACEMENT SCORES AND STATISTICS GRADES FOR FIFTEEN FRESHMEN

Student	Placement Test Score x	Numerical Grade y
1	21	69
2	17	72
3	21	94
4	11	61
5	15	62
6	19	80
7	15	65
8	23	88
9	13	54
10	19	75
11	16	80
12	25	93
13	8	55
14	14	60
15	17	64

Figure 14.1
Scatter diagram for the 15
points.

Figure 14.1
Scatter diagram for the 15
points.

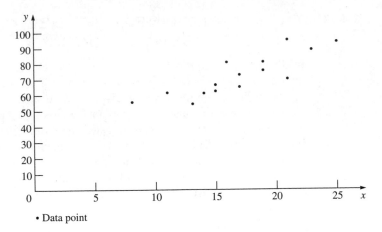

• Data point

It is important to note that the model given in Equation 14.1 describes the relationship between a placement score x and the **average** course grade of all students with this score. To illustrate this, assume temporarily that the unknown y-intercept and slope are actually $\beta_0 = 30$ and $\beta_1 = 2.5$ (in reality, these are unknown and must be estimated). Then for an arbitrarily chosen value of x, say 20, the model states that the mean course grade μ_y of all those who scored 20 on the placement test is given by

$$\mu_y = 30 + 2.5(20) = 80.$$

To write the model so that it describes an individual's course grade y, we must add a **random error component** ϵ to the straight line given in Equation 14.1. The random error component is a random variable that represents the deviation of each person's grade from the mean grade of all who have a placement score of x. Expressing the assumed model in terms of an individual's grade, we have

(14.2A) $$y = \beta_0 + \beta_1 x + \epsilon.$$

To illustrate let's temporarily assume that the unknown y-intercept is $\beta_0 = 30$

Figure 14.2
Assumed model relating
mean grade to test score x.

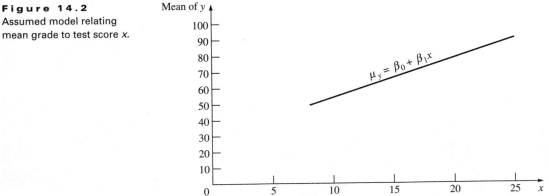

and the unknown slope is $\beta_1 = 2.5$. Then for a randomly selected individual with a placement score of 20, for example, his or her course grade is given by the following.

$$y = 30 + 2.5(20) + \epsilon$$
$$= 80 + \epsilon$$
$$= \text{(the mean grade of all who score 20)} + \text{(a random error component)}$$

Without the random error component in the equation, the model would state that all who score $x = 20$ on the placement test will also obtain the same course grade of 80. The inclusion of ϵ in the model allows for individual grades to vary from the mean grade of 80. For the general model in Equation 14.2A the random error component (ϵ) allows for variation in individual grades (y) from the mean grade ($\beta_0 + \beta_1 x$) of all who score x on the placement test.

(14.2B)
$$y = \overset{\mu_y}{\overbrace{\beta_0 + \beta_1 x}} + \epsilon$$

In Equation 14.2B, $\mu_y = \beta_0 + \beta_1 x$ is called the **deterministic component** of the model, since it always gives the same value for a given score of x (it gave a value of 80 with $x = 20$ for the above illustration). The model in Equation 14.2B is obtained by adding a random error component (ϵ) to the deterministic component ($\beta_0 + \beta_1 x$), and the resulting model is called a **probabilistic model.**

Deterministic   Random error
component   component

$$y = \boxed{\beta_0 + \beta_1 x} + \boxed{\epsilon}$$

The essential distinguishing feature between a deterministic and a probabilistic model is that a deterministic model assumes that the phenomenon of interest can be predicted precisely, while a probabilistic model allows for variation in the predictions through the inclusion of a random error component (see Figure 14.3).

In the sections that follow, inferences will be made in the form of confidence intervals and tests of hypotheses. These inferences require certain assumptions about the behavior of the random error component ϵ in the model. The assumed model and its related assumptions are described in the following box and in Figure 14.4.

Figure 14.3
Probabilistic linear model
$y = \beta_0 + \beta_1 x + \epsilon.$

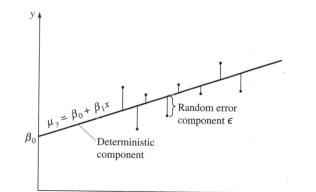

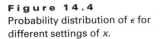

Figure 14.4
Probability distribution of ϵ for different settings of x.

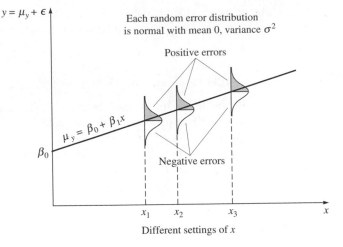

KEY CONCEPT

Simple Linear Probabilistic Model and Assumptions:

(14.3) $$y = \beta_0 + \beta_1 x + \epsilon,$$

y is the **dependent (response) variable** being modeled,

x is the **independent (predictor) variable**,

β_0 is the **y-intercept** of the line,

β_1 is the **slope** of the line,

ϵ is a random variable called the **random error component.**

Assumptions:

1. For each possible setting of x, the random error component ϵ has
   a. a normal probability distribution,
   b. a mean of 0,
   c. a constant variance that is denoted by σ^2.

2. For every possible pair of observations y_i and y_j, the associated random errors ϵ_i and ϵ_j are independent. That is, the error associated with one y value does not affect the error associated with another y value.

Note: The mean of y is given by $\mu_y = \beta_0 + \beta_1 x$, and individual values of y deviate about this straight line with a variance of σ^2.

14.2   FITTING THE MODEL BY THE METHOD OF LEAST SQUARES

In the previous section, we began our investigation of the relationship between placement score x and course grade y by constructing a scatter diagram. Since the pattern of the data points suggested a straight line, we tentatively assumed that the mean course grade μ_y is related to test score x by a straight line. This is the deterministic component of the model, written in Equation 14.1 as

© Day Williams 1990/Photo Researchers, Inc.

$$\mu_y = \beta_0 + \beta_1 x.$$

The coefficients β_0 and β_1 denote the y-intercept and the slope of the regression line, and they are unknown parameters that must be estimated from the sample data. In Chapter 3, we discussed the method of least squares for estimating the slope and y-intercept when fitting a straight line to a set of data points. As discussed there, the **least squares line** is generally considered to be the line of best fit. The formulas for the least squares estimates of β_0 and β_1 are repeated below.

PROCEDURE

The Least Squares Estimates of β_0 and β_1:
The slope is estimated by

(14.4)
$$\hat{\beta}_1 = \frac{SS(xy)}{SS(x)}$$

and the y-intercept by

(14.5)
$$\hat{\beta}_0 = \bar{y} - \hat{\beta}_1 \bar{x},$$

where

$$SS(xy) = \Sigma xy - \frac{(\Sigma x)(\Sigma y)}{n},$$

$$SS(x) = \Sigma x^2 - \frac{(\Sigma x)^2}{n},$$

$$\bar{x} = \frac{\Sigma x}{n}, \quad \text{and}$$

$$\bar{y} = \frac{\Sigma y}{n}.$$

EXAMPLE 14.1 Obtain the estimated model for describing the relationship between test score x and course grade y.

Solution

The assumed model is $\mu_y = \beta_0 + \beta_1 x$, which will be estimated by the least squares line. This estimated model will be denoted by $\hat{y} = \hat{\beta}_0 + \hat{\beta}_1 x$. To find $\hat{\beta}_0$ and $\hat{\beta}_1$, we must first determine $SS(xy)$ and $SS(x)$.

Test Score x	Stat Grade y	xy	x^2	y^2
21	69	1,449	441	4,761
17	72	1,224	289	5,184
21	94	1,974	441	8,836
11	61	671	121	3,721
15	62	930	225	3,844
19	80	1,520	361	6,400
15	65	975	225	4,225
23	88	2,024	529	7,744
13	54	702	169	2,916
19	75	1,425	361	5,625
16	80	1,280	256	6,400
25	93	2,325	625	8,649
8	55	440	64	3,025
14	60	840	196	3,600
17	64	1,088	289	4,096
254	1,072	18,867	4,592	79,026

$$SS(x) = \Sigma x^2 - \frac{(\Sigma x)^2}{n} = 4{,}592 - \frac{(254)^2}{15} = 290.9333$$

$$SS(xy) = \Sigma xy - \frac{(\Sigma x)(\Sigma y)}{n} = 18{,}867 - \frac{(254)(1{,}072)}{15} = 714.4667$$

The slope of the least squares line is

$$\hat{\beta}_1 = \frac{SS(xy)}{SS(x)} = \frac{714.4667}{290.9333} = 2.4558.$$

The y-intercept equals

$$\hat{\beta}_0 = \bar{y} - \hat{\beta}_1 \bar{x} = \frac{1{,}072}{15} - (2.4558)\frac{254}{15} = 29.88.$$

Thus, the estimated model is the least squares line

$$\hat{y} = 29.88 + 2.46x.$$

The least squares line and the 15 data points are shown in Figure 14.5. For each data point, the difference between the observed value y and the predicted value $\hat{y}$ is called the **error** (or **residual**). The size of an error equals the vertical distance from the point to the line. In Chapter 3, the **sum of the squares of the errors,**

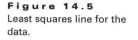

Figure 14.5
Least squares line for the data.

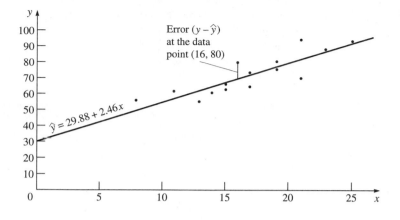

$\Sigma(y - \hat{y})^2$, was denoted by *SSE*, and we will continue with the use of this notation.

In Figure 14.5, the least squares line crosses the *y*-axis at the *y*-intercept of 29.88. The slope of 2.46 indicates that each 1-point increase in placement score *x* will result in a predicted increase of 2.46 in the final course grade.

Using SSE to Estimate the Variance σ^2 of the Random Error Components

The least squares line is usually considered the "best" choice as an estimated model for describing a linear relationship between two variables *x* and *y*. Recall from Chapter 3 that the least squares line is unique in that it is *the line for which the sum of the squares of the errors, SSE, is minimized*. No other line would give a smaller value of *SSE* for the given data points. The importance of *SSE* in regression analysis extends beyond characterizing the least squares line. We will see that *SSE* is an essential part of all hypotheses tests and confidence intervals in this chapter.

In our investigation of the relationship between placement test scores *x* and final statistics grades *y*, we first constructed a scatter diagram to discern what type of model might be appropriate. A plot of the 15 data points suggested that a straight line relationship should be considered, so we assumed the linear model

$$y = \beta_0 + \beta_1 x + \epsilon,$$

where $(\beta_0 + \beta_1 x)$ is the deterministic part and equals μ_y, the mean course grade for a score of *x*. The probabilistic part of the model is ϵ, which is a random error component assumed to have a mean of 0 and a variance of σ^2. We proceeded to obtain as our estimated model the least squares line

$$\hat{y} = 29.88 + 2.46x.$$

Before this can be used to predict course grades, we need to test its utility, that is, how well it describes the relationship between *x* and *y*. Testing the usefulness of the model requires that we first obtain an estimate of σ^2, the variance of the random error components. The best estimate of σ^2 is based on *SSE,* the sum of the squared deviations of the *y*-values from the least squares line. This estimate is obtained by dividing *SSE* by its degrees of freedom *df*, where $df = (n - 2)$. In Chapter 3, *SSE* was calculated by using its definition. However, a much easier shortcut formula is given below.

PROCEDURE

Estimate of the Variance σ^2 of the Random Error Components:

$$(14.6) \qquad \hat{\sigma}^2 = s^2 = \frac{SSE}{df} = \frac{SSE}{n - 2}$$

SSE is most easily calculated by the formula

$$SSE = SS(y) - \hat{\beta}_1 SS(xy).$$

Note:

σ is called the **standard deviation of the model,** and *s* is the **estimated standard deviation of the model.**

In calculating *SSE,* try to retain several significant places in intermediate calculations, since excessive round off in the values of $SS(y)$, $\hat{\beta}_1$, or $SS(xy)$ can seriously affect the accuracy of *SSE.*

EXAMPLE 14.2   Estimate σ, the standard deviation of the model, for the data pertaining to placement test scores (x) and final statistics grades (y).

Solution
In Example 14.1 we obtained

$$SS(xy) = 714.4667 \text{ and } \hat{\beta}_1 = 2.4558.$$

We also found that $\Sigma y^2 = 79,026$ and $\Sigma y = 1,072$. These two quantities are needed to find $SS(y)$.

$$SS(y) = \Sigma y^2 - \frac{(\Sigma y)^2}{n} = 79,026 - \frac{(1,072)^2}{15} = 2,413.7333$$

The error sum of squares can now be computed.

$$\begin{aligned} SSE &= SS(y) - \hat{\beta}_1 SS(xy) \\ &= 2,413.7333 - (2.4558)(714.4667) \\ &= 659.146 \end{aligned}$$

The estimated variance of the random error components is

$$\hat{\sigma}^2 = s^2 = \frac{SSE}{n - 2} = \frac{659.146}{13} = 50.704.$$

Thus, the estimated standard deviation of the model is

$$s = \sqrt{50.704} = 7.12.$$

In the following sections, we will see that s, the estimated standard deviation of the model, is involved in all inferential processes. At this stage, we can give a simple illustration of its significance. The unknown standard deviation of the random error components, σ, measures the variability of y with respect to the regression line $\mu_y = \beta_0 + \beta_1 x$. Similarly, s (which estimates σ) measures the variability of the sample data points with respect to the least squares line (which estimates the regression line). By applying the Empirical rule, we would expect that roughly 95 percent of the sample y values will lie within $2s$ units of the least squares line, that is, few data points would be expected to have an error $(y - \hat{y})$ that is larger than $2s$. For the placement scores example, $2s = 2(7.12) = 14.24$, and a check of Figure 14.5 reveals that each of the 15 points lies within this range of 2 standard deviations from the least squares line.

SECTIONS 14.1 AND 14.2 EXERCISES

The Simple Linear Model and Related Assumptions; Fitting the Model by the Method of Least Squares

14.1   Explain the difference between a deterministic model and a probabilistic model.

14.2   Give an example for which a deterministic model would be more appropriate than a probabilistic model.

14.3   Explain why a probabilistic model will produce different values of y for the same value of x. Is this also true for a deterministic model? Explain.

14.4   For one day's use of a compact car, a rental company charges a fixed fee of $35.00 plus 30 cents per mile. Let x denote the number of miles traveled on a given day, and let y be the rental cost. Give the model that relates y to x, and state whether it is deterministic or probabilistic.

14.5   The service department of an appliance store charges $50 for a service call plus an hourly rate of $45.00 to repair a dishwasher in the home. Let x denote the number of hours required for such a repair and let y denote its cost. Give the model that relates y to x, and state whether it is deterministic or probabilistic.

14.6   Suppose one assumes a simple linear model to describe the relationship between x and y, and the following sample of four data points is observed.

x	2	3	4	5
y	4	6	12	10

a. Construct a scatter diagram.
b. Explain why the model must be probabilistic.
c. If the model were deterministic, what would have to be the value of y when x is 4?

14.7   Consider the following data points:

x	1	3	4	6	9
y	8	9	5	1	0

a. Make a scatter diagram.
b. Find the least squares line for the data, and draw its graph on the scatter diagram. Does the line appear to provide a good fit to the data points?

14.8   Consider the following five data points:

x	0	2	5	8	10
y	1	3	4	6	10

a. Construct a scatter diagram.
b. Find the least squares line for the data, and draw its graph on the scatter diagram. Does the line appear to provide a good fit to the data points?
c. Calculate the sum of the squares of the errors.
d. Determine the estimated variance of the random error components.
e. Obtain the estimated standard deviation of the model.

14.9   For the data in Exercise 14.7, determine the following:
a. SSE, the sum of the squares of the errors;
b. s^2, the estimated variance of the random error components;
c. s, the estimated standard deviation of the model.

14.10   Five data points are such that $SS(y) = 218$ and $SS(xy) = 68$. The least squares line is fit to these points and its slope is 1.7. Determine SSE and s^2.

14.11   The least squares line is fit to 100 data points for which $\Sigma y^2 = 1,750$, $\Sigma y = 280$, $SS(x) = 310$, and $SS(xy) = -155$. Determine SSE and s^2.

14.12   The Jebsen-Taylor Hand Function Test is used to measure the recovery of coordination after traumatic injury. The following are the times after injury (in weeks) and the scores on one subtest for eight patients with similar medial nerve injuries.

Time after injury	x	3	2	5	6	2	4	10	5
Subtest score	y	6	8	5	3	7	6	3	4

a. Construct a scatter diagram of the data.
b. Obtain the least squares line to use as an estimated model of the relationship between x and y.
c. Determine the estimated variance of the random error components.
d. Determine the estimated standard deviation of the model and interpret its value.

14.13   The average prices (in dollars) per ounce of gold and silver for the years 1986 through 1994 are given below. (*Source:* U.S. Bureau of Mines.)

Year	Gold (y)	Silver (x)
1986	368	5.47
1987	478	7.01
1988	438	6.53
1989	383	5.50
1990	385	4.82
1991	363	4.04
1992	345	3.94
1993	361	4.30
1994	389	5.30

For the above data, $\Sigma x = 46.91$, $\Sigma y = 3{,}510$, $\Sigma x^2 = 253.6095$, $\Sigma y^2 = 1{,}383{,}102$, and $\Sigma xy = 18{,}625.9$.

a. Make a scatter diagram of the data points. Does its appearance suggest a linear relationship between x and y?

b. Obtain the least squares line to model the relationship between the price per ounce of silver and gold.

c. Graph the least squares line on your scatter diagram.

14.14   Can physical exercise extend a person's life? This popular belief was supported in a study by Paffenbarger, Hyde, Wing, and Hsieh. They examined the physical activity and other life-style characteristics of 16,936 Harvard alumni for relationships to lengths of life ("Physical Activity, All-Cause Mortality, and Longevity of College Alumni," *New England Journal of Medicine,* 1986, 314). The table below gives estimates of years of added life gained by men expending 2,000 or more kcal per week on exercise, as compared with those expending less than 500 kcal.

Age at the Start of Followup (x)	Estimated Years of Added Life (y)
37	2.51
42	2.34
47	2.10
52	2.11
57	2.02
62	1.75
67	1.35
72	0.72
77	0.42

For the above data, $\Sigma x = 513$, $\Sigma y = 15.32$, $\Sigma x^2 = 30{,}741$, $\Sigma y^2 = 30.298$, and $\Sigma xy = 797.84$.

a. Construct a scatter diagram of the data points, and observe that there appears to be approximately a linear relationship between x and y.

b. Model the relationship between y and x by obtaining the least squares line, and graph this on the scatter diagram.

c. Determine SSE and s^2.

14.15 For the data given in Exercise 14.13, calculate the estimated standard deviation of the model. What is the largest deviation from the least squares line that you would expect in the average price of gold for a given year?

14.16 Consider the data in Exercise 14.14.

a. Calculate the estimated standard deviation of the model.

b. For each of the given x-values, determine the interval $\hat{y} \pm 2s$ and check if the observed value of y lies within this interval.

14.3 INFERENCES FOR THE SLOPE TO ASSESS THE USEFULNESS OF THE MODEL

Before using an estimated model, it is necessary to check how well it describes the relationship between x and y. There are several ways that this can be tested. The procedure to be described is based on the slope β_1 of the regression line. The assumed model is

$$y = \beta_0 + \beta_1 x + \epsilon.$$

The model implies that knowledge of x contributes information for predicting y if the coefficient of x, β_1, is not 0. If β_1 is 0, then this coefficient of x indicates that knowledge of x is irrelevant for predicting y (in which case we would just use $\bar{y}$, the sample mean of the y-values). Consequently, we can test the usefulness of the model by testing the null hypothesis

$$H_0: \beta_1 = 0.$$

The test statistic is based on the least squares estimate $\hat{\beta}_1$ of the slope β_1. Under the assumptions that have been made concerning the random error components, the sampling distribution of $\hat{\beta}_1$ is normal, with a mean of β_1 and a standard deviation of $\dfrac{\sigma}{\sqrt{SS(x)}}$ (see Figure 14.6). By transforming $\hat{\beta}_1$ to standard units, we obtain the test statistic

$$\frac{\hat{\beta}_1 - \beta_1}{\dfrac{\sigma}{\sqrt{SS(x)}}}.$$

Figure 14.6
Sampling distribution of the estimated slope $\hat{\beta}_1$.

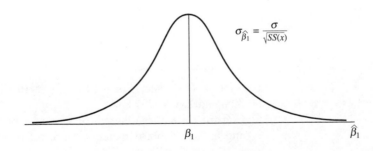

However, σ is unknown and must be replaced by its estimate s, thus resulting in the test summarized below.

PROCEDURE

Testing the Slope for Model Usefulness:

For testing the null hypothesis,

$$H_0: \beta_1 = 0,$$

the test statistic is

(14.7) $$t = \frac{\hat{\beta}_1 - \beta_1}{\dfrac{s}{\sqrt{SS(x)}}} = \frac{\hat{\beta}_1 \sqrt{SS(x)}}{s}.$$

The rejection region is given by the following.

For $H_a: \beta_1 < 0$, H_0 is rejected for $t < -t_\alpha$;

For $H_a: \beta_1 > 0$, H_0 is rejected for $t > t_\alpha$;

For $H_a: \beta_1 \neq 0$, H_0 is rejected for $t < -t_{\alpha/2}$ or $t > t_{\alpha/2}$.

Assumptions:

1. The model is $y = \beta_0 + \beta_1 x + \epsilon$.

2. For each possible setting of x, the random errors ϵ have a normal distribution with mean 0 and variance σ^2. Also, the errors are independent.

Note:

1. The t-distribution is based on $df = n - 2$.

2. $\hat{\beta}_1 = \dfrac{SS(xy)}{SS(x)}$.

3. $s = \sqrt{\dfrac{SSE}{n-2}}$ and $SSE = SS(y) - \hat{\beta}_1 SS(xy)$.

EXAMPLE 14.3 For the placement scores and course grades example, is there sufficient evidence at the 5 percent level to conclude that the assumed linear model is useful and that knowledge of the placement test score x contributes information for predicting the final course grade y?

Solution

The assumed model is $y = \beta_0 + \beta_1 x + \epsilon$, and the estimated model was found to be $\hat{y} = 29.88 + 2.46x$. We also determined that $SS(x) = 290.93$ and $s = 7.12$.

Step 1: Hypotheses.

$$H_0: \beta_1 = 0$$

$$H_a: \beta_1 \neq 0$$

Step 2: Significance level.

$$\alpha = 0.05$$

Step 3: Calculations.

$$t = \frac{\hat{\beta}_1\sqrt{SS(x)}}{s} = \frac{2.46\sqrt{290.93}}{7.12} = 5.89$$

Step 4: Rejection region.

Since H_a is two sided, the rejection region is two tailed. Using Table 4 of Appendix A with $df = n - 2 = 13$, H_0 is rejected for $t < -t_{.025} = -2.16$ and $t > t_{.025} = 2.16$.

Step 5: Conclusion.

Since $t = 5.89$ is in the rejection region, H_0 is rejected and H_a is accepted. Thus, the data indicate that knowledge of the placement test score is useful for predicting the final course grade.

If we conclude, as in the previous example, that the slope of the linear model is not zero, then we may want to estimate it with a confidence interval. The confidence interval formula for β_1 is summarized below.

PROCEDURE

A $(1 - \alpha)$ Confidence Interval for the Slope β_1:

(14.8) $\qquad\qquad\qquad \hat{\beta}_1 \pm t_{\alpha/2}\dfrac{s}{\sqrt{SS(x)}}$

Assumptions:

1. The model is $y = \beta_0 + \beta_1 x + \epsilon$.

2. For each possible setting of x, the random errors ϵ have a normal distribution with mean 0 and variance σ^2. Also, all pairs of errors are independent.

Note:

1. The t-distribution is based on $df = n - 2$.

2. $\hat{\beta}_1 = \dfrac{SS(xy)}{SS(x)}$.

3. $s = \sqrt{\dfrac{SSE}{n-2}}$ and $SSE = SS(y) - \hat{\beta}_1 SS(xy)$.

EXAMPLE 14.4 Obtain a 99 percent confidence interval for the true slope of the regression line for the placement test and course grade example.

Solution

For 99 percent confidence, $t_{\alpha/2} = t_{.005}$. From Table 4 using $df = n - 2 = 13$, this value is found to be 3.012. The 99 percent confidence interval is given by the following.

$$\hat{\beta}_1 \pm t_{\alpha/2} \frac{s}{\sqrt{SS(x)}}$$

$$2.46 \pm (3.012) \frac{7.12}{\sqrt{290.93}}$$

$$2.46 \pm 1.26$$

Thus, with 99 percent confidence we estimate that the true slope of the regression line is between 1.20 and 3.72.

14.4 THE COEFFICIENTS OF CORRELATION AND DETERMINATION TO MEASURE THE USEFULNESS OF THE MODEL

In Chapter 3, the **correlation coefficient** r was considered as a means of evaluating the usefulness of a linear model. It was defined as

(14.9)
$$r = \frac{SS(xy)}{\sqrt{SS(x)SS(y)}}.$$

We saw that r is a measure of the strength of the linear relationship between x and y, and r will have a value near its extremes of -1 or $+1$ when the data points lie near the least squares line (Figure 14.7 A, B). A value of r near 0 indicates the lack of a linear relationship between x and y (see Fig. 14.7C).

The correlation coefficient r between x and y is for the sample of points. The corresponding correlation coefficient for the population from which the points were selected is denoted by the Greek letter ρ (rho). In Chapter 3, we had to leave unanswered the question of how one determines if the sample correlation coefficient r is sufficiently close to -1 or $+1$ to conclude that the linear model is useful. Now that we have a background in hypothesis testing, we could address this issue by testing the following null hypothesis concerning the population's correlation coefficient ρ.

$$H_0: \rho = 0$$

A test statistic can be derived that is based on the sample correlation coefficient r. It is given by

(14.10)
$$t = \frac{r\sqrt{n - 2}}{\sqrt{1 - r^2}}.$$

Figure 14.7
Correlation coefficients for different scatter diagrams. *A*, large positive correlation; *B*, large negative correlation; *C*, lack of correlation.

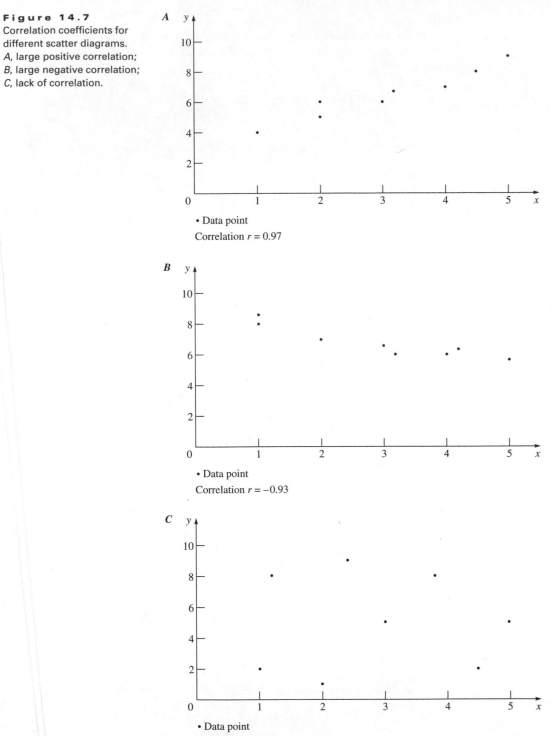

A

• Data point
Correlation $r = 0.97$

B

• Data point
Correlation $r = -0.93$

C

• Data point
Correlation $r = 0.01$

This test is redundant, however, because it is exactly equivalent to the t-test of the preceding section concerning the slope β_1 of the regression line. There we tested

$$H_0: \beta_1 = 0$$

for the placement scores and test grades example. To illustrate the equivalency of the two tests, recall that in Example 14.3 the following value of the test statistic t was obtained.

$$t = \frac{\hat{\beta}_1 \sqrt{SS(x)}}{s} = \frac{2.46\sqrt{290.93}}{7.12} - 5.89$$

We could have found t by performing an equivalent calculation based on the correlation coefficient r. To do this, r must first be determined.

$$r = \frac{SS(xy)}{\sqrt{SS(x)SS(y)}} = \frac{714.47}{\sqrt{(290.93)(2,413.73)}} = 0.853$$

Now using Equation 14.10 to calculate t, we have

$$t = \frac{r\sqrt{n-2}}{\sqrt{1-r^2}} = \frac{0.853\sqrt{15-2}}{\sqrt{1-0.853^2}} - 5.89.$$

Because of the redundancy of this correlation test, we will continue to use the slope test for testing the existence of a linear relationship between x and y.

The Coefficient of Determination

Although the correlation coefficient r is a well-known measure of the strength of the linear relation between x and y, the significance of its value is difficult to interpret. For instance, what does a correlation of $r = 0.8$ signify, and how does it compare to a value of $r = 0.4$? A more meaningful measure of the linear relationship is given by the **coefficient of determination.** This concept can be developed in a manner similar to the partitioning of the total sum of squares that was used in Chapter 13 to introduce analysis of variance.

$SS(y)$ measures the variability in the y-values with respect to their mean $\bar{y}$, since $SS(y)$ equals the sum of the squared deviations of the y-values from $\bar{y}$. In analysis of variance terminology, $SS(y)$ is called the **total sum of squares,** and for our example we found this to be $SS(y) = 2,413.73$.

We have already seen that the **error sum of squares** SSE measures the variability in the y-values with respect to the least squares line, since SSE equals the sum of the squared deviations of the points from the line. For the 15 data points, we obtained a value of $SSE = 659.15$. If knowledge of x could predict y perfectly, then all the data points would lie on the least squares line and SSE would be 0. At the other extreme, if knowledge of x were completely worthless for predicting y, then SSE would equal the total sum of squares $SS(y)$. The difference between these two quantities,

$$SS(y) - SSE,$$

Figure 14.8
Partitioning of the total variation in the y values.

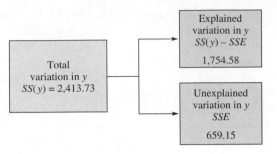

is the reduction in the total sum of squares that can be attributed to the use of x in the model. This reduction is called the **explained variation in y,** that is, the variation in y that is being explained by the model.* For our example,

$$SS(y) - SSE = 2{,}413.73 - 659.15 = 1{,}754.58.$$

This partitioning of the total variation in y is illustrated in Figure 14.8.

If the reduction in the total sum of squares, $SS(y) - SSE$, is divided by the total sum of squares, $SS(y)$, we obtain a ratio that gives the **proportion of the total variation in y that is explained by the model.** This ratio is called the **coefficient of determination.** It is denoted by r^2 because it can be shown to equal the square of the correlation coefficient.

PROCEDURE

The Coefficient of Determination r^2:

$$\textbf{(14.11)} \qquad r^2 = \frac{SS(y) - SSE}{SS(y)} = 1 - \frac{SSE}{SS(y)}$$

Note:

1. r^2 is the proportion of $SS(y)$, the total variation in y, that is explained by the least squares line.

2. r^2 equals the square of the correlation coefficient and, therefore, $0 \leq r^2 \leq 1$.

EXAMPLE 14.5 For the placement test and course grade example, find the coefficient of determination and interpret its value.

Solution

$$r^2 = 1 - \frac{SSE}{SS(y)}$$

$$= 1 - \frac{659.15}{2{,}413.73}$$

$$= 0.73$$

---

* Many statistical computer programs call this **regression sum of squares.**

Thus, the least squares line accounts for 73 percent of the total variation in the y-values.

Note that since the correlation coefficient r had been determined earlier, we could have squared it to obtain the coefficient of determination ($r^2 = 0.853^2 = 0.73$).

At the beginning of the section we posed the question of how a correlation coefficient of $r = 0.8$ compares to a value of $r = 0.4$. We can now address this question by obtaining the corresponding coefficients of determination, which are $0.8^2 = 0.64$ and $0.4^2 = 0.16$. Thus, a least squares line with $r = 0.8$ accounts for 64 percent of the total variation in the y-values, while only 16 percent of the variation in the y-values is explained by a least squares line with $r = 0.4$.

SECTIONS 14.3 and 14.4 EXERCISES

Inferences for the Slope to Assess the Usefulness of the Model; The Coefficients of Correlation and Determination to Measure the Usefulness of the Model

14.17   In Exercise 14.7, a scatter diagram was constructed for the following data. The graph suggested a linear relation between x and y.

x	1	3	4	6	9
y	8	9	5	1	0

a. Measure the strength of the linear relation by calculating the correlation coefficient r.
b. Measure the strength of the linear relation by calculating the coefficient of determination r^2.
c. Interpret the value of r^2 obtained in part b.

14.18   Consider the following five data points:

x	7	2	6	1	5
y	8	2	5	0	4

a. Construct a scatter diagram.
b. Obtain the least squares line, and draw its graph on the scatter diagram.
c. Find the value of r^2.
d. Is there evidence of a linear relationship between x and y? Test at the 0.05 significance level.

14.19   The tread depth (in hundredths of an inch) and the number of miles of usage (in thousands) are given on page 660 for a sample of 10 tires of the same brand. Determine the coefficients of correlation and determination.

Miles (x)	39	37	35	15	25	38	18	60	36	40
Tread (y)	10	15	16	30	21	14	34	3	10	14

14.20   Construct a 90 percent confidence interval to estimate the slope of the assumed model in Exercise 14.18.

14.21   With reference to Exercise 14.19, is there sufficient evidence at the 1 percent level to conclude that x and y are negatively correlated? (Formulate the hypotheses in terms of the slope β_1.)

14.22   For 100 data points, $\Sigma x = 387$, $\Sigma y = 588$, $\Sigma x^2 = 1,981$, $\Sigma y^2 = 5,023$, and $\Sigma xy = 2,530$. Find the coefficient of determination and interpret its value.

14.23   Find the approximate p-value for the test in Exercise 14.21.

14.24   The least squares line was fit to a sample of data points for which $SS(y) = 952$ and $SSE = 207$. Determine r^2 and interpret its value.

14.25   The Jebsen-Taylor Hand Function Test is used to measure the recovery of coordination after traumatic injury. The following are the times after injury (in weeks) and the scores on one subtest for eight patients with similar medial nerve injuries.

Time after injury	x	3	2	5	6	2	4	10	5
Subtest score	y	6	8	5	3	7	6	3	4

For the above data, $\Sigma x = 37$, $\Sigma y = 42$, $\Sigma x^2 = 219$, $\Sigma y^2 = 244$, and $\Sigma xy = 165$. A scatter diagram and the least squares line were obtained in Exercise 14.12, and the scatter diagram suggested a linear relationship between x and y.
   a. Measure the strength of the linear relationship by calculating the correlation coefficient r.
   b. What is signified by the fact that r is negative?
   c. Measure the strength of the linear relation between x and y by calculating the coefficient of determination.
   d. Interpret the value of r^2 in part c.

14.26   With reference to Exercise 14.25, is there sufficient evidence to conclude that knowledge of x contributes information for predicting y? Test at the 5 percent level.

14.27   Construct a 99 percent confidence interval to estimate the slope of the assumed model in Exercise 14.25.

14.28   Obtain the approximate p-value for Exercise 14.26.

14.29   The average prices (in dollars) per ounce of gold and silver for the years 1986 through 1994 are given below. (*Source:* U.S. Bureau of Mines.)

Year	Gold (y)	Silver (x)
1986	368	5.47
1987	478	7.01
1988	438	6.53
1989	383	5.50
1990	385	4.82
1991	363	4.04
1992	345	3.94
1993	361	4.30
1994	389	5.30

For the above data, $\Sigma x = 46.91$, $\Sigma y = 3{,}510$, $\Sigma x^2 = 253.6095$, $\Sigma y^2 = 1{,}383{,}102$, and $\Sigma xy = 18{,}625.9$. A scatter diagram and the least squares line were obtained in Exercise 14.13, and the scatter diagram suggested a linear relationship between x and y.

a. Calculate r^2 and interpret its value.

b. Is there sufficient evidence to conclude that knowledge of the silver price contributes information for predicting the price of gold? Test with $\alpha = 0.01$.

14.30   Construct a 95 percent confidence interval to estimate the slope of the assumed model in Exercise 14.29.

14.31   Find the approximate p-value for the test in Exercise 14.29.

14.32   Many baseball experts consider pitching the key to a successful season. A frequently used measure of a pitching staff's effectiveness is its earned run average (ERA), which a pitcher strives to keep as low as possible. The following table gives the number of wins and the earned run average for each American League team for the 1995 season.

Team	ERA (x)	Wins (y)
Cleveland	3.84	100
Boston	4.40	86
Seattle	4.52	79
New York	4.55	79
California	4.52	78
Texas	4.67	74
Baltimore	4.32	71
Kansas City	4.49	70
Chicago	4.85	68
Oakland	4.97	67
Milwaukee	4.83	65
Detroit	5.50	60
Toronto	4.90	56
Minnesota	5.77	56

For the above data, $\Sigma x = 66.13$, $\Sigma y = 1,009$, $\Sigma x^2 = 315.4055$, $\Sigma y^2 = 74,589$, and $\Sigma xy = 4,702.35$.

a. Construct a scatter diagram of the 14 data points.

b. Model the relationship between the number of wins and the earned run average by obtaining the least squares line.

c. Calculate the coefficient of determination and interpret its value.

d. Is there sufficient evidence at the 1 percent level to conclude that knowledge of a team's ERA contributes information for predicting total wins?

14.33   Construct a 95 percent confidence interval to estimate the slope of the assumed model in Exercise 14.32.

14.34   Obtain the approximate p-value for part d of Exercise 14.32.

14.5   USING THE MODEL TO ESTIMATE AND PREDICT

We began our study of placement test scores x and final statistics grades y by constructing a scatter diagram to determine what type of mathematical relationship might exist between x and y. The graph of the 15 data points suggested that a straight line model might be appropriate. We then assumed the linear model

$$y = \beta_0 + \beta_1 x + \epsilon$$

and we proceeded to obtain an estimated model of this by finding the least squares line

$$\hat{y} = 29.88 + 2.46x.$$

Before this estimated model could be used, we needed to check its usefulness by determining how well it describes the relation between x and y for the sample data. A test of the model's utility was conducted in Section 14.3 by performing a hypothesis test concerning the slope β_1. Further support of the model's usefulness was obtained in the previous section when the coefficient of determination was found to be $r^2 = 0.73$. We saw that this value indicates that 73 percent of the total variation in the y-values is being accounted for by the least squares line. We are now at the stage where we can use the estimated model to make inferences about final course grades. There are two types of inferences for which we want to use the least squares line:

1. **Estimating the mean value of y, μ_y, for a given value of x.**
   For example, we might want to estimate the mean course grade for all students who achieve a score of 20 on the math placement test.

2. **Predicting a single value of y for a given value of x.**

As an illustration of this situation, suppose we have a student who scored 20 on the placement test, and we want to predict his or her final statistics grade.

In each of the two examples above, the least squares line is utilized by substituting $x = 20$ into it and evaluating $\hat{y}$. This gives the value

$$\hat{y} = 29.88 + 2.46x$$
$$= 29.88 + 2.46(20)$$
$$= 79.08$$

The same value of 79.08 is used in both cases to estimate the mean grade of all students who have a test score of 20, as well as to predict what a single student's course grade will be. The two types of problems will differ in regard to their maximum error. In other words, if a 95 percent confidence interval is used to estimate the mean grade, and a 95 percent prediction interval is used to predict the single grade, the two intervals will have different widths. Which interval do you think will be wider? Before answering, keep in mind that the size of the maximum error increases as the width of the interval increases. In Chapter 8, we saw that sample means fluctuate less than individual values and, consequently, the error associated with a mean score will be smaller than the associated error for an individual score. We can estimate more precisely the mean grade of all students with a score of 20 and predict less precisely the grade of a single student who achieves this score. A general confidence interval and a prediction interval for these two situations are given below.

PROCEDURE

A $(1 - \alpha)$ Confidence Interval for Estimating the Mean of y at $x = x_0$:

(14.12)
$$\hat{y} \pm t_{\alpha/2} s \sqrt{\frac{1}{n} + \frac{(x_0 - \bar{x})^2}{SS(x)}}$$

Note:

1. The t-distribution is based on $df = n - 2$.

2. x_0 is the given value of x, and $\hat{y}$ is calculated by substituting x_0 into the least squares equation.

3. n is the number of data points, and $\bar{x}$ is the mean of their x values.

4. $s = \sqrt{\dfrac{SSE}{n-2}}$ where $SSE = SS(y) - \hat{\beta}_1 SS(xy)$.

Assumptions:

These are the same as stated in the box for "Simple Linear Probabilistic Model and Assumptions," page 644.

PROCEDURE

A $(1 - \alpha)$ Prediction Interval for Predicting One y-Value at $x = x_0$:

$$(14.13) \qquad \hat{y} \pm t_{\alpha/2}s \sqrt{1 + \frac{1}{n} + \frac{(x_0 - \bar{x})^2}{SS(x)}}$$

Note:

1. The t-distribution is based on $df = n - 2$.

2. x_0 is the given value of x, and $\hat{y}$ is calculated by substituting x_0 into the least squares equation.

3. n is the number of data points, and $\bar{x}$ is the mean of their x-values.

4. $s = \sqrt{\dfrac{SSE}{n - 2}}$, where $SSE = SS(y) - \hat{\beta}_1 SS(xy)$.

Assumptions:

These are the same as stated in the box for "Simple Linear Probabilistic Model and Assumptions," page 644.

Before illustrating the preceding formulas, we wish to point out that the term **estimate** is used in reference to a **parameter**, while the term **predict** is used when relating to a **random variable.** The value of a parameter is estimated, while the value of a random variable is predicted. Formula 14.12 is used to estimate a mean (parameter) and is called a confidence interval. Formula 14.13 is used to predict a single value of y (random variable) and is referred to as a prediction interval. Notice that the only difference in the formulas is the inclusion of a 1 under the radical in the error bound for the prediction interval.

EXAMPLE 14.6 Obtain a 95 percent confidence interval to estimate the mean statistics grade for all students who achieve a score of 20 on the placement exam.

Solution

The confidence interval for estimating the mean of y at a given value x_0 is

$$\hat{y} \pm t_{\alpha/2}s \sqrt{\frac{1}{n} + \frac{(x_0 - \bar{x})^2}{SS(x)}}.$$

The least squares line was found to be $\hat{y} = 29.88 + 2.46x$. For $x = x_0 = 20$, $\hat{y} = 29.88 + 2.46(20) = 79.08$. For 95 percent confidence, $t_{\alpha/2} = t_{.025} = 2.16$. This is obtained from Table 4 using $df = n - 2 = 13$. Earlier, we obtained $s = 7.12$, $SS(x) = 290.93$, and $\bar{x} = 254/15 = 16.93$.

$$79.08 \pm (2.16)(7.12) \sqrt{\frac{1}{15} + \frac{(20 - 16.93)^2}{290.93}}$$

$$79.08 \pm 4.84$$

Thus, with 95 percent confidence we estimate that the mean course grade for all students who score 20 on the placement test is between 74.24 and 83.92.

EXAMPLE 14.7 Determine a 95 percent prediction interval for the purpose of predicting the final course grade of a student who obtains a placement test score of 20.

Solution

For predicting the course grade of a single student, the prediction interval in Formula 14.13 is used.

$$\hat{y} \pm t_{\alpha/2}s \sqrt{1 + \frac{1}{n} + \frac{(x_0 - \bar{x})^2}{SS(x)}}$$

$$79.08 \pm (2.16)(7.12) \sqrt{1 + \frac{1}{15} + \frac{(20 - 16.93)^2}{290.93}}$$

$$79.08 \pm 16.12$$

Therefore, we predict that the student's course grade will be between 62.96 and 95.20. The large width of this prediction interval renders it of little value from a practical point of view.

In the last example, note that the error bound for the prediction interval is 16.12, while the error bound for the confidence interval obtained in Example 14.6 is only 4.84. For the same amount of confidence and the same x_0, the error bound of the confidence interval for the mean of y will always be smaller than that of the prediction interval for a single y.

By examining Formulas 14.12 and 14.13 used to obtain confidence and prediction intervals, we see that in each case the error bound is smallest when the given value of x_0 is $\bar{x}$. At $x_0 = \bar{x}$ the term $\frac{(x_0 - \bar{x})^2}{SS(x)}$ in the error bound becomes 0. However, this term increases, and thus does the error of estimation and prediction, as x_0 is chosen farther away from $\bar{x}$. For instance, if we had chosen x_0 to be 24 instead of 20, then the error bound would increase from 4.84 to 7.51 for the mean grade and from 16.12 to 17.12 for an individual's grade. For this reason, one should be cautious in using the least squares line to estimate and predict at values of x_0 that are distant from the mean of the sample x values. Another reason for caution is that while a straight line might be a good model near $\bar{x}$, it could be a very poor choice for values of x outside the extremes of the x values for which the model was developed. Outside the extremes of x, curvature might be present in the actual relationship between x and y. For instance, suppose the true (but unknown) model relating y to x was the parabolic curve in Figure 14.9, but only data points with x values between a and b were used to obtain an estimated model. Although a least squares line might suffice for values of x within this interval, the straight line would be a very poor choice for values of x that are distant from the interval.

Figure 14.9
Excessive curvature outside
the range of the data points.

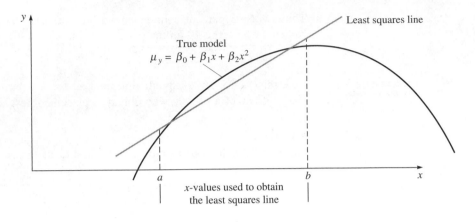

SECTION 14.5 EXERCISES

Using the Model to Estimate and Predict

14.35 Consider the following data points.

x	-2	0	3	4	5
y	0	2	3	3	4

 a. Construct a scatter diagram.
 b. Fit the least squares line to the data.
 c. Calculate the coefficient of determination and interpret its value.
 d. Test at the 0.05 level if knowledge of x contributes information for predicting y.
 e. Obtain a 95 percent confidence interval to estimate the mean of y when x is 3.
 f. Obtain a 95 percent prediction interval for a single value of y when x is 3.
 g. Compare the widths of the intervals obtained in parts e and f.

14.36 A sample of 25 data points resulted in the following values: $\Sigma x = 100$, $\Sigma y = 375$, $\Sigma x^2 = 500$, $\Sigma y^2 = 6{,}125$, and $\Sigma xy = 1{,}700$.
 a. Determine the least squares line.
 b. Calculate the coefficient of determination and interpret its value.
 c. Does knowledge of x contribute information for predicting y? Test at the 0.05 level.
 d. Predict with 90 percent probability the value of y when x is 6.
 e. Estimate with 90 percent confidence the mean value of y when x is 6.

14.37 A least squares line was fit to 20 data points, and the following results were obtained.

Least squares line $\hat{y} = 21.7 + 10.8x$

$$SS(x) = 22.2, \qquad \bar{x} = 20, \qquad SSE = 44.8$$

a. Obtain a 90 percent confidence interval for the average value of y when $x = 25$.
b. If x had been chosen as 21 in part a, would the confidence interval be wider or narrower? Why?
c. What choice of x in part a would result in a confidence interval of smallest width?

14.38 Refer to Exercise 14.37 and obtain a 90 percent prediction interval for y when $x = 25$.

14.39 Exercise 14.32 gave the number of wins and the earned run average for each American League team for the 1995 season. These figures are repeated below.

Team	ERA (x)	Wins (y)
Cleveland	3.84	100
Boston	4.40	86
Seattle	4.52	79
New York	4.55	79
California	4.52	78
Texas	4.67	74
Baltimore	4.32	71
Kansas City	4.49	70
Chicago	4.85	68
Oakland	4.97	67
Milwaukee	4.83	65
Detroit	5.50	60
Toronto	4.90	56
Minnesota	5.77	56

For the above data, $\Sigma x = 66.13$, $\Sigma y = 1,009$, $\Sigma x^2 = 315.4055$, $\Sigma y^2 = 74,589$, and $\Sigma xy = 4,702.35$. From Exercise 14.32, the estimated standard deviation of the model is $s = 6.651$, and the least squares line is

$$\hat{y} = 171.24 - 20.995x$$

Estimate with 95 percent confidence the mean number of wins for teams with an earned run average of 4.67.

14.40 In Exercises 14.13 and 14.29, we gave the following average prices (in dollars) per ounce of gold and silver for the years 1986 through 1994.

Year	Gold (y)	Silver (x)
1986	368	5.47
1987	478	7.01
1988	438	6.53
1989	383	5.50
1990	385	4.82
1991	363	4.04
1992	345	3.94
1993	361	4.30
1994	389	5.30

For the above data, $\Sigma x = 46.91$, $\Sigma y = 3,510$, $\Sigma x^2 = 253.6095$, $\Sigma y^2 = 1,383,102$, and $\Sigma xy = 18,625.9$. The estimated standard deviation of the model is $s = 17.60$, and the least squares line is

$$\hat{y} = 200.499 + 36.357x$$

For a year in which the average cost of silver is $5.00 per ounce, predict with probability 0.95 the cost of an ounce of gold.

14.41   Refer to Exercise 14.39 and obtain a 95 percent prediction interval for the number of wins by a team whose earned run average is 4.67.

14.42   With reference to Exercise 14.40, estimate with 95 percent confidence the mean cost of an ounce of gold for years in which the average cost of silver is $5.00 per ounce.

14.43   The time required for a factory worker to install a certain component in a video camcorder appears to be related to the number of days of experience with this procedure. Installment times (in seconds) and the numbers of days of experience appear below for a sample of 10 workers.

Experience	x	6	8	10	8	1	3	2	5	5	7
Time required	y	32	30	25	28	39	35	40	30	33	38

   a. Construct a scatter diagram of the data points.
   b. Obtain the least squares line to approximate the relationship between y and x.
   c. Calculate the coefficient of determination and interpret its value.
   d. Test at the 5 percent level the utility of the model.
   e. For a randomly selected worker with 5 days experience, predict with probability 0.95 the time required to install the component.

14.44   In Exercise 14.43, estimate with 95 percent confidence the average installation time for all workers with 5 days' experience.

14.6 CONDUCTING A REGRESSION ANALYSIS WITH MINITAB

In this section, we will show how MINITAB can relieve the burden of performing many tedious and laborious calculations associated with the regression analysis of the preceding sections. In addition, the illustrations will serve to review and tie together the various steps involved in the model building process, from the initial formulation of the model through its ultimate use.

In our investigation of the relationship between placement test scores x and final statistics grades y, the first step was to discern what type of mathematical relationship might exist between the two variables. This was done by making a scatter diagram of the 15 data points. To have MINITAB accomplish this, the x- and y-values of the points are first stored in columns C1 and C2. The **PLOT** command is then applied to the two columns. In the **PLOT** command, the column of y-values is specified first, followed by the column that contains the x-values.

```
READ C1 C2
21  69
17  72
21  94
11  61
15  62
19  80
15  65
23  88
13  54
19  75
16  80
25  93
 8  55
14  60
17  64
END
NAME C1 'X' C2 'Y'
GSTD    # SWITCHES MINITAB TO STANDARD GRAPHICS MODE
PLOT C2 C1
```

Menu Commands:
Graph ➤
Plot
In **Graph variables**
Y-box enter **C2**
In **Graph variables**
X-box enter **C1**
Select OK

In response to the above commands, MINITAB will produce the output that appears in Exhibit 14.1.

The scatter diagram in Exhibit 14.1 suggests that test score x and course grade y might be approximately linearly related. Therefore, we tentatively assume the simple linear model

$$y = \beta_0 + \beta_1 x + \epsilon,$$

where ϵ is a random error component.

The second step is to estimate the assumed model by obtaining the least squares

■■■■■ **Exhibit 14.1**

(a) Character scatter diagram.

```
MTB > NAME C1 'X' C2 'Y'    # this command is optional
MTB > GSTD                  # puts minitab in std graphics mode
MTB > PLOT C2 C1            # the location of y values is specified first
```

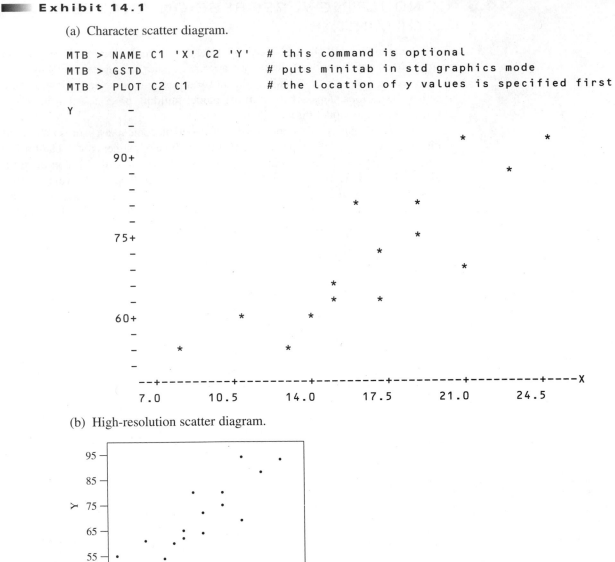

(b) High-resolution scatter diagram.

Menu Commands:
Stat ➤
Regression ➤
Regression
In **Response** *box*
enter **C2**
In **Predictors** *box*
enter **C1**
Select OK

line. To have MINITAB produce the estimated model, type the command **REGRESS** as follows.

```
REGRESS C2 1 C1
```

In the above, the **REGRESS** command is followed by the column of y-values; then the number 1 (the number of predictor variables); and lastly, by the column of x-values. The output produced by the **REGRESS** command appears in Exhibit 14.2. In Exhibit 14.2, the least squares line is labeled as the regression equation and is

$$y = 29.9 + 2.46x.$$

If desired, additional decimal digits in the coefficients can be obtained from the "Coef" column just below the regression equation. If these values are used, then the estimated model is

$$y = 29.882 + 2.4558x.$$

The estimate of σ, the standard deviation of the random error components (standard deviation of the model), is given as $s = 7.121$. The value of s^2 also appears in the analysis of variance table. Recall that $s^2 = SSE/(n - 2)$. From the "SS" (sum of squares) column for the "Error" row, we have $SSE = 659.2$. Next to this in the "MS" (mean square) column appears the mean square for error, which is $s^2 = 50.7$ (note that this equals 7.121^2).

After constructing the scatter diagram (to assist in formulating an assumed model) and obtaining an estimated model (the least squares line), the next step in the model-building process is to assess the usefulness of the model. The coefficient of determination appears in Exhibit 14.2 as "R-sq," and its value is $r^2 = 72.7$ percent. This can also be determined from the analysis of variance table by calculating

$$r^2 = \frac{\text{Regression } SS}{\text{Total } SS} = \frac{1,754.6}{2,413.7} = 0.727.$$

Exhibit 14.2

```
MTB > REGRESS C2 1 C1

The regression equation is
Y = 29.9 + 2.46 X

Predictor       Coef        Stdev      t-ratio         p
Constant      29.882        7.304         4.09     0.001
X             2.4558       0.4175         5.88     0.000

s = 7.121        R-sq = 72.7%      R-sq(adj) = 70.6%

Analysis of Variance

SOURCE        DF          SS          MS          F         p
Regression     1      1754.6      1754.6      34.60     0.000
Error         13       659.2        50.7
Total         14      2413.7
```

Since nearly 73 percent of the variation in the y values is being explained by the model, r^2 suggests that the fit of the least squares line to the data points is quite good. However, we can formally check this by testing the slope β_1 of the assumed model. For testing

$$H_0: \beta_1 = 0$$

$$H_a: \beta_1 \neq 0,$$

the value of the test statistic is given in the "t-ratio" column for the row labeled "X". This is $t = 5.88$. The p-value appears to the right and equals 0.000. Since this is less than 0.05, we would reject H_0 at the 5 percent level and conclude that a straight line model is useful for describing the relationship between placement test score x and final course grade y.*

Concluding that the assumed linear model is useful, we can now utilize the estimated model (the least squares line) to estimate the mean of y or to predict an individual y for a given value of x. By using the subcommand **PREDICT** with the **REGRESS** command, MINITAB will construct a 95 percent confidence interval for μ_y and a 95 percent prediction interval for y. For example, suppose as in Section 14.5 we want to obtain a 95 percent confidence interval for the mean statistics grade of all students who achieve a score of 20 on the placement exam. In addition, we want a 95 percent prediction interval to predict the final course grade of a student who obtains a placement test score of 20. These intervals are obtained by the following commands.

> **Menu Commands:**
> **Stat** ➤
> **Regression** ➤
> **Regression**
> *In* **Response** *box*
> *enter* **C2**
> *In* **Predictors** *box*
> *enter* **C1**
> *Select* Options
> *In* **Pred. intervals**
> **for new obs.** *box*
> *enter* **20**
> *In* **Confidence level**
> *box enter* **95**
> *Select* OK
> *Select* OK

```
REGRESS C2 1 C1;
PREDICT 20.
```

The main command is the same as used earlier, except that it ends with a semicolon since a subcommand will appear on the next line. The subcommand **PREDICT** is followed by the value of x. As many as 10 **PREDICT** subcommands can be used with the same **REGRESS** command. In addition to the output given earlier in Exhibit 14.2, MINITAB will also produce 95 percent confidence and prediction intervals for each of the specified x values. The output for the above commands appears in Exhibit 14.3.

The value of 79.00 in the "Fit" column is obtained by substituting $x = 20$ into the estimated model

$$y = 29.882 + 2.4558x.$$

In Section 14.5, we obtained a value of 79.08 since we used 29.88 and 2.46 for the coefficients. Because of this rounding off, MINITAB's confidence interval (74.16 to 83.84) and prediction interval (62.87 to 95.13) differ slightly from those obtained earlier.

MINITAB has also produced some output in Exhibit 14.3 that pertains to topics that are discussed in Chapter 16.

There are three levels of output that can be obtained with the **REGRESS** command. The maximum amount will be given by typing the following command prior to using the **REGRESS** command.

BRIEF 3

---

* The F-ratio in the analysis of variance table offers an equivalent alternative for testing the usefulness of the model. In Chapter 13, it was pointed out that the square of the value of a t-statistic equals the value of an F-statistic ($5.88^2 = 34.6$). Since $F = 34.6$ has a p-value of 0.000, we would reject H_0 and conclude that the assumed model is useful.

The minimum amount is achieved by using 1 in place of 3 in the command above. The output that appears in our exhibits corresponds to the default level of 2.

Exhibit 14.3

```
MTB > REGRESS C2 1 C1;
SUBC> PREDICT 20.
```

```
The regression equation is
Y = 29.9 + 2.46 X
```

Predictor	Coef	Stdev	t-ratio	p
Constant	29.882	7.304	4.09	0.001
X	2.4558	0.4175	5.88	0.000

s = 7.121   R-sq = 72.7%   R-sq(adj) = 70.6%

Analysis of Variance

SOURCE	DF	SS	MS	F	p
Regression	1	1754.6	1754.6	34.60	0.000
Error	13	659.2	50.7		
Total	14	2413.7			

Fit	Stdev.Fit	95% C.I.	95% P.I.
79.00	2.24	(74.16, 83.84)	(62.87, 95.13)

SECTION 14.6 EXERCISES

Conducting a Regression Analysis with MINITAB

14.45 Exercise 14.14 referred to a study of the effects of physical exercise on the length of a person's life. The table below gives estimates of years of added life gained by men expending 2,000 or more kcal per week on exercise, as compared with those expending less than 500 kcal.

Age at the Start of Followup (x)	Estimated Years of Added Life (y)
37	2.51
42	2.34
47	2.10
52	2.11
57	2.02
62	1.75
67	1.35
72	0.72
77	0.42

MINITAB was used to fit a least squares line to the data, and the computer output appears on the following page.

```
MTB > NAME C1 'X' C2 'Y'
MTB > REGRESS C2 1 C1;
SUBC> PREDICT 50.

The regression equation is
Y = 4.57 - 0.0503 X

Predictor        Coef        Stdev      t-ratio          p
Constant       4.5674       0.3739        12.21      0.000
X            -0.050267     0.006398        -7.86      0.000

s = 0.2478      R-sq = 89.8%      R-sq(adj) = 88.4%

Analysis of Variance

SOURCE          DF           SS            MS          F          p
Regression       1        3.7901        3.7901      61.72      0.000
Error            7        0.4298        0.0614
Total            8        4.2200

    Fit    Stdev.Fit        95% C.I.           95% P.I.
 2.0541       0.0940    ( 1.8318, 2.2763)  ( 1.4272, 2.6809)
```

a. Give the estimated model, that is, the least squares line.
b. Give the coefficient of determination and interpret its value.
c. For testing the utility of the model, give the value of the test statistic and its p-value. Is there sufficient evidence at the 0.05 level to conclude that the model is useful?
d. Give and interpret the 95 percent confidence interval for the mean number of years of added life for men who are 50 at the start of followup.

M▶ 14.46 In Exercise 14.40, we gave the following average prices (in dollars) per ounce of gold and silver for the years 1986 through 1994.

Year	Gold (y)	Silver (x)
1986	368	5.47
1987	478	7.01
1988	438	6.53
1989	383	5.50
1990	385	4.82
1991	363	4.04
1992	345	3.94
1993	361	4.30
1994	389	5.30

MINITAB was used to fit a least squares line to the data, and the computer output appears on the following page. Use this to solve the following.

a. What is the least squares line?
b. What is the coefficient of determination? Do you think this value indicates that the model fits the data well?
c. For testing the utility of the model, give the value of the test statistic and its p-value. Is there sufficient evidence at the 0.05 level to conclude that the model is useful?
d. Find and interpret the 95 percent prediction interval for the price of an ounce of gold for a year in which the average cost of silver is $5.00.
e. Why would the interval obtained in part d probably be of little practical value?

```
MTB > REGRESS C1 1 C2;
SUBC> PREDICT 5.

The regression equation is
Y = 200 + 36.4 X
```

Predictor	Coef	StDev	T	P
Constant	200.50	30.96	6.48	0.000
X	36.357	5.832	6.23	0.000

```
S = 17.60        R-Sq = 84.7%     R-Sq(adj) = 82.6%
```

Analysis of Variance

Source	DF	SS	MS	F	P
Regression	1	12034	12034	38.86	0.000
Error	7	2168	310		
Total	8	14202			

Fit	StDev Fit	95.0% C.I.	95.0% PI
382.28	6.00	(368.10, 396.46)	(338.31, 426.76)

M▶ 14.47 Refer to the MINITAB output in Exercise 14.45.
a. Use the analysis of variance table to determine the coefficient of determination, and compare the result with the value from part b of Exercise 14.45.
b. Find the estimated standard deviation of the model and give an interpretation of its value.
c. Find the value of SSE. What can be said about this value when compared to the values of SSE for other straight lines that could be fit to the data?

M▶ 14.48 Refer to the MINITAB output in Exercise 14.46.
a. What is the estimated standard deviation of the model?
b. Use the analysis of variance table to determine s, and check your result with that obtained in part a.

M▶ 14.49 Refer to the MINITAB output in Exercise 14.46.
a. Obtain the coefficient of determination and interpret its value.
b. Use the analysis of variance table to determine r^2, and compare your result with that obtained in part a.

M▸ 14.50   Exercise 14.32 gave the number of wins and the earned run average for each American League team for the 1995 season. The data are repeated below.

Team	ERA (x)	Wins (y)
Cleveland	3.84	100
Boston	4.40	86
Seattle	4.52	79
New York	4.55	79
California	4.52	78
Texas	4.67	74
Baltimore	4.32	71
Kansas City	4.49	70
Chicago	4.85	68
Oakland	4.97	67
Milwaukee	4.83	65
Detroit	5.50	60
Toronto	4.90	56
Minnesota	5.77	56

Use MINITAB to solve the following.
a. Construct a scatter diagram of the 14 data points.
b. Model the relationship between the number of wins and earned run average by obtaining the least squares line.
c. Find the coefficient of determination and interpret its value.
d. Test at the 5 percent level if there is sufficient evidence to conclude that knowledge of a team's ERA contributes information for predicting total wins.
e. Obtain a 95 percent prediction interval for the number of wins by a team whose earned run average is 4.67.

M▸ 14.51   *Money* magazine (March, 1990) had 50 tax professionals complete a 1040 Federal income tax return for a hypothetical family. The results were very surprising and also somewhat discouraging. Only 2 of the 50 pros came up with the correct amount of tax due. The tax assessments of the 50 preparers ranged from a low of $9,806 to a high of $21,216 (the correct tax was $12,038). In addition to considerable differences in the amounts of tax due, there was also a great deal of variability in the fees charged by the preparers. The fee for each preparer and the amount by which their result differed from the correct tax are given below. All figures are in dollars. (Adapted from Topolnicki, D. M. "The Pros Flub Our Third Annual Tax-Return Test." *Money,* Vol 19, No. 3, March 1990, p. 90.)

Fee	Error	Fee	Error	Fee	Error	Fee	Error
990	2,232	300	53	1,015	1,484	750	4,228
795	2,019	450	189	600	1,659	2,500	4,307
750	1,665	950	256	1,100	1,831	276	5,064
1,950	1,400	800	457	422	1,852	900	5,188
400	1,345	1,450	494	550	2,051	280	5,454
960	1,111	1,425	609	1,200	2,074	4,000	5,473
1,300	875	650	614	2,000	2,098	750	5,518
850	651	770	737	1,500	2,198	970	5,973
1,450	618	750	808	1,100	2,788	1,150	8,103
1,360	215	975	912	1,100	2,867	400	8,288
1,285	0	720	1,124	1,750	3,079	520	9,178
750	0	960	1,300	1,350	3,240		
271	54	450	1,407	640	3,338		

One might expect that the performance of a tax preparer as measured by the accuracy of the return is related to the fee charged. To check on this, use MINITAB to

a. Construct a scatter diagram of the preparation fees (x) and the errors (y).

b. Does the appearance of the scatter diagram suggest that fee and error size are correlated? Support your opinion by using MINITAB to obtain the coefficients of correlation and determination.

c. Formally check your opinion in part (b) by testing at the 5 percent level if a linear relationship exists between the fee charged and the size of the error.

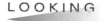

LOOKING BACK

This chapter is concerned with investigating the relationship between two variables x and y. The ultimate goal is to obtain a mathematical model (equation) that will use knowledge of x to make inferences and predictions concerning y. To achieve this, a sample of n pairs of x and y values is obtained, and then we proceed systematically as follows to fit a model to the data.

First, a **scatter diagram** is constructed to discern what type of relationship might exist between x and y. We have only considered relationships that can be modeled by a straight line. If the scatter diagram indicates that a more complicated model is required, then one could consider a multiple regression analysis, which is an extension of the methodology of this chapter.

If the scatter diagram suggests that the data points tend to lie near a straight line, then we assume that x and y are related by the **simple linear model**

$$y = \beta_0 + \beta_1 x + \epsilon.$$

We then proceed to obtain an estimated model of this by finding the **least squares line.**

PROCEDURE

$$\hat{y} = \hat{\beta}_0 + \hat{\beta}_1 x$$

where

$$\hat{\beta}_1 = \frac{SS(xy)}{SS(x)}, \quad \text{and} \quad \hat{\beta}_0 = \bar{y} - \hat{\beta}_1 \bar{x}.$$

The next step is to check how well the estimated model describes the relationship between x and y. The **coefficient of determination,** r^2, gives an informal indication of this, since it is the proportion of the total variation in y that is explained by the least squares line. This is given by the following.

PROCEDURE

$$r^2 = \frac{SS(y) - SSE}{SS(y)} = 1 - \frac{SSE}{SS(y)}$$

A formal test of the usefulness of the model can be accomplished by performing a hypothesis test concerning the slope of the line. To test H_0: $\beta_1 = 0$, the test statistic is

PROCEDURE

$$t = \frac{\hat{\beta}_1 \sqrt{SS(x)}}{s},$$

where $s = \sqrt{\dfrac{SSE}{n - 2}}$, $SSE = SS(y) - \hat{\beta}_1 SS(xy)$, and the t-distribution has $df = n - 2$.

If H_0 is rejected, then we conclude that the linear model is useful for describing the relation between x and y.

After concluding that the model is useful, it can then be utilized to do the following.

1. Estimate the mean of y, μ_y, for a given value x_0 of x.

PROCEDURE

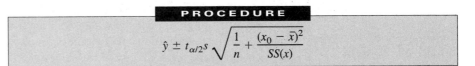

$$\hat{y} \pm t_{\alpha/2} s \sqrt{\frac{1}{n} + \frac{(x_0 - \bar{x})^2}{SS(x)}}$$

2. Predict a single value of y for a given value x_0 of x.

PROCEDURE

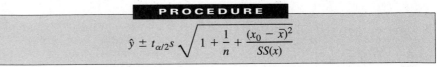

$$\hat{y} \pm t_{\alpha/2} s \sqrt{1 + \frac{1}{n} + \frac{(x_0 - \bar{x})^2}{SS(x)}}$$

Key Words

In reviewing this chapter, you should be able to define, explain, and illustrate each of the following.

simple linear model *(page 640)*

regression line *(page 641)*

scatter diagram *(page 641)*

random error component (ϵ) *(page 642)*

deterministic component *(page 643)*

probabilistic model *(page 643)*

least squares line *(page 645)*

error (residual) *(page 646)*

error sum of squares (*SSE*) *(page 646)*

standard deviation of the model (σ) *(page 648)*

estimated standard deviation of the model (*s*) *(page 648)*

correlation coefficient (*r*) *(page 655)*

coefficient of determination (r^2) *(page 657)*

explained variation in *y* *(page 658)*

confidence interval for the mean of *y* *(page 663)*

prediction interval for an individual *y* *(page 664)*

M▶ MINITAB Commands

READ _ _ *(page 669)*

END *(page 669)*

GSTD *(page 669)*
PLOT _ _ *(page 669)*

NAME _ _ *(page 669)*

BRIEF _ *(page 672)*

REGRESS _ _ _ *(page 671)*

REGRESS _ _ _; *(page 672)*
PREDICT _.

REVIEW EXERCISES

14.52   The following data are given.

x	10	13	8	15	5
y	14	10	15	4	20

  a. Construct a scatter diagram.
  b. Obtain the least squares line.
  c. Calculate the coefficients of correlation and determination.
  d. Does knowledge of *x* contribute information for the prediction of *y*? Test at the 5 percent level.
  e. Determine the estimated standard deviation of the model.
  f. Estimate the slope of the model with a 90 percent confidence interval.
  g. Construct a 95 percent confidence interval to estimate the mean of *y* when *x* is 11.

(Continued on next page)

    h. Predict the value of y when x is 11 by constructing a 95 percent prediction interval.

14.53   A local automobile dealership pays each salesperson a weekly salary of \$135 plus \$75 for each car that he/she sells that week. Let x denote the number of cars sold in a week, and let y denote the salesperson's salary for that week.
a. What is the model that gives the relation between y and x?
b. Is the model deterministic or probabilistic? Explain.

14.54   The least squares line was fit to 40 points, and the following results were obtained: $\Sigma y = 680$, $\Sigma y^2 = 12{,}560$, and $r^2 = 0.76$. Determine:
a. the error sum of squares,
b. the estimated variance of the random error components,
c. the estimated standard deviation of the model.

14.55   The following figures give the daily attendance and the number of hot dog sales for a sample of 10 games of a minor league baseball team.

Attendance (x)	Hot Dog Sales (y)
8,747	6,845
5,857	4,168
8,360	5,348
6,945	5,687
8,688	6,007
4,534	3,216
7,450	5,018
5,874	4,652
9,821	7,002
5,873	3,897

For the data, $\Sigma x = 72{,}149$, $\Sigma y = 51{,}840$, $SS(x) = 25{,}378{,}309$, $SS(y) = 13{,}893{,}508$, and $SS(xy) = 17{,}604{,}101$.
a. Construct a scatter diagram.
b. Obtain the least squares line to model the relationship between hot dog sales and attendance.
c. Measure the strength of the linear relationship by finding the coefficient of determination, and interpret its value.

14.56   For Exercise 14.55, test the utility of the model at the 5 percent significance level, and determine the p-value for the test.

14.57   Refer to Exercise 14.55 and obtain a 95 percent confidence interval to estimate mean hot dog sales for days when the attendance is 6,000.

14.58   Refer to Exercise 14.55 and predict with probability 0.95 hot dog sales for a day when the attendance is 6,000.

14.59   The following data are the heights (inches) and weights (pounds) of a sample of 12 members of a high school football team.

Player	Height (x)	Weight (y)
1	62	135
2	68	182
3	69	168
4	73	198
5	70	174
6	68	159
7	75	221
8	72	197
9	71	182
10	70	170
11	66	154
12	77	234

For the above data, $\Sigma x = 841$, $\Sigma y = 2{,}174$, $\Sigma x^2 = 59{,}117$ $\Sigma y^2 = 402{,}480$, and $\Sigma xy = 153{,}547$.

a. Construct a scatter diagram of the data points.

b. Model the relationship between weight and height by obtaining the least squares line.

c. Calculate the coefficient of determination and interpret its value.

d. Test at the 5 percent level the utility of the model.

e. For a randomly selected player with a height of 70 inches, predict his weight with a 95 percent prediction interval.

14.60 In Wall Street terminology, the "January effect" refers to a historic pattern in which low-priced stocks such as those traded on the NASDAQ exchange often outperform the Dow Jones Industrial Average during January (*The Wall Street Journal,* December 4, 1989). The table below gives the percentage change during January for the DJIA and the NASDAQ composite during the years 1980 to 1989.

Year	DJIA % Change (x)	NASDAQ % Change (y)
80	4.4	7.0
81	−1.7	−2.2
82	−0.4	−3.8
83	2.8	6.9
84	3.0	−3.6
85	6.2	12.7
86	1.6	3.3
87	13.8	12.4
88	1.0	4.3
89	7.2	4.7

a. Construct a scatter diagram of the data points.

b. Obtain the least squares line to model the relationship between the January percentage changes in the DJIA and the NASDAQ composite.

(Continued on next page)

c. Calculate the correlation coefficient for x and y.

d. Test at the 0.05 level the usefulness of the model.

e. What proportion of the total variation in the y-values is being explained by the model?

14.61 Refer to Exercise 14.60 and estimate with 95 percent confidence the average percentage change in the NASDAQ composite for Januarys in which the DJIA increases by 2 percent.

14.62 Refer to Exercise 14.60 and predict with 95 percent probability the percentage change in the NASDAQ composite for a January in which the DJIA increases by 2 percent.

MINITAB Assignments

M▶ 14.63 Exercise 14.32 gave the number of wins and the earned run average for each of the 14 American League teams during the 1995 season. The data suggest that the number of games won by a major league team is related to the team's earned run average. Is there a relationship between the number of wins and a team's batting average? These figures are given below for the American League teams during the 1995 season.

Team	Batting Average (x)	Wins (y)
Cleveland	.291	100
Boston	.280	86
Seattle	.275	79
New York	.276	79
California	.277	78
Texas	.265	74
Baltimore	.262	71
Kansas City	.260	70
Chicago	.280	68
Oakland	.264	67
Milwaukee	.266	65
Detroit	.247	60
Toronto	.260	56
Minnesota	.279	56

Use MINITAB to solve the following.

a. Construct a scatter diagram.

b. Fit a least squares line to the 14 points.

c. Find the coefficient of determination. Do you think its value indicates that the model is useful?

d. Test the utility of the model at the 5 percent significance level.

M▶ 14.64 The owner of a very expensive gift shop suspects that her weekly sales are related to the performance of the stock market for that week. To explore this possibility, she determines the amount of sales and the mean Dow Jones

Industrial Average for each of 30 randomly selected weeks. These figures are given below, with the sales figures expressed in units of $1,000. Refer to the MINITAB output to solve the following.

Week	DJIA (x)	Sales (y)	Week	DJIA (x)	Sales (y)
1	7015	58.3	16	6679	52.7
2	7318	62.9	17	6513	39.3
3	6581	46.3	18	6922	58.7
4	6623	48.2	19	7146	60.9
5	6917	58.2	20	6429	40.5
6	7503	65.8	21	7409	70.3
7	6223	36.7	22	6315	39.1
8	7332	62.3	23	6487	45.9
9	7197	59.8	24	6715	55.1
10	7513	71.8	25	7543	70.2
11	7703	66.9	26	7619	70.3
12	6523	37.9	27	6125	39.1
13	6332	30.6	28	6987	45.9
14	7197	69.8	29	7215	55.1
15	7713	65.2	30	7707	70.2

```
MTB > NAME C1 'X' C2 'Y'
MTB > REGRESS C2 1 C1;
SUBC> PREDICT 7600.

The regression equation is
Y = - 108 + 0.0233 X

Predictor        Coef        StDev           T          P
Constant       -107.71        12.57       -8.57      0.000
X             0.023318     0.001795       12.99      0.000

S = 4.771        R-Sq = 85.8%      R-Sq(adj) = 85.3%

Analysis of Variance
Source          DF          SS            MS          F          P
Regression       1        3842.1        3842.1      168.77     0.000
Error           28         637.4          22.8
Total           29        4479.5

     Fit  StDev Fit         95.0% C.I.                95.0% PI
  69.512      1.409     ( 66.626,  72.398)   (  59.319,  79.705)
```

a. What is the estimated model for relating sales to the DJIA?

b. Give the coefficient of determination and interpret its value in the context of this problem.

M▶ 14.65 Refer to Exercise 14.64 and test at the 1 percent level if the model is useful. Give the p-value for the test.

M▶ 14.66 With reference to Exercise 14.64, give and interpret the 95 percent confidence interval for the mean amount of sales for weeks when the average DJIA is 7600.

M▶ 14.67 The following are the median prices of existing one-family homes in 45 metropolitan areas for the years 1990 and 1994. The figures are in units of $1,000.

Metropolitan Area	1990 (x)	1994 (y)
Albany-Schenectady-Troy, NY	106.9	112.0
Anaheim-Santa Ana, CA	242.4	211.0
Baltimore, MD	105.9	115.4
Birmingham, AL	80.8	100.2
Buffalo-Niagara Falls, NY	77.2	82.3
Chicago, IL	116.8	144.1
Cincinnati, OH-KY-IN	79.8	96.5
Cleveland, OH	80.6	98.5
Columbus, OH	81.6	94.8
Dallas, TX	89.5	95.0
Denver, CO	86.4	116.8
Detroit, MI	76.7	87.0
Ft. Worth-Arlington, TX	76.7	82.5
Hartford, CT	157.3	133.4
Honolulu, HI	352.0	360.0
Houston, TX	70.7	80.5
Indianapolis, IN	74.8	90.7
Jacksonville, FL	72.4	81.9
Kansas City, MO-KS	74.1	87.1
Las Vegas, NV	93.0	110.5
Louisville, KY-IN	60.8	80.5
Memphis, TN-AR-MS	78.1	86.3
Miami-Hialeah, FL	89.3	103.2
Milwaukee, WI	84.4	109.0
Minneapolis-St. Paul, MN-WI	88.7	101.5
Nashville, TN	81.8	96.5
New Orleans, LA	67.8	76.9
New York-N. New Jersey-Long Island, NY-	174.9	173.2
Oklahoma City, OK	53.2	66.7
Orlando, FL	82.8	90.7
Philadelphia, PA-NJ	108.7	119.5
Phoenix, AZ	84.0	91.4
Pittsburgh, PA	70.1	80.7
Portland, OR	79.5	116.9
Richmond-Petersburg, VA	87.5	95.4
Rochester, NY	89.8	85.6

(*Continued on next page*)

(*Continued*)		
St. Louis, MO-IL	76.7	85.0
Sacramento, CA	137.5	124.5
Salt Lake City-Ogden, UT	69.4	98.0
San Antonio, TX	63.6	78.2
San Diego, CA	183.2	176.0
San Francisco, CA	259.3	255.6
Seattle-Tacoma, WA	142.0	155.9
Tampa-St. Petersburg-Clearwater, FL	71.4	76.2
Washington, DC-MD-VA	150.5	157.9

Source: *Statistical Abstract of the United States,* 1995.

Use MINITAB to obtain an estimated model for relating 1994 prices to 1990 prices, and test the utility of the model at the 5 percent significance level.

CHAPTER 15

NONPARAMETRIC TESTS

© George Semple

MONEY *magazine (March, 1990) had 50 tax professionals complete a 1040 Federal income tax return for a hypothetical family. The results were very surprising and somewhat discouraging. Only 2 of 50 pros came up with the correct amount of tax due. The tax assessments of the 50 preparers ranged from a low of $9,806 to a high of $21,216. The correct tax was $12,038. In addition to considerable differences in the amounts of tax due, there was also a great deal of variability in the fees charged by the preparers. The smallest fee was $271, while the largest charge was a whopping $4,000.*

The fee for each preparer and the amount by which their result differed from the correct tax are given in Exercise 15.94. You will be asked to use the method presented in Section 15.7 to assess if the performance of a tax preparer, as measured by the accuracy of the return, is related to the fee charged.

LOOKING
▶
A H E A D

Most of the statistical tests in the previous chapters are based on the assumption that the sampled populations have normal distributions. Some have additional requirements pertaining to the variances of the populations. For instance, the pooled two-sample t-test for comparing two population means is quite restrictive, since it assumes that each population has a normal distribution and that the variances of the two populations are equal. In situations where little is known about the nature of a population, a researcher may be reluctant to use a test that is based on such assumptions. As an alternative, one may prefer to employ a procedure from an area of methodology that is referred to as **nonparametric statistics**. These techniques usually require few assumptions about the sampled populations. Moreover, compared to their parametric counterparts from the previous chapters, they involve simpler and fewer calculations. In fact, in many nonparametric tests the sample values are replaced by their relative ranks, which are generally whole numbers, and may thus be easier to work with than the original measurements. Some nonparametric tests, such as the sign test discussed in Sections 15.1 and 15.2, are even simpler and require little in the way of calculations.

Recall that the **power of a test** pertains to its ability to correctly reject a false null hypothesis (correctly accept a true alternative hypothesis). The major drawback to the use of nonparametric tests is that they are less powerful than their parametric counterparts in situations for which the assumptions of the latter are true. Loosely speaking, a nonparametric test requires more evidence to "prove" the alternative hypothesis. However, the difference in power is often small. Furthermore, when the assumptions of a parametric test are not satisfied, a nonparametric test may actually be more powerful.

In this chapter, we will discuss nonparametric tests that can be used in place of the one-sample test for a population mean, the independent samples test for two means, the paired-sample test for two means, and the analysis of variance for a completely randomized design and for a randomized block design. In addition, we will consider a nonparametric correlation coefficient that offers an alternative to Pearson's product-moment correlation coefficient. We will also discuss a test for checking whether a sequence of observations is random.

15.1 THE ONE-SAMPLE SIGN TEST

The sign test for a single sample is a nonparametric alternative to the one-sample z- and t-tests that were discussed in Sections 10.2 and 10.3. The sign test does not require that the sampled population have a normal distribution. Its derivation only assumes that the sample is random and that the population has a continuous distribution (in practice, it is often applied to discrete distributions as well). While the one-sample z- and t-tests are used to test a hypothesis concerning a population's mean, the one-sample sign test applies to its median.* The basis for the test is the binomial distribution, and it is one of the easiest nonparametric tests to apply. The one-sample sign test is illustrated in the following example.

EXAMPLE 15.1

For the purpose of conducting future experiments, a psychologist is designing a maze for laboratory mice. The contention is made that the maze can be traversed in an average time of less than 8 seconds. To test this belief, 11 test runs are conducted, and the following times in seconds are recorded.

7.8	7.7	7.8	8.1	7.6	7.8	7.9	8.2	7.6	8.0	7.7

Is there sufficient evidence at the 0.05 level to support the hypothesis that the median time required to traverse the maze is less than 8 seconds?

Solution

The median of a population will be denoted by $\tilde{\mu}$ (read "mu tilde"). For this problem, $\tilde{\mu}$ represents the true median time required to complete the maze.

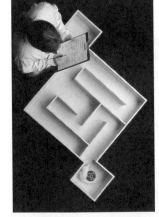

David M. Dennis/Tom Stack & Associates

Step 1: Hypotheses.

$$H_0: \tilde{\mu} = 8$$

$$H_a: \tilde{\mu} < 8$$

Step 2: Significance level.

$$\alpha = 0.05$$

Step 3: Calculations.

The test statistic is based on x, the number of values in the sample that fall above the assumed median value of 8 in the null hypothesis. To determine x, each sample value is replaced by a plus sign $(+)$ if it exceeds 8 and by a minus sign $(-)$ if it is below 8. If a value equals 8, it is discarded and the sample size n is reduced by 1. The sample of 11 measurements is thus replaced by the following signs (we have used * to indicate that the measurement has been deleted).

---

* If the population has a symmetric distribution, then the mean and the median are equal. In this case, the hypotheses in the sign test can be stated in terms of the population mean.

Sample value:	7.8	7.7	7.8	8.1	7.6	7.8	7.9	8.2	7.6	8.0	7.7
Sign replacement:	−	−	−	+	−	−	−	+	−	*	−

The number of plus signs is $x = 2$, and the sample size is reduced to $n = 10$ since there is 1 value that equals 8.

Step 4: Rejection region.

Under the assumption that H_0 is true, the random variable x has a binomial distribution with $p = 0.5$ and $n = 10$. Since H_a involves the $<$ relation, an unusually small value of x will provide evidence against H_0 and in support of H_a. Therefore, the rejection region is left tailed, and it is determined from the table of binomial probabilities (Table 1, Appendix A). The rejection region is shown in Figure 15.1, and it consists of the values of x in the left tail for which the cumulative probability is as close as possible to $\alpha = 0.05$. From Table 1, using $n = 10$ and $p = 0.5$, the values $x \leq 2$ have a total probability of 0.055.

From Figure 15.1, we should reject H_0 for $x \leq 2$ since this will result in a value of $\alpha = 0.055$. This value is very close to the desired value of 0.05, and it is preferred to a rejection region of $x \leq 1$ for which α would be only 0.011. Thus, we are actually conducting the test at the $\alpha = 0.055$ significance level.

Step 5: Conclusion.

Since the value of the test statistic in Step 3 is $x = 2$, and this is in the rejection region, H_0 is rejected. Therefore, there is sufficient evidence at the $\alpha = 0.055$ significance level to conclude that the median time required to run the maze is less than 8 seconds.

For sample sizes larger than 20, Table 1 cannot be used to determine the rejection region. However, for n this large, the normal approximation to the binomial distribution can be used to produce the following test statistic.

$$z = \frac{x - \mu}{\sigma} = \frac{x - np}{\sqrt{npq}} = \frac{x - n(0.5)}{\sqrt{n(0.5)(0.5)}} = \frac{x - 0.5n}{0.5\sqrt{n}}$$

The large-sample test is summarized on the following page.

Figure 15.1
Rejection region for Example 15.1.

Probability 0.055

PROCEDURE

Large-Sample Sign Test for a Population Median $\tilde{\mu}$:

For testing the null hypothesis H_0: $\tilde{\mu} = \tilde{\mu}_0$ ($\tilde{\mu}_0$ is a constant), the test statistic is

(15.1)
$$z = \frac{x - 0.5n}{0.5\sqrt{n}}.$$

The rejection region is given by the following.

For H_a: $\tilde{\mu} < \tilde{\mu}_0$, H_0 is rejected when $z < -z_\alpha$;

For H_a: $\tilde{\mu} > \tilde{\mu}_0$, H_0 is rejected when $z > z_\alpha$;

For H_a: $\tilde{\mu} \neq \tilde{\mu}_0$, H_0 is rejected when $z < -z_{\alpha/2}$ or $z > z_{\alpha/2}$.

Assumption: The sample is random from a continuous population.

Note:

1. x is the number of sample observations that exceed $\tilde{\mu}_0$ (x is the number of + signs).

2. Sample values equal to $\tilde{\mu}_0$ are discarded, and n is the number of values that are retained.

EXAMPLE 15.2

In Example 10.6, we were given a sample of 40 dining charges per person at an expensive restaurant. These amounts (in dollars) are repeated below.

38.00	36.25	37.75	38.62	38.50	40.50	41.25	43.75	44.50	42.50
43.00	42.25	42.75	42.88	44.25	45.75	45.50	35.75	35.12	39.95
35.88	35.50	36.75	35.50	36.25	36.62	37.25	37.75	37.62	36.80
38.50	38.25	37.88	38.88	38.25	46.00	46.50	45.25	34.60	39.50

Is there sufficient evidence at the 1 percent level to conclude that the median cost per person exceeds $38.00?

Solution
Step 1: Hypotheses.

$$H_0: \tilde{\mu} = 38.00$$

$$H_a: \tilde{\mu} > 38.00$$

Step 2: Significance level.

$$\alpha = 0.01$$

Step 3: Calculations.

Each sample observation is replaced by + if it is greater than 38 and by − if it falls below 38.

*	−	−	+	+	+	+	+	+	+
+	+	+	+	+	+	+	−	−	+
−	−	−	−	−	−	−	−	−	−
+	+	−	+	+	+	+	+	−	+

The number of plus signs is $x = 23$, and $n = 39$ since the first sample value equals 38 and is discarded.

$$z = \frac{x - 0.5n}{0.5\sqrt{n}} = \frac{23 - 0.5(39)}{0.5\sqrt{39}} = 1.12$$

Step 4: Rejection region.

The rejection region is right tailed since H_a involves the $>$ relation. From the z-table (Table 3, Appendix A), H_0 is rejected for $z > z_{.01} = 2.326$.

Step 5: Conclusion.

$z = 1.12$ is not in the rejection region, and H_0 is not rejected. Therefore, there is insufficient evidence at the 0.01 level to conclude that the median cost exceeds $38.00 per person.

It is interesting to note that the use of the sign test in the previous example did not result in the rejection of H_0. However, when the large-sample z-test was applied in Example 10.6 to the data, there was sufficient evidence to reject the null hypothesis. This reinforces the statement made earlier that nonparametric tests usually require more evidence than their parametric counterparts to reject H_0. In this example, the sign test only considers whether a measurement falls above or below $38.00 and it ignores the magnitude of the difference. Thus, the sign test is not fully using all the relevant information contained in the sample. In a sense, this is the "price" that we are paying for the simplicity and fewer assumptions associated with this nonparametric test.

15.2 THE TWO-SAMPLE SIGN TEST: PAIRED SAMPLES

The sign test can also be applied as a nonparametric alternative to the paired t- and z-tests that were considered in Section 11.4. Although the tested hypotheses do not involve means, the sign test can be used to determine if one population tends to yield larger or smaller values than the other population. When the sign test is applied to data consisting of paired samples, the difference is calculated for each pair, and this is replaced by either a $+$ or $-$ sign, depending on whether the difference is positive or negative. If the difference equals zero, then that pair is discarded and n, the number of pairs, is reduced by one. The test procedure is then the same as in the previous section, except that it now tests the null hypothesis that the median of the population of differences is zero.

EXAMPLE 15.3 In Example 11.11, 10 infants were involved in a study to compare the effectiveness of 2 medications for the treatment of diaper rash. For each baby, 2 areas of approximately the same size and rash severity were selected, and 1 area was treated with medication A and the other with B. The number of hours until the rash disappeared was recorded for each medication and each infant. The results are repeated below.

Infant	Medication A	Medication B	Difference $D = A - B$
1	46	43	3
2	50	49	1
3	46	48	−2
4	51	47	4
5	43	40	3
6	45	40	5
7	47	47	0
8	48	44	4
9	46	41	5
10	48	45	3

Is there sufficient evidence at the 5 percent level to conclude that a difference exists in the effectiveness of the 2 medications, that is to say, that one of the medications tends to heal more quickly than the other?

Solution

The difference for each of the 10 pairs has been calculated and listed above. We will let $\tilde{\mu}_D$ denote the median of the hypothetical population of all possible differences. A value of $\tilde{\mu}_D$ that is different from 0 is supportive of the statement that 1 of the medications tends to heal more quickly than the other.

Step 1: Hypotheses.

$$H_0: \tilde{\mu}_D = 0$$

$$H_a: \tilde{\mu}_D \neq 0$$

Step 2: Significance level.

$$\alpha = 0.05$$

Step 3: Calculations.

The differences for the 10 pairs are replaced by the following signs:

+	+	−	+	+	+	*	+	+	+

The value of the test statistic is $x = 8$, since there are 8 plus signs. The number of pairs is reduced to $n = 9$, because the difference for pair 7 is 0, and thus is discarded.

Step 4: Rejection region.

Since the alternative hypothesis is 2 sided, the rejection region is 2 tailed. H_0 is rejected for values of x that are unusually small or large. The rejection region is

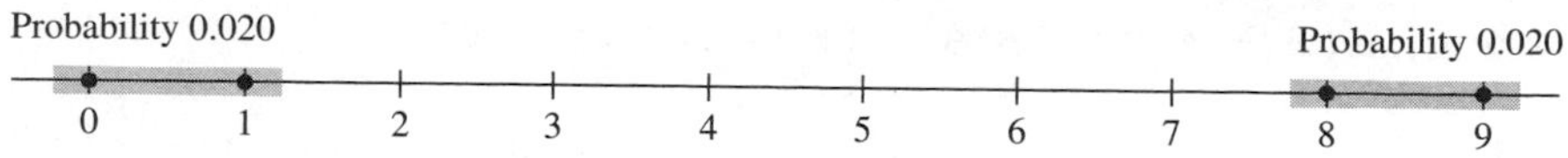

Figure 15.2
Rejection region for Example
15.3.

Probability 0.020 ... Probability 0.020

0 1 2 3 4 5 6 7 8 9

shown in Figure 15.2, and is obtained from Table 1 using $n = 9$ and $p = 0.5$. It consists of those values of x in the 2 tails for which the cumulative probability is as close as possible to $\alpha = 0.05$. From Figure 15.2, by choosing 0, 1, 8, and 9 for the rejection region, α will equal 0.04, and the test will be conducted at this significance level.

Step 5: Conclusion.

In Step 3, the value of the test statistic is $x = 8$, and this lies in the rejection region. H_0 is thus rejected, and we conclude with $\alpha = 0.04$ that there is a difference in the healing effectiveness of the 2 medications. From an inspection of the data, it appears that medication B tends to heal more quickly.

MINITAB EXAMPLE 15.4

Use MINITAB to perform the sign test for the data in Example 15.3.

Solution
The two samples are stored in columns C1 and C2, and the **LET** command is used to determine the column of differences.

```
READ C1 C2
46 43
50 49
46 48
51 47
43 40
45 40
47 47
48 44
46 41
48 45
END
LET C3 = C1 - C2
```

Menu Commands:
Stat ➤
Nonparametrics ➤
1-Sample Sign
In **Variables** *box*
enter **C3**
Select **Test mean**
button and enter **0**
In **Alternative** *box*
select **not equal**
Select OK

We now apply the **STEST** command to the differences that are stored in column C3.

```
STEST 0 C3;      # 0 is the value specified in Ho
ALTERNATIVE 0.   # 0 is the code for not equal in Ha
```

The output generated from the above commands is shown in Exhibit 15.1.

The p-value in Exhibit 15.1 is 0.0391. Since this is less than the specified significance value of 0.05, the null hypothesis would be rejected and, thus, the alternative hypothesis accepted.

Exhibit 15.1

```
MTB > LET C3 = C1 - C2
MTB > STEST 0 C3;      # 0 is the value specified in Ho
SUBC> ALTERNATIVE 0.  # 0 is the code for not equal in Ha

SIGN TEST OF MEDIAN = 0.00000 VERSUS not = 0.00000

          N   BELOW  EQUAL  ABOVE        P     MEDIAN
C3       10     1      1      8     0.0391     3.000
```

SECTIONS 15.1 AND 15.2 EXERCISES

The One-Sample Sign Test; The Two-Sample Sign Test: Paired Samples

15.1   A products testing laboratory wanted to compare the tread life of two brands of tires for light trucks. The rear wheels for each of 11 trucks were randomly assigned 1 tire of each brand. The tires were driven until their treadwear indicators showed, and the numbers of miles accumulated are displayed below.

Truck	Brand 1	Brand 2
1	37,500	37,500
2	53,900	52,200
3	44,900	45,300
4	27,400	26,300
5	32,900	30,600
6	58,800	56,800
7	19,600	18,900
8	31,500	30,700
9	38,600	36,600
10	42,600	42,000
11	45,700	44,500

Use the sign test to determine if the data provide sufficient evidence at the 5 percent level to conclude that the two brands of tires differ in their average length of life.

15.2   A random sample of size $n = 15$ produced the following values.

352	354	365	358	338	358	364	341
369	353	363	360	342	358	376	

Use the sign test with $\alpha = 0.05$ to determine if sufficient evidence exists to say that the median of the sampled population differs from 350.

15.3   An administrator of a large community hospital claims that the average age of blood donors during the last 12 months has fallen below the previous value of 40 years. To check on this belief, 20 records were randomly selected from those of all donors during the last 12 months. The ages of the 20 donors are given below.

32	43	51	43	21	42	23	29	35	37
39	28	27	64	27	31	36	28	31	22

Use the sign test to determine if there is sufficient evidence at the 5 percent level to support the administrator's claim.

15.4   A consumer prefers a particular brand of potato chips that is packaged with a label weight of 15 ounces. However, he believes that the bags tend to be underweight. In an attempt to prove this belief, 15 bags were purchased over a period of several weeks. The resulting weights are given below.

14.1	14.0	14.3	15.1	14.2	14.7	15.3	14.9
15.1	14.2	13.7	14.1	14.2	14.3	14.2	

Use the sign test to determine if the data present sufficient evidence at the 0.05 level to support the consumer's belief.

15.5   Determine the p-value for the test in Exercise 15.3.

15.6   The balances of a random sample of 10 savings accounts at a credit union appear below.

$1,500.89	$995.98	$1,258.87	$598.32	$1,282.58
$1,929.64	$303.76	$1,372.56	$429.76	$1,075.12

At the 10 percent level, use the sign test to determine if sufficient evidence exists to conclude that the average balance of all savings accounts at this institution exceeds $1,000.00.

15.7   A manufacturer of air pollution controls believes that a chemical treatment of its air filters will increase the number of airborne particles collected by the filters. To test this theory, a treated and an untreated filter are installed at 16 sites. The following data are the number of airborne particles collected over a one-hour period.

Site	Treated	Untreated
1	58	50
2	27	19
3	38	29
4	45	34
5	27	22
6	16	13
7	87	70
8	97	82
9	25	29
10	36	30
11	45	29
12	39	39
13	56	43
14	43	29
15	47	49
16	54	51

Is there sufficient evidence that the chemical treatment is effective? Use the sign test at the 0.01 significance level.

15.8   Determine the p-value for the test in Exercise 15.6.

15.9   A manufacturer of platinum-tipped spark plugs believes that they last longer than conventional spark plugs. To show this, one platinum plug and one conventional plug were installed in each of six-2-cylinder engines. The effective life in hours was determined for each plug and appears below.

Engine	1	2	3	4	5	6
Platinum plug	640	570	530	410	600	580
Conventional plug	470	370	460	490	380	410

Is there sufficient evidence to support the manufacturer's contention that the platinum plugs do tend to last longer? Use the sign test and a significance level of 10 percent.

15.10   To test the claim that a certain herb lowers blood pressure, 36 people participated in an experiment. The systolic blood pressure of each was recorded initially and 90 minutes after digesting the herb. The observed blood pressures are given on the following page.

Subject	Initial Blood Pressure	90 Minutes Later
1	117	114
2	125	129
3	108	106
4	145	143
5	139	140
6	156	151
7	143	144
8	127	124
9	143	141
10	132	129
11	103	104
12	124	125
13	142	140
14	187	182
15	105	103
16	117	114
17	121	119
18	142	139
19	165	160
20	134	131
21	111	116
22	125	121
23	118	118
24	138	139
25	132	130
26	142	135
27	128	134
28	120	117
29	154	149
30	141	140
31	119	115
32	121	117
33	153	149
34	138	139
35	167	171
36	182	175

Use the sign test and a 0.01 level of significance to determine if there is sufficient evidence to conclude that the herb is effective in lowering blood pressure.

15.11   A certain candy bar is supposed to weigh 75 grams. Fifty bars were randomly selected from a production run, and the following weights in grams were recorded.

75.4	68.8	71.3	74.4	77.0	73.0	70.8	75.2	70.4	73.6
73.4	75.2	73.7	77.2	74.9	73.1	68.8	73.5	77.9	76.2
75.2	74.1	73.9	75.8	74.1	74.3	70.5	74.0	73.5	71.6
75.6	76.1	71.9	71.1	75.4	74.0	77.4	76.4	78.8	74.2
73.9	74.5	72.5	71.6	73.0	75.2	71.0	75.9	73.5	74.1

Use the sign test and $\alpha = 5$ percent to determine if there is sufficient evidence to conclude that the median weight of all bars in this production run differs from 75 grams.

15.12   Determine the p-value for the test in Exercise 15.10.

15.13   Determine the p-value for the test in Exercise 15.11.

15.14   A shoe manufacturer has developed a new running shoe that purportedly enables an athlete to run faster. Eight adults participated in an experiment in which each ran a mile with regular track shoes and then ran a mile the next day with the new shoes. Their running times in seconds appear below.

Runner	Track Shoe	New Shoe
1	321	318
2	307	299
3	397	401
4	269	260
5	285	285
6	364	363
7	295	289
8	302	296

Is there sufficient evidence to support the manufacturer's claim? Use the sign test with $\alpha = 0.05$.

15.15   As part of its admissions evaluation process, an actuarial science department administers an examination to each applicant. To determine if there is a significant difference in the grading of the exams by 2 faculty members, 10 exams were graded by both professors and the following scores were assigned:

Applicant	1	2	3	4	5	6	7	8	9	10
Professor A's score	75	87	89	63	93	54	83	71	88	71
Professor B's score	69	84	80	57	95	49	79	65	88	67

Do the data provide sufficient evidence to indicate that one professor tends to assign higher grades? Use the sign test at the 0.05 significance level.

15.16   Find the approximate *p*-value for the test in Exercise 15.14.

MINITAB Assignments

M▶ 15.17   The following are the heights in inches of 36 blue spruce trees randomly selected from a 10-acre tree farm.

84	79	67	86	75	89	76	91	83
74	87	78	86	90	84	79	73	88
87	79	73	76	87	86	80	82	74
85	87	69	84	83	74	81	85	79

Use MINITAB to perform the sign test to determine if sufficient evidence exists at the 0.05 level to conclude that the median height of all trees on the farm exceeds 80 inches.

M▶ 15.18   A fish wholesaler has a catch of several thousand lobsters. A prospective buyer selected 50 at random and obtained the following weights in ounces.

21.3	21.1	21.4	18.9	20.2	19.3	19.1	18.3	19.9	22.0
20.6	20.7	21.9	20.1	17.1	19.3	21.2	18.4	21.0	21.6
16.5	18.9	17.4	20.8	18.5	18.1	21.1	19.3	21.5	20.1
21.8	20.2	19.7	18.9	19.5	20.0	18.7	21.6	20.9	21.5
17.5	16.1	20.1	21.8	19.4	21.6	23.1	20.5	22.0	20.6

The prospective buyer will purchase the entire catch if it can be shown that the average weight exceeds 19.9 ounces. Formulate a suitable set of hypotheses, and use MINITAB to perform the sign test at the 1 percent significance level.

M▶ 15.19   Use MINITAB to perform the sign test for the data in Exercise 15.10.

M▶ 15.20   To satisfy a senior colloquium requirement, a student designed an experiment to assess the awareness of freshmen at her university concerning beginning salaries of graduating seniors in the teaching and social work professions. Ten students were interviewed and asked to estimate the average starting salaries for the two professional areas. The figures in the following table are the percentages that the students' estimates were off from the actual values reported by the university's career development center.

	Profession	
Student	Teaching	Social Work
1	11.7	11.8
2	5.3	4.3
3	23.8	21.7
4	25.6	25.6
5	4.3	5.8
6	14.5	15.2
7	1.5	6.4
8	7.6	12.5
9	10.4	13.2
10	13.6	14.2

Is there sufficient evidence to indicate a difference in the accuracies of the estimates for the two professions? Use MINITAB to perform the sign test at the 5 percent significance level.

15.3 The Mann-Whitney U-Test: Independent Samples

The pooled two-sample t-test for comparing two population means was discussed in Section 11.2. The test is quite restrictive because it assumes that each population has a normal distribution and the variances of the populations are equal. The Mann-Whitney U-test offers a nonparametric alternative that is less restrictive in that the populations do not have to be normal. The test does assume, however, that the populations have continuous distributions that are identical in shape. (Like the sign test, the populations are assumed to be continuous, but in practice the Mann-Whitney U-test is frequently applied to discrete distributions as well.) The Mann-Whitney U-procedure is used to test the null hypothesis that the populations also have identical locations. The alternative hypothesis is that one of the distributions is shifted to the left or right of the other (Figure 15.3).

Like the pooled t-test, the Mann-Whitney U-test also requires independent random samples. The test, and all the remaining procedures in this chapter, are based

Figure 15.3
Identical distributions with shifted locations.

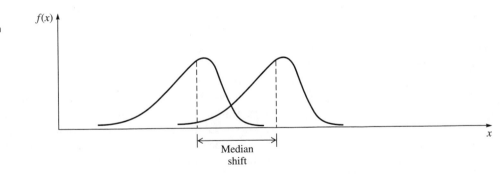

on replacing the sample measurements with their relative ranks. To illustrate this concept, consider the following samples of the actual weights of "eight-ounce" bags of two brands of potato chips (the units are ounces).

Brand One: $n_1 = 10$					Brand Two: $n_2 = 11$				
8.47	8.32	8.20	8.41	8.38	8.13	8.28	8.31	8.24	8.17
8.42	8.29	8.45	8.45	8.29	8.10	8.27	8.26	8.04	8.21
					8.29				

The two samples are combined into a single group of $n_1 + n_2 = 21$ values, and these are arranged in order of size from smallest to largest (to distinguish between the samples, we have underlined the values from sample one). The 21 measurements are then ranked from 1 (smallest value) to 21 (largest value). The assigned ranks appear in parentheses below each measurement.

8.04	8.10	8.13	8.17	8.20	8.21	8.24
(1)	(2)	(3)	(4)	(5)	(6)	(7)
8.26	8.27	8.28	8.29	8.29	8.29	8.31
(8)	(9)	(10)	(12)	(12)	(12)	(14)
8.32	8.38	8.41	8.42	8.45	8.45	8.47
(15)	(16)	(17)	(18)	(19.5)	(19.5)	(21)

When there are ties, the measurements are assigned the mean of the ranks that they jointly share. For instance, there are 3 values of 8.29, and these values occupy positions 11, 12, and 13 in the array. Consequently, each value of 8.29 is assigned a mean rank of $(11 + 12 + 13)/3 = 12$. There are also 2 values of 8.45, which occupy positions 19 and 20. Each value is thus assigned a mean rank of $(19 + 20)/2 = 19.5$.

For each sample, the sum of the ranks (**rank sum**) is determined. For sample one (its measurements are underlined) the rank sum is as follows.

$$R_1 = 5 + 12 + 12 + 15 + 16 + 17 + 18 + 19.5 + 19.5 + 21$$
$$= 155$$

The rank sum of sample two is as follows.

$$R_2 = 1 + 2 + 3 + 4 + 6 + 7 + 8 + 9 + 10 + 12 + 14$$
$$= 76$$

If the two samples actually come from populations with the same location, then we would expect that the ranks as determined above should be randomly dispersed among the two samples. However, if one of the rank sums is unusually small compared to the other, then this is construed as evidence that one population is shifted relative to the other and that the medians of the populations are not the same. A test statistic can be developed that is based on R_1 and R_2. This is accomplished by utilizing the following quantities.

$$U_1 = n_1 n_2 + \frac{n_2(n_2 + 1)}{2} - R_2$$

$$U_2 = n_1 n_2 + \frac{n_1(n_1 + 1)}{2} - R_1$$

Either U_1 or U_2 can be used as a test statistic. Often the test statistic is chosen to be the smaller of these two, and it is denoted by U. Special tables have been constructed that give the values of U for which the null hypothesis should be rejected. However, by standardizing U to its z-value, one can use a simpler approximate test based on the z-distribution. It works well even for values of n_1 and n_2 as small as 10. Consequently, we shall use this procedure, which is summarized below. Also, when using a z-statistic there is no advantage in calculating both U_1 and U_2 and then selecting the smaller. For the sake of simplicity, we will only calculate U_1 and use its z-value as the test statistic.

PROCEDURE

Mann-Whitney U-Test for $(\tilde{\mu}_1 - \tilde{\mu}_2)$ $(n_1 \geq 10$ and $n_2 \geq 10)$:

For testing the null hypothesis H_0: $(\tilde{\mu}_1 - \tilde{\mu}_2) = 0$, the test statistic is

(15.2)
$$z = \frac{U_1 - \frac{n_1 n_2}{2}}{\sqrt{\frac{n_1 n_2 (n_1 + n_2 + 1)}{12}}}.$$

The rejection region is given by the following.

For H_a: $(\tilde{\mu}_1 - \tilde{\mu}_2) < 0$, H_0 is rejected when $z < -z_\alpha$;

For H_a: $(\tilde{\mu}_1 - \tilde{\mu}_2) > 0$, H_0 is rejected when $z > z_\alpha$;

For H_a: $(\tilde{\mu}_1 - \tilde{\mu}_2) \neq 0$, H_0 is rejected when $z < -z_{\alpha/2}$ or $z > z_{\alpha/2}$.

Assumptions:

1. Independent random samples.

2. The sampled populations have continuous probability distributions with the same shape.

Note:

$U_1 = n_1 n_2 + \frac{n_2(n_2 + 1)}{2} - R_2$, where R_2 is the rank sum for sample 2.

EXAMPLE 15.5   Use the two samples of weights considered at the beginning of this section to determine if a difference exists in the median weights of the two brands of potato chips. Test at the 0.05 significance level.

Solution

Let $\tilde{\mu}_1$ and $\tilde{\mu}_2$ denote the population medians for brands 1 and 2, respectively.

Step 1: Hypotheses.

$$H_0: (\tilde{\mu}_1 - \tilde{\mu}_2) = 0$$

$$H_a: (\tilde{\mu}_1 - \tilde{\mu}_2) \neq 0$$

Step 2: Significance level.

$$\alpha = 0.05$$

Step 3: Calculations.

The two samples, after they have been combined and ranked, are repeated below. The underlined values are from sample 1, $n_1 = 10$, and $n_2 = 11$.

8.04	8.10	8.13	8.17	8.20	8.21	8.24
(1)	(2)	(3)	(4)	(5)	(6)	(7)
8.26	8.27	8.28	8.29	8.29	8.29	8.31
(8)	(9)	(10)	(12)	(12)	(12)	(14)
8.32	8.38	8.41	8.42	8.45	8.45	8.47
(15)	(16)	(17)	(18)	(19.5)	(19.5)	(21)

As determined earlier, the rank sum for the second sample is $R_2 = 76$. Next, we need the value of U_1.

$$U_1 = n_1 n_2 + \frac{n_2(n_2 + 1)}{2} - R_2$$

$$= 10(11) + \frac{11(11 + 1)}{2} - 76 = 176 - 76 = 100$$

The z-value for U_1 is used as the test statistic.

$$z = \frac{U_1 - \frac{n_1 n_2}{2}}{\sqrt{\frac{n_1 n_2(n_1 + n_2 + 1)}{12}}}$$

$$= \frac{100 - \frac{10(11)}{2}}{\sqrt{\frac{10(11)(10 + 11 + 1)}{12}}} = \frac{45}{14.201} = 3.17$$

Step 4: Rejection region.

Since H_a contains the $\neq$ relation, the rejection region is two tailed and consists of those values for which $z < -z_{\alpha/2} = -z_{.025} = -1.96$, and those values for which $z > z_{\alpha/2} = z_{.025} = 1.96$.

Step 5: Conclusion.

$Z = 3.17$ is in the rejection region. Therefore, H_0 is rejected, and we conclude that the median weights of the 2 brands of potato chips do differ.

MINITAB EXAMPLE 15.6   For the two samples in Example 15.5, have MINITAB perform the Mann-Whitney
U-test.

Solution
We begin by storing samples one and two in columns C1 and C2, respectively.

```
SET C1
8.47 8.32 8.20 8.41 8.38 8.42 8.29 8.45 8.45 8.29
END
SET C2
8.13 8.28 8.31 8.24 8.17 8.29 8.10 8.27 8.26 8.04 8.21
END
```

Now type the following commands.

```
MANN-WHITNEY C1 C2;
ALTERNATIVE 0.        # 0 is the code for ≠ in alt hyp
```

> **Menu Commands:**
> **Stat ➤**
> **Nonparametrics ➤**
> **Mann-Whitney**
> *In* **First Sample** *box*
> *enter* **C1**
> *In* **Second Sample**
> *box enter* **C2**
> *In* **Alternative** *box*
> *select* **not equal**
> *Select* OK

The resulting output appears in Exhibit 15.2.
      MINITAB uses as the test statistic the rank sum R_1, and this is labeled W in
Exhibit 15.2 (W = 155). In addition to calculating the *p*-value for the test statistic,
MINITAB also provides a 95 percent (approximately) confidence interval for the
difference between the population medians (MINITAB denotes the population medians by ETA1 and ETA2). Since the *p*-value of 0.0017 is less than $\alpha = 0.05$, the
null hypothesis would be rejected. We conclude that there is a difference in the
median weights of the two brands.

■■■ **Exhibit 15.2**

```
MTB > MANN C1 C2;
SUBC> ALTE 0.

Mann-Whitney Confidence Interval and Test

C1            N = 10      Median =        8.3950
C2            N = 11      Median =        8.2400
Point estimate for ETA1-ETA2 is       0.1600
95.5 pct c.i.  for ETA1-ETA2 is (0.0700,0.2400)
W = 155.0
Test of ETA1 = ETA2 vs. ETA1 not = ETA2 is significant at 0.0017
The test is significant at 0.0017 (adjusted for ties)
```

      For the **MANN-WHITNEY** command used in the above example, 95 percent
is the default value that MINITAB uses for the confidence interval. A different value

Menu Commands:
Stat ➤
Nonparametrics ➤
Mann-Whitney
In **First Sample** *box*
enter **C1**
In **Second Sample**
box enter **C2**
In **Confidence Level**
box enter **90**
In **Alternative** *box*
select **not equal**
Select OK

can be used by specifying it after **MANN-WHITNEY**. For instance, a 90 percent confidence interval could have been obtained in Exhibit 15.2 by typing the following.

```
MANN 90 C1 C2;
ALTE 0.
```

As we mentioned at the beginning of this section, the Mann-Whitney *U*-test assumes that the populations have an identical shape (this implies that the standard deviations must be the same). If this common shape is that of a normal distribution, then the pooled *t*-test could be used instead, and it would be slightly more powerful than the Mann-Whitney *U*-test. However, in situations for which the populations have distributions with the same nonnormal shape, the Mann-Whitney test may be considerably more powerful.

SECTION 15.3 EXERCISES

The Mann-Whitney *U*-Test: Independent Samples

15.21   By the year 1998, food manufacturers are required to add folic acid to grain-based products such as bread, flour, and pasta. The FDA mandated this action in response to studies suggesting that the vitamin might lower the risk of certain birth defects. One natural source of folic acid is orange juice. Suppose the owner of an orange grove believes that juice made with oranges from a competitor's grove contains less folic acid than his product. An independent testing agency measured the amount of folic acid in one-ounce samples of orange juice from the two groves, and the results are shown below.

Orange Grove	Amount of Folic Acid (mcg)						
Owner's	17.0	17.2	17.1	17.0	16.9	17.0	17.1
	17.0	17.1	16.7	17.1	16.7	17.1	
Competitor's	17.0	16.8	16.9	16.9	17.0	16.8	16.9
	17.0	16.9	16.7	17.1	16.8		

Apply the Mann-Whitney *U*-test to determine if the average amount of folic acid is higher for juice made with oranges from the owner's grove. Use a significance level of 5 percent.

15.22   Independent random samples from 2 populations produced the following observations. Use the Mann-Whitney *U*-test to determine if the median of population 1 exceeds that of population 2. Use $\alpha = 0.10$.

Sample 1	Sample 2
158	150
127	170
138	129
145	134
127	122
116	113
187	119
197	182
150	167
127	155
149	137

15.23   Apply the Mann-Whitney U-test to the following independent random samples to determine if the medians of the sampled populations differ. Test at the 0.05 significance level.

Sample 1	56	47	52	67	58	59	51	62	63	64	70	57
Sample 2	45	63	51	48	49	51	69	57	46	71	65	

15.24   To determine if a difference exists in the average weight of apples from 2 orchards, 11 apples were randomly selected from Orchard One and 10 were selected at random from Orchard Two. Their weights in ounces are given below. Use the Mann-Whitney U-test to determine if the medians of the sampled populations differ. Test with $\alpha = 0.05$.

Orchard	Weights in Ounces										
One	4.7	5.3	5.9	4.8	5.1	6.2	6.1	6.1	5.3	6.1	4.9
Two	6.3	5.7	5.8	4.9	6.9	6.8	7.2	6.9	6.8	7.3	

15.25   Some automobile dealers are able to sell a new car at a higher price than that advertised by adding extra charges that are often called "documentary fees." In 1990, Illinois became the first state to limit the amount of these extra charges. Suppose a state's attorney general wants to compare documentary fees charged by dealers in her state with those charged by dealers in a neighboring state. Random samples of 16 and 14 dealers were selected from the 2 states and produced the following values (all figures are in dollars).

Home State				Neighboring State			
125	188	190	175	185	150	99	110
190	135	180	195	149	125	120	139
125	125	120	225	175	145	165	
225	200	150	265	100	160	173	

Use the Mann-Whitney *U*-test to determine if there is sufficient evidence at the 5 percent level to conclude that a difference exists in the median documentary fees charged by dealers in the 2 states.

15.26   Determine the *p*-value for the test in Exercise 15.24.

15.27   Determine the *p*-value for the test in Exercise 15.25.

15.28   To determine if a new additive improves the mileage performance of gasoline, 12 test runs were conducted with the additive, and 10 runs were made without it. The test results appear below, with all figures in miles per gallon (mpg).

With Additive	Without Additive
32.6	31.3
30.1	29.7
29.8	29.1
34.6	30.3
33.5	30.9
29.6	29.9
33.8	28.7
30.0	30.1
29.7	27.6
29.5	28.6
32.7	
33.1	

Is there sufficient evidence to conclude that the additive increases gasoline mileage? Use the Mann-Whitney *U*-test and a significance level of 5 percent.

15.29   Pellet stoves use wood pellets made from industrial lumber waste as fuel. A stove manufacturer believes that its pellet stoves are more efficient than those of a competing brand. An independent testing laboratory measured the efficiency ratings for 10 of the manufacturer's stoves and for 12 of the competitor's stoves. The results are given below.

Manufacturer	Competitor
80	74
83	75
79	79
85	80
79	72
82	74
77	80
81	77
84	78
76	78
	70
	76

Is there sufficient evidence at the 0.01 level to support the manufacturer's belief? Use the Mann-Whitney U-test.

15.30   Find the p-value for the test in Exercise 15.28.

15.31   Find the p-value for the test in Exercise 15.29.

15.32   To compare the protein content of two brands of animal feed, bags of each brand were randomly selected and analyzed. The percentages of protein for each sample are given below.

Brand 1	Brand 2
27.3	21.6
25.7	23.9
29.2	22.8
28.2	21.9
22.9	22.1
26.2	22.2
28.4	23.3
29.2	20.9
31.9	22.7
28.0	23.0
24.0	22.5
	22.4
	20.9
	21.1

Use the Mann-Whitney U-test to determine if the brands differ in their median protein content. Use a significance level of 0.05.

MINITAB Assignments

M▶ 15.33   Use MINITAB to apply the Mann-Whitney U-test to the samples in Exercise 15.32.

M▶ 15.34   Use MINITAB to apply the Mann-Whitney U-test to the samples in Exercise 15.28.

M▶ 15.35   Use MINITAB to apply the Mann-Whitney U-test to the samples in Exercise 15.25.

M▶ 15.36   An economist wanted to compare the hourly labor rates of television repairmen in two states. Service shops were randomly selected from each state, and the following hourly charges in dollars were obtained.

First State	Second State
50.00	45.00
48.00	47.00
48.00	41.00
47.00	49.00
46.00	41.50
49.00	45.00
51.50	42.50
48.00	44.00
49.50	49.00
47.50	46.00
45.00	58.00
50.00	55.00
60.00	50.00
54.50	
55.00	
49.50	

Use MINITAB to apply the Mann-Whitney U-test to determine whether a difference exists in the median hourly rates for these two states. Test with $\alpha = 0.05$.

15.4 THE KRUSKAL-WALLIS *H*-TEST FOR A COMPLETELY RANDOMIZED DESIGN

The Kruskal-Wallis H-test is a nonparametric alternative to the one-way analysis of variance for analyzing a completely randomized design. Chapter 13 discusses the one-way analysis of variance that is used to test the null hypothesis that k (2 or more) population means are equal. In addition to independent random samples, it assumes that the k populations have normal distributions with the same variance. The Kruskal-Wallis H-test is less restrictive. It assumes independent random samples, and like the Mann-Whitney U-test, it also assumes that the sampled populations have continuous distributions that are identical in shape. The Kruskal-Wallis H-procedure tests the null hypothesis that the k-distributions all have the same location. The alternative hypothesis is that at least one distribution is shifted relative to the others, and thus, the k populations do not all have the same central location (see Figure 15.4).

In conducting the Kruskal-Wallis test, there are many similarities to the Mann-Whitney test in the preceding section. The samples are combined into a single group of $n = n_1 + n_2 + \cdots + n_k$ values, and the measurements are arranged in order of size from smallest to largest. The measurements are then ranked from 1 (smallest value) to n (largest value). In the event of ties, the tied measurements are assigned mean ranks in the same manner as described for the Mann-Whitney test. Next, the rank sum, R_i, is determined for each of the k samples, and a test statistic is calculated based on the values of the k rank sums. The statistic is

Figure 15.4
$k = 3$ identical distributions with shifted locations.

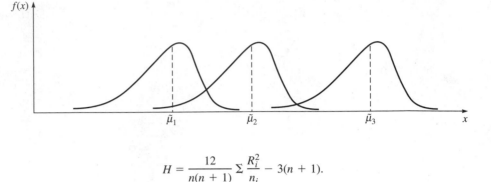

$$H = \frac{12}{n(n + 1)} \Sigma \frac{R_i^2}{n_i} - 3(n + 1).$$

It can be shown that if H_0 is true and the sample sizes are sufficiently large (each $n_i \geq 5$), the test statistic H has approximately a chi-square distribution with $df = k - 1$. H, moreover, measures the extent to which the ranks are randomly dispersed among the k samples. If there are great differences in the average ranks for the k samples, then H will be large, suggesting that the populations differ in their central locations. On the other hand, a small value of H indicates that the average ranks are nearly equal and thus supportive of the null hypothesis. Consequently, the rejection region is always an upper tail, and H_0 is rejected for $H > \chi_\alpha^2$. The Kruskal-Wallis H-test is summarized below.

PROCEDURE

Kruskal-Wallis *H*-Test for a Completely Randomized Design:

For testing the hypotheses

$$H_0: \tilde{\mu}_1 = \tilde{\mu}_2 = \cdots = \tilde{\mu}_k,$$

H_a: at least two populations have different medians,

the test statistic is

(15.3) $$H = \frac{12}{n(n + 1)} \Sigma \frac{R_i^2}{n_i} - 3(n + 1).$$

The null hypothesis is rejected for $H > \chi_\alpha^2$.

Assumptions:

1. Independent random samples.

2. The sampled populations have continuous probability distributions with the same shape.

3. Each sample size n_i is at least 5.

Note:

1. The chi-square distribution has $df = k - 1$.

2. R_i is the rank sum of sample i, and $n = n_1 + n_2 + \cdots + n_k$.

EXAMPLE 15.7

*© Greg Vaughn/Tom Stack
& Associates*

A commercial farm that grows and sells fish for consumption is testing three types of feed. Each food is fed for six weeks to one of three large pools of newly born fish of the same type. Eight fish are then randomly selected from each pool and weighed. Their weights in ounces are given below.

Diet 1	Diet 2	Diet 3
56	61	53
51	66	49
52	60	41
48	59	47
54	68	45
49	71	43
57	59	45
59	69	40

Use the Kruskal-Wallis *H*-test to determine whether one or more of the feeds tend to produce larger fish. Test at the 0.05 significance level.

Solution

Step 1: Hypotheses.

Let $\tilde{\mu}_1$, $\tilde{\mu}_2$, and $\tilde{\mu}_3$ denote the medians of the weight distributions for the three feeds.

$$H_0: \tilde{\mu}_1 = \tilde{\mu}_2 = \tilde{\mu}_3$$

H_a: At least two of the median weights differ.

Step 2: Significance level.

$$\alpha = 0.05$$

Step 3: Calculations.

Think of the 3 samples as being combined into one group of $n = n_1 + n_2 + n_3 = 24$ measurements, and then assign the smallest value a rank of 1, the next in size a rank of 2, and so on up to the largest, which gets a rank of 24 (ties are averaged as explained previously). The resulting assigned ranks are indicated in parentheses and given below.

Diet 1	Diet 2	Diet 3
56 (14)	61 (20)	53 (12)
51 (10)	66 (21)	49 (8.5)
52 (11)	60 (19)	41 (2)
48 (7)	59 (17)	47 (6)
54 (13)	68 (22)	45 (4.5)
49 (8.5)	71 (24)	43 (3)
57 (15)	59 (17)	45 (4.5)
59 (17)	69 (23)	40 (1)

$$R_1 = 95.5 \qquad R_2 = 163 \qquad R_3 = 41.5$$

$$n_1 = 8 \qquad n_2 = 8 \qquad n_3 = 8$$

For each feed the ranks are added, and the resulting rank sums are $R_1 = 95.5$, $R_2 = 163$, and $R_3 = 41.5$. (As a check, you can use the fact that the rank sums must add to $n(n + 1)/2 = 24(25)/2 = 300$.) The test statistic H equals the following.

$$H = \frac{12}{n(n + 1)} \Sigma \frac{R_i^2}{n_i} - 3(n + 1)$$

$$= \frac{12}{24(25)} \left(\frac{95.5^2}{8} + \frac{163^2}{8} + \frac{41.5^2}{8} \right) - 3(25)$$

$$= 18.53$$

Step 4: Rejection region.

H has approximately a chi-square distribution with $df = k - 1 = 2$. H_0 is rejected for $H > \chi_\alpha^2 = \chi_{.05}^2 = 5.99$. This value is obtained from Table 5.

Step 5: Conclusion.

Since $H = 18.53$ is in the rejection region, the null hypothesis is rejected. We therefore conclude that at least one of the feeds tends to produce larger fish.

MINITAB EXAMPLE 15.8 Use MINITAB to perform the Kruskal-Wallis H-test for the three samples of weights in Example 15.7.

Solution
The MINITAB command is **KRUSKAL-WALLIS** (or just **KRUS**, since only the first four letters of a command are needed). **KRUSKAL-WALLIS** requires a method of storing the data that differs from that for the analysis of variance command **AOVONEWAY**. All samples must be stacked in a single column. Consequently, we will enter in C1 the values of sample one, followed by the sample two values, and then the values of sample three. We also need to create a column C2 that identifies the sample associated with each value in C1.

```
SET C1
56 51 52 48 54 49 57 59
61 66 60 59 68 71 59 69
53 49 41 47 45 43 45 40
END
SET C2
1 1 1 1 1 1 1 1
2 2 2 2 2 2 2 2
3 3 3 3 3 3 3 3
END
```

Menu Commands:
Stat ➤
Nonparametrics ➤
Kruskal-Wallis
In **Response** *box*
enter **C1**
In **Factor** *box*
enter **C2**
Select OK

A faster way of placing the indices 1, 2, and 3 in column C2 is to type the following.

```
SET C2
(1:3)8
END
```

Now type the following command.

```
KRUS C1 C2   # Samples are in C1 and indices are in C2
```

The output produced by MINITAB is displayed in Exhibit 15.3.

Exhibit 15.3

```
MTB > KRUS C1 C2 # Samples are in C1 and indices are in C2

LEVEL       NOBS      MEDIAN   AVE. RANK    Z VALUE
    1          8       53.00        11.9      -0.28
    2          8       63.50        20.4       3.86
    3          8       45.00         5.2      -3.58
OVERALL       24                    12.5

H = 18.53   d.f. = 2   p = 0.000
H = 18.58   d.f. = 2   p = 0.000 (adj. for ties)
```

The value of the test statistic, $H = 18.53$, agrees with the value that we obtained in Example 15.7. Beside the *H*-value appears its *p*-value of 0.000, which indicates that the null hypothesis of equal medians must be rejected.

When there are ties in the data, MINITAB also computes an adjusted value of *H* (18.58 in Exhibit 15.3), which involves a slight modification of Formula 15.3.

Exhibit 15.3 also gives a *z*-value for the average rank of each sample. This is an indication of how that average rank differs from the overall mean rank (12.5) for all 24 observations.

SECTION 15.4 EXERCISES

The Kruskal-Wallis *H*-Test for a Completely Randomized Design

15.37   A completely randomized design produced the following sample results.

	Sample 1	Sample 2	Sample 3
	69	63	51
	73	64	56
	70	60	59
	68	61	56
	74	65	60
		67	

Use the Kruskal-Wallis H-test to determine if at least 2 of the sampled populations have different medians. Use $\alpha = 10$ percent.

15.38   Consider the following independent random samples.

Sample 1	Sample 2	Sample 3	Sample 4
17.9	12.3	11.1	13.3
14.2	13.8	10.1	13.9
15.7	10.9	14.2	10.8
12.7	12.7	10.3	16.5
14.9	10.2	12.1	13.3
14.0		10.5	

Test to determine if the sampled populations differ in location. Use the Kruskal-Wallis H-test and a significance level of 5 percent.

15.39   Exercise 15.32 considered a comparison of two brands of animal feed. The Mann-Whitney U-test was used on the following samples to determine if a difference exists in the median protein content of the two brands. Use the Kruskal-Wallis H-test and $\alpha = 0.05$ to perform the analysis.

Brand 1		Brand 2	
27.3	28.4	21.6	20.9
25.7	29.2	23.9	22.7
29.2	31.9	22.8	23.0
28.2	28.0	21.9	22.5
22.9	24.0	22.1	22.4
26.2		22.2	20.9
		23.3	21.1

15.40   A particular camcorder model is sold by mail order companies, camera shops, and general merchandise outlets. The manufacturer is interested in determining if the average selling price differs for these three sources. Seven stores of each type were randomly selected, and the following prices were obtained.

Type of Store	Selling Price in Dollars						
Mail order	899	929	900	979	925	950	959
Camera shop	995	935	950	979	979	995	929
General merch.	979	999	950	995	925	975	989

Use the Kruskal-Wallis H-test and $\alpha = 0.05$ to determine if a difference exists in the median prices for the types of stores.

15.41   Determine the approximate p-value for the test in Exercise 15.39.

15.42   A consumer products evaluation magazine tested three brands of flashlight batteries. Each brand was used in five flashlights, and the lights were left on until the batteries failed. The life lengths in hours are given below.

Brand	Length of Life in Hours				
A	7.3	6.9	5.8	7.9	8.2
B	6.7	7.1	6.0	6.5	5.9
C	7.5	8.3	7.9	8.4	8.3

Do the data provide sufficient evidence to indicate that at least two of the brands differ in their average length of life? Use the Kruskal-Wallis *H*-test and $\alpha = 0.01$.

15.43   A study was conducted to compare the four leading computer word processing programs. Each of four groups of students received training in one program for a four-week period. At the conclusion of the instruction, all participants were rated on their ability to perform a given task. Their ratings are given below.

Program 1	Program 2	Program 3	Program 4
24	29	24	25
22	27	38	23
21	21	25	24
20	25	24	25
21	22	29	23
	28		

Is there sufficient evidence to indicate differences in performance ratings for at least two of the four word processing programs? Use the Kruskal-Wallis *H*-test and a significance level of 5 percent.

15.44   Determine the approximate *p*-value for the test in Exercise 15.42.

15.45   In a state for which the price of milk is not regulated, 8 stores were randomly selected in the western, central, and eastern parts of the state. The prices charged by the 24 stores for a quart of milk appear below. Do the data provide sufficient evidence to conclude that there is a difference in the median prices for a quart of milk in the 3 geographical regions? Test at the 0.05 significance level, and use the Kruskal-Wallis *H*-test.

Region	Cost in Cents for a Quart of Milk							
Western	63	64	63	60	55	62	60	62
Central	67	62	69	68	65	65	65	66
Eastern	68	68	73	64	69	72	69	68

15.46   Radon, the second largest cause of lung cancer after smoking, is a radioactive gas produced by the natural decay of radium in the ground. Radon can seep into a home through openings in the foundation, and the EPA recommends that corrective measures be taken if levels reach 4 or more picocuries per liter (pc/l). To investigate radon levels in four public schools, an official took a sample of six readings at each school. The results, in pc/l, appear below.

School 1	School 2	School 3	School 4
1.7	5.3	5.1	1.2
1.3	4.2	4.1	2.9
0.8	3.9	5.2	1.5
1.5	5.7	3.3	2.3
0.9	3.8	2.6	2.1
1.1	5.2	4.9	1.8

Do the data suggest that radon levels tend to be higher at one or more of the schools? Use the Kruskal-Wallis H-test and a significance level of 5 percent.

15.47   Determine the approximate p-value for the test in Exercise 15.45.

MINITAB Assignments

M▶ 15.48   Use MINITAB to conduct the test in Exercise 15.46.

M▶ 15.49   A manufacturer of bond writing paper conducted a study to compare the quality of paper for 3 different sources of pulp used in the manufacturing process. For each type of pulp, 16 sheets of paper were randomly selected and assigned a quality rating. The results were as follows.

Pulp A		Pulp B		Pulp C	
78	74	73	62	79	91
69	89	65	80	87	90
75	71	78	76	84	87
89	80	76	72	98	92
76	88	73	56	87	85
78	84	65	69	92	78
65	82	69	61	91	86
78	91	74	71	87	92

Is there sufficient evidence to conclude that a difference exists in the average quality ratings for the 3 sources of pulp? Use MINITAB to apply the Kruskal-Wallis H-test and let $\alpha = 0.05$.

M▶ 15.50   With reference to Exercise 15.49, use MINITAB to apply the Kruskal-Wallis H-test to determine if there is a difference in the average quality ratings for sources A and C. Test at the 0.05 significance level.

M▶ 15.51   Use MINITAB to apply the Kruskal-Wallis test to the samples in Exercise 15.43.

15.5 THE FRIEDMAN F_r-TEST FOR A RANDOMIZED BLOCK DESIGN

In the preceding section, we discussed a nonparametric alternative to the one-way analysis of variance for analyzing a completely randomized design. We now introduce the corresponding nonparametric test for the randomized block design that was considered in Section 13.5. Similar in several respects to the Kruskal-Wallis H-test, it is called the Friedman F_r-test and was developed by the economist Milton Friedman, currently a popular columnist and former Nobel prize winner. As is the usual case for nonparametric tests, the required assumptions are less restrictive. We assume that the sampled populations have continuous distributions with the same shape, and that the k experimental units within each block are randomly assigned to the k treatments. Like the Kruskal-Wallis H-test, the Friedman F_r-procedure tests the null hypothesis that the distributions for the k treatments all have the same central location.

The Friedman F_r-test is conducted by first arranging the k measurements within each block according to size. For each of the b blocks, the k measurements are ranked from 1 for the smallest value to k for the largest. Tied measurements are assigned mean ranks in the same manner as discussed in the preceding sections. After assigning ranks, we determine the rank sum R_i for each of the k treatments. Then we calculate the test statistic

$$F_r = \frac{12}{bk(k + 1)} \Sigma R_i^2 - 3b(k + 1).$$

As with the Kruskal-Wallis H-statistic, a large value for the test statistic F_r provides evidence that a difference exists in the average values of the k treatments. Another similarity is that the test statistic has approximately a chi-square distribution with $df = k - 1$, provided that the sample sizes are sufficiently large (either k or b exceeds 5). The test procedure is outlined in the box on the following page and illustrated in Example 15.9.

PROCEDURE

Friedman F_r-Test for a Randomized Block Design:

For testing the hypotheses

$$H_0: \tilde{\mu}_1 = \tilde{\mu}_2 = \cdots = \tilde{\mu}_k,$$

$$H_a: \text{at least two treatment medians differ,}$$

the test statistic is

$$(15.4) \qquad F_r = \frac{12}{bk(k+1)} \Sigma R_i^2 - 3b(k+1).$$

The null hypothesis is rejected for $F_r > \chi_\alpha^2$.

Assumptions:

1. The sampled populations have continuous distributions with the same shape.

2. The k experimental units within each block are randomly assigned to the k treatments.

3. Either the number of treatments k or the number of blocks b exceeds 5.

Note:

1. The chi-square distribution has $df = k - 1$.

2. R_i is the rank sum of the ith treatment for $i = 1, 2, \ldots, k$.

EXAMPLE 15.9   In Example 13.5, we considered a study involving 10 infants who participated in a comparison of 3 medications for the treatment of diaper rash. For each baby, three areas of approximately the same size and rash severity were selected, and one area was treated with medication A, one with B, and one with C. The treatments were randomly assigned for each infant. The number of hours for the rash to disappear was recorded for each medication and each infant, and the values obtained appear below.

	Medication		
Infant	A	B	C
1	46	43	42
2	50	49	47
3	46	48	41
4	51	47	46
5	43	40	40
6	45	40	41
7	47	47	42
8	48	44	41
9	46	41	42
10	48	45	43

In Chapter 13, an analysis of variance was used to test if a difference exists in the average healing times for the 3 medications. We will now use the Friedman F_r-test to analyze this randomized block design and determine if there is a significant difference at the 1 percent level.

Solution

Let $\tilde{\mu}_A$, $\tilde{\mu}_B$, and $\tilde{\mu}_C$ denote the median number of hours for the diaper rash to disappear for medications A, B, and C, respectively.

Step 1: Hypotheses.

$$H_0: \tilde{\mu}_A = \tilde{\mu}_B = \tilde{\mu}_C$$

H_a: Not all the medians are equal.

Step 2: Significance level.

$$\alpha = 0.01$$

Step 3: Calculations.

The first step in the analysis is to rank the measurements within each block. In this experiment, the infants serve as the blocks. The measurements 46, 43, and 42 in the first block are replaced by their relative ranks 3, 2, and 1, respectively. This replacement of measurements with ranks is conducted for each of the 10 blocks, and the assigned ranks are indicated in parentheses below.

	Medication		
Infant	A	B	C
1	46 (3)	43 (2)	42 (1)
2	50 (3)	49 (2)	47 (1)
3	46 (2)	48 (3)	41 (1)
4	51 (3)	47 (2)	46 (1)
5	43 (3)	40 (1.5)	40 (1.5)
6	45 (3)	40 (1)	41 (2)
7	47 (2.5)	47 (2.5)	42 (1)
8	48 (3)	44 (2)	41 (1)
9	46 (3)	41 (1)	42 (2)
10	48 (3)	45 (2)	43 (1)
Rank sum	$R_1 = 28.5$	$R_2 = 19$	$R_3 = 12.5$

The rank sums for the 3 treatments A, B, and C are $R_1 = 28.5$, $R_2 = 19$, and $R_3 = 12.5$, respectively. (You can perform an easy check by using the fact that the rank sums must add to $bk(k + 1)/2 = 10(3)(4)/2 = 60$.) Next, we calculate the value of the test statistic.

$$F_r = \frac{12}{bk(k + 1)} \Sigma R_i^2 - 3b(k + 1)$$

$$= \frac{12}{10(3)(4)} (28.5^2 + 19^2 + 12.5^2) - 3(10)(4)$$

$$= 12.95$$

Step 4: Rejection region.

Since the number of blocks exceeds 5, the test statistic has approximately a chi-square distribution with $df = k - 1 = 2$. Using Table 5 to determine the critical value, we find that H_0 is rejected if $F_r > \chi_{.01}^2 = 9.21$.

Step 5: Conclusion.

The calculated value $F_r = 12.95$ exceeds the table value 9.21, and consequently the null hypothesis is rejected. Thus, there is sufficient evidence at the 1 percent level to conclude that at least 2 of the treatments differ in the average time required to eliminate the rash.

The next example illustrates how MINITAB can be used to perform a Friedman F_r-test for the randomized block design.

MINITAB EXAMPLE 15.10   Use MINITAB to perform the Friedman F_r-test for the medication data in Example 15.9.

Solution

The MINITAB command is **FRIEDMAN**. With this command all values of the response variable (healing time) are stacked in a single column C1, and two auxiliary columns C2 and C3 are used to indicate for each response value the associated treatment and block numbers.

```
NAME C1 'HealTime' C2 'Treatmnt' C3 'Infant'
READ C1 C2 C3
46 1  1
50 1  2
46 1  3
51 1  4
43 1  5
45 1  6
 .  .  .
 .  .  .
41 3  8
42 3  9
43 3 10
END
```

Menu Commands:
Stat ➤
Nonparametrics ➤
Friedman
In **Response** box
enter **C1**
In **Treatment** box
enter **C2**
In **Blocks** box
enter **C3**
Select OK

Note that the manner in which the data are stored is identical to that used in Example 13.6 in which we had MINITAB obtain the analysis of variance table. You may find it helpful to review that example.

Next we type the following command.

```
FRIEDMAN C1 C2 C3   # Data in C1, Trt & Blk indices in C2 & C3
```

MINITAB will produce the output in Exhibit 15.4. The value of the test statistic F_r is denoted by S and has the value 12.95, the same as that obtained in the previous example. Its p-value is given as 0.002, and thus the null hypothesis would be rejected since this is less than $\alpha = 0.01$.

As with the Kruskal-Wallis H-test, when some ranks are averaged because of ties in the data, MINITAB will also compute an adjusted value of the test statistic. Also appearing in Exhibit 15.4 are the rank sums and estimates of the population medians for the 3 treatments (the estimates are not necessarily the same as the sample medians).

Exhibit 15.4

```
MTB > FRIEDMAN C1 C2 C3   # Data in C1, Trt & Blk indices in C2 & C3

Friedman Test
Friedman test of HealTime by Treatmnt blocked by Infant

S = 12.95   d.f. = 2   p = 0.002
S = 13.63   d.f. = 2   p = 0.001 (adjusted for ties)

                       Est.    Sum of
Treatmnt      N      Median    RANKS
       1     10      46.500     28.5
       2     10      43.500     19.0
       3     10      42.000     12.5
Grand median =       44.000
```

SECTION 15.5 EXERCISES

The Friedman F_r-Test for a Randomized Block Design

15.52   The following data pertain to a randomized block design that was used to compare the averages of treatments I, II, and III. Use the Friedman F_r-test to determine if there is a significant difference at the 1 percent level.

	Treatment		
Block	**I**	**II**	**III**
1	21	17	16
2	25	24	19
3	19	18	16
4	24	19	15
5	20	18	18
6	21	22	20

15.53   A randomized block design was used to compare four treatment medians. The data appear below. Does a significant difference exist between the treatment medians? Use $\alpha = 0.05$ and the Friedman F_r-test.

	Treatment			
Block	**A**	**B**	**C**	**D**
1	11.8 2.5	10.9 1	12.8 4	11.8 2.5
2	13.0 3.5	11.1 1	13.0 3.5	12.1 2
3	11.1 2	10.1 1	11.9 4	11.8 3
4	11.9 2	10.5 1	12.9 4	12.0 3
5	11.4 2	10.3 1	12.3 4	11.9 3
6	11.7 2	10.7 1	12.7 4	12.0 3
7	10.9 2	10.1 1	11.6 3	11.7 4
8	12.4 3	11.4 1	13.5 4	12.1 2

15.54   As part of its admissions evaluation process, an actuarial science department administers an examination to each applicant. To determine if there is a significant difference in the grading of the exams by 2 faculty members, 10 exams were graded by each professor, and the following grades were assigned.

Applicant	1	2	3	4	5	6	7	8	9	10
Professor A's Grade	75	87	89	63	93	54	83	71	88	71
Professor B's Grade	69	84	80	57	95	49	79	65	88	67

Use the Friedman F_r-test to determine if the data provide sufficient evidence of a difference in the median grades assigned by the two professors. Test at the 0.01 significance level.

15.55   The products testing laboratory for a national magazine wanted to compare the tread life of two brands of snow tires. The 2 drive wheels for each of 12 cars were randomly assigned 1 tire of each brand. The tires were driven until their treadwear indicators showed, and the numbers of miles accumulated are displayed below.

Car	Brand 1	Brand 2
1	47,300	51,600
2	63,700	62,000
3	54,700	55,100
4	37,200	36,100
5	42,700	40,400
6	68,600	66,600
7	29,800	28,700
8	41,700	40,900
9	48,800	46,800
10	52,800	52,200
11	55,100	55,100
12	51,300	50,400

Do the data provide sufficient evidence that the two brands of tires differ in their average length of life? Use the Friedman F_r-test and $\alpha = 10$ percent.

15.56   Determine the approximate p-value for the test in Exercise 15.54.

15.57   Determine the approximate p-value for the test in Exercise 15.55.

15.58   Three light bulb manufacturers, A, B, and C, claim that their brands have an expected life of 750 hours. An independent laboratory tested a 25-watt, 40-watt, 60-watt, 75-watt, 100-watt, and a 150-watt bulb of each brand, and the following life-lengths in hours were obtained.

Manufacturer	Wattage of Bulb					
	25	40	60	75	100	150
A	835	798	801	743	753	738
B	903	817	813	792	788	763
C	798	806	803	758	717	762

Is there a significant difference in the average life-lengths for the three brands? Use the Friedman F_r-test and $\alpha = 0.05$.

15.59   An investigative reporter planned to do an article on jewelry appraisals. To determine if there was a significant difference in diamond appraisals by 4 large jewelry companies, the reporter had anonymous individuals submit 6 particular diamonds for appraisal by the 4 companies. The quoted offers are shown below.

	Jewelry Company			
Diamond	1	2	3	4
First	$7,400	$7,500	$7,000	$8,000
Second	$1,200	$1,150	$950	$1,350
Third	$2,500	$2,400	$2,200	$2,800
Fourth	$400	$450	$400	$400
Fifth	$1,500	$1,500	$1,100	$1,750
Sixth	$5,800	$5,500	$4,800	$5,900

Formulate a suitable set of hypotheses and use the Friedman F_r-test with a significance level of 10 percent.

15.60　To satisfy a senior colloquium requirement, a student designed an experiment to assess the awareness of freshmen at her university concerning beginning salaries of graduating seniors in five professions. Ten students were interviewed and asked to estimate the average starting salaries for the professional areas displayed below. The figures in the following table are the percentages that the students' estimates were off from the actual values reported by the university's career development center.

	Profession				
Student	Teaching	Accounting	Social Work	Insurance	Government
1	11.7	12.9	11.8	19.7	10.8
2	5.3	4.9	4.3	6.5	2.8
3	23.8	29.7	21.7	30.6	20.5
4	25.6	36.7	25.6	37.4	23.8
5	4.3	5.9	5.8	10.6	6.1
6	14.5	20.4	15.2	21.7	15.1
7	1.5	5.6	6.4	8.9	9.4
8	7.6	12.5	12.5	8.5	8.5
9	10.4	11.5	13.2	17.8	14.6
10	13.6	19.8	14.2	18.6	15.7

Is there sufficient evidence to indicate differences in the accuracies of the estimates for the five professions? Use the Friedman F_r-test and a significance level of 5 percent.

15.61　Determine the approximate p-value for the test in Exercise 15.59.

MINITAB Assignments

M▶ 15.62　Use MINITAB to perform the test in Exercise 15.60.

M▶ 15.63   Total compensation for CEOs at the nation's 350 largest companies averaged $1,779,663 in 1994. In several instances, the CEO's bonus exceeded his/her salary. *The Wall Street Journal/William M. Mercer CEO Compensation Survey* gave the following compensations for 29 utilities (*The Wall Street Journal,* April 12, 1995). The figures are in thousands of dollars.

Utilities Company	Compensation ($1,000's)	
	Salary	Bonus
American Electric Power	620.0	209.4
Ameritech	709.6	754.2
Bell Atlantic	831.2	778.7
BellSouth	588.5	765.0
Centerior Energy	360.0	74.5
Central & South West	599.8	162.7
Columbia Gas System	682.0	50.0
Con Edison	603.3	353.6
Detroit Edison	540.0	33.5
Dominion Resources	571.1	273.7
Duke Power	558.5	103.5
FLP	795.8	650.0
General Public Utilities	573.8	292.5
GTE	784.6	1,219.5
MCI Communications	850.0	900.0
Niagra Mohawk	457.9	0.0
Nynex	800.0	885.0
Ohio Edison	461.7	178.8
PSEG	652.5	265.3
Pacific Enterprises	641.0	428.6
Pacific Gas & Electric	550.0	134.2
Pacific Telesis	541.5	391.7
PacifiCorp	504.1	302.1
Peco Energy	485.3	305.7
SCEcorp	664.0	0.0
Southwestern Bell	762.0	1,190.0
Sprint	863.9	1,085.6
U S West	700.0	560.0
Union Electric	400.0	120.0

Use MINITAB to apply the Friedman F_r-test to determine if the difference in average salary and average bonus is statistical significant at the 5 percent level.

M▶ 15.64   A national supermarket chain conducted a survey of its three largest competitors, primarily for the purpose of comparing prices of store-brand products. Prices were obtained for 20 popular items of the same size, and the costs appear on the following page.

Store-Brand Product	Supermarket Chain			
	Investigating Chain	**First Competitor**	**Second Competitor**	**Third Competitor**
Paper Towels	$0.79	$0.69	$0.85	$0.57
Trash Bags	$2.98	$3.19	$2.99	$3.02
Skim Milk	$2.05	$2.19	$2.11	$2.01
Lettuce	$0.79	$0.89	$0.79	$0.74
Jam	$1.59	$1.65	$1.69	$1.47
Coffee	$3.49	$3.59	$3.59	$3.29
Corn Flakes	$2.19	$2.25	$2.29	$2.15
Frozen Peas	$0.79	$0.85	$0.79	$0.65
Orange Juice	$1.75	$1.99	$1.89	$1.71
Bananas	$0.39	$0.39	$0.35	$0.37
Dish Detergent	$0.99	$1.15	$1.19	$0.95
Canned Dog Food	$0.57	$0.65	$0.59	$0.49
Vegetable Soup	$0.67	$0.69	$0.55	$0.73
Butter	$1.39	$1.45	$1.49	$1.19
Potatoes	$2.19	$2.39	$2.09	$2.25
Bread	$0.99	$0.99	$0.99	$0.99
Ground Beef	$2.25	$2.59	$2.45	$2.31
Onions	$0.49	$0.59	$0.53	$0.44
Tissues	$1.19	$1.35	$1.29	$1.05
Ice Cream	$2.59	$2.79	$2.75	$2.47

Use MINITAB to test if a significance difference exists in the median prices charged by the four supermarket chains. Let $\alpha = 0.05$.

M▶ 15.65 Refer to Exercise 15.64 and test at the 5 percent level if a significance difference exists in the median prices charged by the three competing chains.

15.6 THE RUNS TEST FOR RANDOMNESS

Throughout this book, whenever reference has been made to a sample for the purpose of making an inference, we have always assumed that the sample was random. The assumption of randomness is an integral part of the inferential process. In this section, we will consider a very simple method for detecting a lack of randomness in a sequence of sample observations. It is based on the **theory of runs**, where a **run** is defined below.

KEY CONCEPT

Consider a sequence of data, each of which is classified into one of two types. A **run** is a succession of occurrences of the same type. It is preceded and followed by either a different type or nothing at all.

To illustrate a run, consider a professor who constructs a quiz consisting of 12 true-false questions. The correct answers to questions one through twelve are as follows.

$$\underline{T} \quad \underline{F \ F} \quad \underline{T \ T} \quad \underline{F} \quad \underline{T \ T} \quad \underline{F} \quad \underline{T \ T}$$

This sequence of the two types of letters T and F contains seven runs, each of which is underlined. The theory of runs utilizes the number of runs in a sequence to check on the assumption of randomness in the data. A lack of randomness can be exhibited by too few runs that might indicate a tendency of the data to group together. For example, the smallest possible number of runs for a sequence containing seven T's and five F's is two. An example of this follows.

$$\underline{T \ T \ T \ T \ T \ T} \quad \underline{F \ F \ F \ F \ F}$$

At the other extreme, a lack of randomness can be indicated by too many runs, which could suggest that the data have been altered or tampered with. For our illustration, the maximum number of possible runs is 11. One way that this occurs is in the following sequence.

$$\underline{T} \quad \underline{F} \quad \underline{T} \quad \underline{F} \quad \underline{T} \quad \underline{F} \quad \underline{T} \quad \underline{F} \quad \underline{T} \quad \underline{F} \quad \underline{T \ T}$$

It is important to note that the test for randomness that we are about to describe is based on the order in which the observations are obtained and not on the frequency of occurrence of each type. For example, the sequence of true-false answers could consist of nine T's and three F's, and in spite of this imbalance, it might be considered random, depending on the number of runs in the sequence.

PROCEDURE

Runs Test for Randomness:

For testing the hypotheses

H_0: the sequence of sample observations is random,

H_a: the sequence is not random,

the test statistic is

$$x = \text{number of runs in the sequence.}$$

The null hypothesis is rejected if $x \le L_{\alpha/2}$ or if $x \ge R_{\alpha/2}$, where $L_{\alpha/2}$ and $R_{\alpha/2}$ are values from Table 7 that determine the left and right tails of the rejection region.

Note:

1. In Table 7, n_1 and n_2 denote the number of occurrences of the two types of elements in the sequence.

2. A blank space in the table indicates that n_1 or n_2 is not sufficiently large for a rejection region in that tail.

EXAMPLE 15.11 Determine if there is sufficient evidence at the 5 percent significance level to indicate a lack of randomness in the following sequence of true-false answers:

> T F F F T T F T T F T T

Solution
Step 1: Hypotheses.

$$H_0: \text{The sequence of T's and F's is random.}$$

$$H_a: \text{The sequence of T's and F's is not random.}$$

Step 2: Significance level.

$$\alpha = 0.05$$

Step 3: Calculations.

> T F F F T T F T T F T T

The arrangement contains $x = 7$ runs, consisting of $n_1 = 7$ T's and $n_2 = 5$ F's.

Step 4: Rejection region.

From Table 7, using the values of $n_1 = 7$, $n_2 = 5$, and a tail area of $\alpha/2 = 0.05/2 = 0.025$, we obtain $L_{\alpha/2} = L_{.025} = 3$ and $R_{\alpha/2} = R_{.025} = 11$. Therefore, the null hypothesis is rejected if $x \le 3$ or $x \ge 11$.

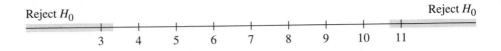

Reject H_0 Reject H_0

        3   4   5   6   7   8   9   10   11

Step 5: Conclusion.

Since $x = 7$ runs is not in the rejection region, H_0 is not rejected. Thus, there is insufficient evidence to indicate a lack of randomness in the sequence.

When n_1 or n_2 exceeds the range of values given in Table 7, an approximate z-statistic can be used. If both n_1 and n_2 are at least 10, the sampling distribution of the following statistic is approximately normal.

PROCEDURE

Large-Sample Runs Test for Randomness:
For testing the hypotheses

H_0: the sequence of sample observations is random,

H_a: the sequence is not random,

the test statistic is

(15.5) $$z = \frac{x - \mu}{\sigma}.$$

The null hypothesis is rejected if $x < -z_{\alpha/2}$ or if $x > z_{\alpha/2}$.

Assumptions:
$n_1 \geq 10$ and $n_2 \geq 10$.

Note:

1. x is the number of runs in the sequence, and

$$\mu = \frac{2n_1 n_2}{n_1 + n_2} + 1, \qquad \sigma = \sqrt{\frac{2n_1 n_2 (2n_1 n_2 - n_1 - n_2)}{(n_1 + n_2)^2 (n_1 + n_2 - 1)}}.$$

2. n_1 and n_2 denote the number of occurrences of the 2 types of elements in the sequence.

EXAMPLE 15.12   The author of a mathematics text claims that the pronouns "he" and "she" are used randomly in the exercises. An examination of 40 consecutive problems in which these pronouns were used revealed the following sequence.

H	H	H	S	S	S	S	S	H	H	H	H	H	S	H	H	H	H	H	H
S	H	S	S	S	S	S	S	S	H	S	S	H	H	H	S	S	H	H	H

Is there sufficient evidence at the 0.05 level to conclude that the author is not using the pronouns randomly?

Solution
Step 1: Hypotheses.

H_0: The sequence of H's and S's is random.

H_a: The sequence is not random.

Step 2: Significance level.

$$\alpha = 0.05$$

Step 3: Calculations.

H H H S S S S S H H H H H S H H H H H H

S H S S S S S S S H S S H H H S S H H H

The sequence contains $n_1 = 22$ H's and $n_2 = 18$ S's. The number of runs is $x = 13$.

$$\mu = \frac{2n_1 n_2}{n_1 + n_2} + 1$$

$$= \frac{2(22)(18)}{22 + 18} + 1$$

$$= 20.8$$

$$\sigma = \sqrt{\frac{2n_1 n_2 (2n_1 n_2 - n_1 - n_2)}{(n_1 + n_2)^2 (n_1 + n_2 - 1)}}$$

$$= \sqrt{\frac{2(22)(18)[2(22)(18) - 22 - 18]}{(22 + 18)^2 (22 + 18 - 1)}} = 3.09$$

$$z = \frac{x - \mu}{\sigma}$$

$$= \frac{13 - 20.8}{3.09}$$

$$= -2.52$$

Step 4: Rejection region.

The rejection region is two tailed, and H_0 is rejected for $z < -z_{\alpha/2} = -z_{.025} = -1.96$ and $z > 1.96$.

Step 5: Conclusion.

From Step 3, $z = -2.52$, which is in the rejection region. Therefore, H_0 is rejected, and we conclude at the 5 percent significance level that the use of "he" and "she" is not random.

The runs test is not restricted to just a sequence of qualitative observations. It can be used to test the randomness of quantitative data by recording for each sample value whether it lies above (A) or below (B) the median. If a sample observation equals the median, it is disregarded.

EXAMPLE 15.13   To illustrate the Mann-Whitney U-test in Example 15.5, two samples of weights of bags of potato chips were considered. The weights of brand two are repeated on the following page.

8.13, 8.28, 8.31, 8.24, 8.17, 8.29, 8.10, 8.27, 8.26, 8.04, 8.21

If the weights were obtained in the order above, is there sufficient evidence at the 0.05 significance level to reject the assumption of randomness?

Solution
Step 1: Hypotheses.

$$H_0: \text{The sample is random.}$$

$$H_a: \text{The sample is not random.}$$

Step 2: Significance level.

$$\alpha = 0.05$$

Step 3: Calculations.

The median of the sample is easily shown to be 8.24. The sample values are now classified as above (A) or below (B) 8.24 (a sample value of 8.24 is discarded).

8.13	8.28	8.31	8.24	8.17	8.29	8.10	8.27	8.26	8.04	8.21
B	A	A	—	B	A	B	A	A	B	B

The sequence of A's and B's contains $x = 7$ runs.

B A A B A B A A B B

Step 4: Rejection region.

Table 7 can be used with $n_1 = 5$ A's and $n_2 = 5$ B's. H_0 is rejected for $x \leq L_{\alpha/2} = L_{.025} = 2$, and for $x \geq R_{\alpha/2} = R_{.025} = 10$.

Reject H_0 2   3   4   5   6   7   8   9   10   Reject H_0

Step 5: Conclusion.

$x = 7$ runs is not in the rejection region. Thus, there is insufficient evidence at the 5 percent level to reject the assumption that the sample is random.

MINITAB EXAMPLE 15.14   Use MINITAB to determine the number of runs above or below the median for the weights in Example 15.13.

Menu Commands:
Calc ➤
Column Statistics
For **Statistic**
select **Median**
In **Input var.** *box*
enter **C1**
Select OK

Stat ➤
Nonparametrics ➤
Runs Test
In **Variables** *box*
enter **C1**
Select **Above and below**
button and enter **8.24**
Select OK

Solution

First, store the data in column C1 and use the **MEDIAN** command to determine the median of the sample.

```
SET C1
8.13 8.28 8.31 8.24 8.17 8.29 8.10 8.27 8.26 8.04 8.21
END
MEDIAN C1
```

MINITAB will give the median value of 8.24. To obtain the number of runs above or below the median, type the following.

```
RUNS 8.24 C1      # 8.24 is the median of the data in C1
```

MINITAB will produce the output in Exhibit 15.5.

The number of runs, 7, appears in Exhibit 15.5, and MINITAB also gives its approximate *p*-value based on a normal approximation. When there are too few observations to assure a good approximation, a warning message is printed, as has occurred in this illustration. The approximation is generally good if at least 10 observations are below the median and at least 10 are above it.

MINITAB's output for the **RUNS** test also gives the number of observations above the median (5 for the above example) and the number of observations at or below the median (6 in the above data). The latter is labeled "BELOW" in Exhibit 15.5.

■■■ Exhibit 15.5

```
MTB > MEDIAN C1
    MEDIAN =      8.2400
MTB > RUNS 8.24 C1

    C1

    K =      8.2400

    THE OBSERVED NO. OF RUNS =    7
    THE EXPECTED NO. OF RUNS =    6.4545
     5 OBSERVATIONS ABOVE K    6 BELOW
  * N SMALL--FOLLOWING APPROX. MAY BE INVALID
            THE TEST IS SIGNIFICANT AT   0.7265
            CANNOT REJECT AT ALPHA = 0.05
```

SECTION 15.6 EXERCISES

The Runs Test for Randomness

For each of the following exercises that involve more than one row of data, the sequence of observations is row-wise.

15.66   Determine the number of runs in the following sequence.

| X X X Y Y Y X X Y Y Y X X X Y X Y X Y |

15.67   Determine the number of runs above or below the median for the following sequence of 11 numbers.

| 15   76   45   34   26   54   21   67   49   18   34 |

15.68   Construct a sequence of 15 elements consisting of the letters A and B, where the sequence contains 7 runs.

15.69   Construct a sequence of 13 numbers for which the median is 25 and there are 5 runs above or below the median.

15.70   A calculator with a random number key generated the following sequence of random digits.

| 9   5   6   7   2   9   5   7   0   2   4   6   3   5 |

Classify each digit as odd or even, and test if sufficient evidence exists to conclude a lack of randomness in the occurrence of even and odd integers. Use $\alpha = 5$ percent.

15.71   A coin was tossed 25 times, and the following sequence was observed.

H T T H H T T T H T T T T
H H T H T H T H H H T H

Do the data suggest that heads and tails are occurring in a nonrandom manner? Use Table 7 and a significance level of 5 percent.

15.72   Use the large-sample runs test to do Exercise 15.71.

15.73   A social club claims that it does not discriminate with regard to membership in its organization. During the last two years, the club has extended membership offers to 38 people, and the sexes of those who have received invitations are listed below in the order in which they were given.

M M M M M F M M M F M M M M F F M M M
M M F M M F M M M M F F F M M M M M F

Does there appear to be a lack of randomness in the sequence of sexes? Use the large-sample runs test and a significance level of 1 percent.

15.74   Two professors, A and B, are scheduled to teach a section of English literature during the same period next semester. The registrar's office claims that students are randomly assigned to one of the sections as they register for the course. Forty students signed up for the course, and the section assignments in the order of registration are given below.

A	A	A	B	A	A	B	B	B	B	B	A	A	A	A	B	B	B	B
A	A	A	B	A	A	A	B	B	B	A	B	B	B	B	B	A	A	A

Is there sufficient evidence at the 5 percent significance level to reject the claim that section assignments were made randomly? Use Table 7 to perform the test.

15.75   For Exercise 15.74, use the large-sample runs test and a significance level of 0.05.

15.76   A self-service gasoline station sells two national brands of motor oil, B_1 and B_2. The sequence of sales for the last 30 oil purchases appears below.

B_1	B_2	B_1	B_1	B_2	B_1	B_2	B_1	B_2	B_1	B_2	B_1	B_2	B_2	B_1
B_2	B_1	B_2	B_1	B_2	B_2	B_1	B_1	B_1	B_2	B_2	B_1	B_1	B_2	B_1

Does there appear to be a lack of randomness in the order in which the brands are selected? Test at the 5 percent level.

15.77   Fifty hamburgers were purchased from a drive-in restaurant in the chronological order listed below. The following figures give the amount of fat in grams for each hamburger.

22.6	23.1	20.8	20.9	22.7	23.6	20.4	19.9	22.7	20.5
20.0	18.7	21.6	20.9	21.5	21.8	20.2	19.7	18.9	19.5
19.3	21.2	18.4	21.0	21.6	20.6	20.7	21.9	20.1	17.1
18.1	21.1	19.3	21.5	20.1	16.5	18.9	17.4	20.8	18.5
21.6	23.1	20.5	22.0	20.6	17.5	16.1	20.1	21.8	19.4

Do the data suggest nonrandomness in the fat contents? Test at the 5 percent level of significance (the median of the sample is 20.55).

MINITAB Assignments

M▶ 15.78   Use MINITAB to perform the test in Exercise 15.77.

M▶ 15.79   Use MINITAB to determine the number of runs above or below the median for the data in Exercise 15.70.

M▶ 15.80   The following are the ages of 70 adults who recently responded to an ad for life insurance designed for older citizens. The ages are listed in the order of response.

69	82	68	82	76	74	69	76	59	62
72	58	68	90	58	67	59	56	62	71
63	82	64	59	56	65	67	69	64	55
74	78	75	71	67	74	81	59	67	63
62	64	63	59	68	67	62	64	59	57
67	68	62	64	59	56	57	62	63	64
91	75	87	56	65	64	69	79	57	75

Use MINITAB to test if the ages occur in a nonrandom pattern. Use a 5 percent significance level.

15.7   SPEARMAN'S RANK CORRELATION COEFFICIENT

The correlation coefficient that was introduced in Chapter 3 is called Pearson's product-moment correlation and is based on the assumption that x and y have a bivariate normal distribution. In the early part of this century, Charles Spearman developed a nonparametric correlation coefficient that is applicable under very general conditions. It is called **Spearman's rank correlation coefficient** and, interestingly, it is calculated with the same formula used for finding Pearson's product-moment correlation coefficient. Spearman's correlation coefficient, however, uses the ranks of the samples, rather than the actual measurements.

To illustrate the calculation of Spearman's rank correlation coefficient, r_s, suppose that a consumer testing group evaluated seven brands of peanut butter on the basis of several factors such as taste, texture, aroma, color, and so on. The total number of points assigned to each brand and its retail price are listed below.

Brand	Price	Rating
1	$1.69	65
2	$1.89	83
3	$1.85	89
4	$1.29	52
5	$1.49	79
6	$1.25	58
7	$1.45	67

We will calculate r_s to measure the strength of the relation between a brand's rating and its price. To determine the rank correlation coefficient, first replace the prices by their relative ranks, and then do the same for the ratings. Ties are handled by assigning mean ranks to tied values as described in Section 15.3. The two ranks for each brand are given below, where the ranks are denoted by x and y.

Brand	Price (x)	Rating (y)	xy	x^2	y^2
1	5	3	15	25	9
2	7	6	42	49	36
3	6	7	42	36	49
4	2	1	2	4	1
5	4	5	20	16	25
6	1	2	2	1	4
7	3	4	12	9	16
	28	28	135	140	140

The rank order correlation coefficient r_s can be calculated with the same formula that was used in Chapter 3 for r.

$$r_s = \frac{SS(xy)}{\sqrt{SS(x)SS(y)}} = \frac{\Sigma xy - \dfrac{(\Sigma x)(\Sigma y)}{n}}{\sqrt{\left[\Sigma x^2 - \dfrac{(\Sigma x)^2}{n}\right]\left[\Sigma y^2 - \dfrac{(\Sigma y)^2}{n}\right]}}$$

$$= \frac{135 - \dfrac{(28)(28)}{7}}{\sqrt{\left[140 - \dfrac{(28)^2}{7}\right]\left[140 - \dfrac{(28)^2}{7}\right]}} = \frac{23}{\sqrt{(28)(28)}} = 0.821$$

Note that $\Sigma x = \Sigma y$, $\Sigma x^2 = \Sigma y^2$, and $SS(x) = SS(y)$. When there are no ties in the values of each sample, these relations always hold. In fact, in this case there is a much simpler formula that can be used to calculate r_s. This is given below, and it is recommended that you always use this formula except in situations for which there are several ties.

PROCEDURE

Spearman's Rank Correlation Coefficient:

$$r_s = \frac{SS(xy)}{\sqrt{SS(x)SS(y)}},$$

where

$$SS(xy) = \Sigma xy - \frac{(\Sigma x)(\Sigma y)}{n},$$

$$SS(x) = \Sigma x^2 - \frac{(\Sigma x)^2}{n},$$

$$SS(y) = \Sigma y^2 - \frac{(\Sigma y)^2}{n}.$$

When there are no ties in the values of each sample, the following shortcut formula is algebraically equivalent.

$$r_s = 1 - \frac{6\Sigma d_i^2}{n(n^2 - 1)},$$

where

$$d_i = x_i - y_i.$$

Note:

1. x_i is the rank of the ith observation in sample 1, y_i is the rank of the ith observation in sample 2, n is the number of pairs.

2. Unless there are several ties, it is recommended that the shortcut formula be used.

EXAMPLE 15.15 Use the shortcut formula to calculate the rank correlation coefficient for the peanut butter ratings and prices.

Solution

The relative ranks for the prices and the ratings are copied below. We have also added a column of differences in the ranks and a column of their squares.

Brand	Price (x)	Rating (y)	$d = x - y$	d^2
1	5	3	2	4
2	7	6	1	1
3	6	7	-1	1
4	2	1	1	1
5	4	5	-1	1
6	1	2	-1	1
7	3	4	-1	1
				$\Sigma d^2 = 10$

$$r_s = 1 - \frac{6\Sigma d^2}{n(n^2 - 1)} = 1 - \frac{6(10)}{7(49 - 1)} = 0.821$$

The rank correlation coefficient r_s can be used as the test statistic for testing the null hypothesis that there is no association between two variables. Letting ρ_s denote the true rank correlation between the variables, the null hypothesis is $H_0: \rho_s = 0$, and this is rejected if the value of r_s differs significantly from 0. Critical values that define the rejection region have been tabulated, and some of these values are given in Table 8 of Appendix A. The hypothesis test is described below and is illustrated in Example 15.16.

PROCEDURE

Spearman's Test for Rank Correlation:
For testing the hypothesis

$$H_0: \rho_s = 0 \quad \text{(there is no association between the variables)}$$

the test statistic is

(15.6) $$r_s = 1 - \frac{6\Sigma d_i^2}{n(n^2 - 1)}.$$

The rejection region is given by the following.

For $H_a: \rho_s < 0$, H_0 is rejected when $r_s < -r_\alpha$;

For $H_a: \rho_s > 0$, H_0 is rejected when $r_s > r_\alpha$;

For $H_a: \rho_s \neq 0$, H_0 is rejected when $r_s < -r_{\alpha/2}$ or $r_s > r_{\alpha/2}$.

Assumption:
The sample of pairs is random.

Note:

1. The number of pairs is n, and d_i is the difference between ranks for the two sample values in the ith pair.

2. The value of r_α or $r_{\alpha/2}$ that determines the rejection region is found from Table 8.

EXAMPLE 15.16   In a 1988 special report on "America's Best Colleges," *U.S. News & World Report* gave the following tuition charges and the average annual pay (1987–88) for full professors at the following large universities. With $\alpha = 5$ percent, use Spearman's test to determine if the sample correlation coefficient is significantly different from zero.

School	Tuition	Avg. Pay
Calif. Inst. of Tech.	$11,789	$69,900
Columbia University	$12,878	$64,800
Georgetown University	$11,990	$64,300
Harvard University	$13,665	$73,200
Princeton University	$13,380	$67,800
Stanford University	$12,564	$70,800
University of Chicago	$12,200	$64,200
Univ. of Cal./Berkeley	$ 6,037	$64,200
Univ. of Pennsylvania	$12,750	$64,300
Yale University	$12,960	$67,700

Solution

Step 1: Hypotheses.

$$H_0: \rho_s = 0 \quad \text{(tuition and salary are not associated)}$$

$$H_a: \rho_s \neq 0 \quad \text{(tuition and salary are associated)}$$

Step 2: Significance level.

$$\alpha = 0.05$$

Step 3: Calculations.

The 10 tuitions are replaced by their ranks, and the same is done for the salaries.

School	Tuition(x)	Pay(y)	d	d^2
Calif. Inst. of Tech.	2	8	−6	36
Columbia University	7	5	2	4
Georgetown University	3	3.5	−0.5	0.25
Harvard University	10	10	0	0
Princeton University	9	7	2	4
Stanford University	5	9	−4	16
University of Chicago	4	1.5	2.5	6.25
Univ. of Cal./Berkeley	1	1.5	−0.5	0.25
Univ. of Pennsylvania	6	3.5	2.5	6.25
Yale University	8	6	2	4
				$\Sigma d^2 = 77$

$$r_s = 1 - \frac{6\Sigma d^2}{n(n^2 - 1)} = 1 - \frac{6(77)}{10(100 - 1)} = 0.533$$

Step 4: Rejection region.

The rejection region is two tailed because the relation in H_a is $\neq$. Table 8 is used with $n = 10$ and a tail area of $\alpha/2 = 0.025$. The value of $r_{.025}$ is 0.648. Therefore, H_0 is rejected for $r < -0.648$ and for $r > 0.648$.

Reject H_0 Reject H_0

-1 -0.648 0   0.648   1

Step 5: Conclusion.

Since $r_s = 0.533$ is not in the rejection region, the null hypothesis is not rejected. There is insufficient evidence at the 0.05 level to conclude that there is an association between salary and tuition.

MINITAB EXAMPLE 15.17

MINITAB does not have an explicit command for calculating Spearman's rank correlation coefficient. However, it can easily be determined by using the **RANK** and **CORRELATION** commands. We will illustrate by using the tuition and salaries data presented in Example 15.16. We begin by entering the data in columns C1 and C2.

```
READ C1 C2
11789 69900
12878 64800
11990 64300
13665 73200
13380 67800
12564 70800
12200 64200
 6037 64200
12750 64300
12960 67700
END
```

Menu Commands:
Manip ➤
Rank
In **Rank data in** *box*
enter **C1**
In **Store ranks in** *box*
enter **C3**
Select **OK**

Manip ➤
Rank
In **Rank data in** *box*
enter **C2**
In **Store ranks in** *box*
enter **C4**
Select **OK**

Stat ➤
Basic Statistics ➤
Correlation
In **Variables** *box enter*
C3 C4
Select **OK**

We now need to obtain the relative ranks for the numbers in C1 and C2. The **RANK** command accomplishes this.

```
RANK C1 C3    # ranks nos. in C1 and puts ranks in C3
RANK C2 C4
```

The ranks that have been created in columns C3 and C4 are shown in Exhibit 15.6. With these columns of ranks, Spearman's rank correlation can be obtained by applying the **CORRELATION** command to C3 and C4. The results appear in Exhibit 15.6. Notice that the rank correlation of 0.530 differs slightly from the value of 0.533 that was obtained in Example 15.16. This is because there are ties in the column of salaries, and in Example 15.16 the shortcut formula was used. Since MINITAB uses the exact formula, there is a small discrepancy in the results.

▨ Exhibit 15.6

```
MTB > PRINT C3 C4

    ROW    C3     C4

      1     2    8.0
      2     7    5.0
      3     3    3.5
      4    10   10.0
      5     9    7.0
      6     5    9.0
      7     4    1.5
      8     1    1.5
      9     6    3.5
     10     8    6.0

MTB > CORRELATION C3 C4

Correlation of C3 and C4 = 0.530
```

SECTION 15.7 EXERCISES

Spearman's Rank Correlation Coefficient

15.81   Use Table 8 to determine the rejection region for testing the hypotheses

$$H_0: \rho_s = 0$$

$$H_a: \rho_s \neq 0$$

where $\alpha = 0.05$ and $n = 24$.

15.82   Use Table 8 to determine the rejection region for detecting a positive rank correlation at the significance level $\alpha = 0.01$. Assume there are 20 pairs of data.

15.83   Use Table 8 to determine the rejection region for detecting a negative rank correlation at the significance level $\alpha = 0.05$, where $n = 13$ points.

15.84   Calculate Spearman's rank correlation coefficient for the following data.

x	37	32	36	31	35
y	88	82	85	80	84

15.85   The table on the following page gives the populations of the 8 most populous metropolitan areas in the United States in 1990. Also given is each area's population rank in 1980 (*The Wall Street Journal,* March 3, 1990). Rank the 1990 populations from high (assign a rank of 1) to low (assign a rank of 8), and calculate the correlation coefficient of the population ranks for 1980 and 1990.

Metropolitan Area	1990 Population (in thousands)	1980 Population Rank
L.A./Long Beach	8,771	2
New York	8,625	1
Chicago	6,308	3
Philadelphia	4,973	4
Detroit	4,409	5
Washington, DC	3,710	7
Houston	3,509	8
Boston	2,837	6

15.86   Do the following data provide sufficient evidence of an association between x and y? Use Spearman's rank correlation coefficient, and test at the 10 percent significance level.

x	y
185	65
437	32
287	57
930	21
487	43
431	39
846	25

15.87   The following sample gives the tread depth (in hundredths of an inch) and the number of miles of usage (in thousands) for a sample of 10 tires of the same brand. In Exercise 14.19, the reader was asked to calculate Pearson's correlation coefficient r for the data ($r = -0.909$). Calculate Spearman's rank correlation coefficient r_s.

Miles (x)	39	37	35	15	25	38	18	60	36	40
Tread (y)	10	15	16	30	21	14	34	3	10	14

15.88   Two computer magazines performed independent evaluations of 9 personal computer systems. Each magazine ranked the systems from best to worst, and their rankings appear in the table on the following page. Is there sufficient evidence of a positive correlation between the rankings of the 2 magazines? Test at the 5 percent significance level.

Computer System	First Magazine's Ratings	Second Magazine's Ratings
1	6	7
2	2	3
3	8	9
4	3	2
5	1	1
6	4	5
7	9	8
8	5	4
9	7	6

15.89   Use the results of Exercise 15.87 to determine if there is sufficient evidence to conclude that mileage and tread depth are negatively correlated. Test with $\alpha = 5$ percent.

15.90   Determine the approximate p-value for the test in Exercise 15.88.

15.91   A professor conducted a study to determine if there is an association between the final examination scores in her course and the times required to complete the exam. Exam scores and times (in minutes) appear below for 15 students.

Student	Exam Score	Time
1	86	108
2	93	100
3	73	115
4	78	113
5	54	118
6	93	99
7	69	110
8	78	109
9	84	111
10	82	117
11	41	120
12	67	116
13	98	89
14	74	112
15	71	110

Use Spearman's rank correlation coefficient to determine if there is an association between the exam score and the time required to complete the exam. Test at the 1 percent significance level.

MINITAB Assignments

M▶ 15.92   The Dow Jones Travel Index tracks the cost of hotel and car rental rates in the 20 major cities listed below. For its May 7, 1996, survey, the following Tuesday through Thursday daily room rates and car rental daily rates were given (*The Wall Street Journal,* May 10, 1996). All figures are in dollars.

City	Hotel	Car	City	Hotel	Car
Atlanta	152	52	Minneapolis	114	52
Boston	180	46	New Orleans	133	52
Chicago	167	51	New York	225	63
Cleveland	119	43	Orlando	104	38
Dallas	115	49	Phoenix	149	33
Denver	113	37	Pittsburgh	124	38
Detroit	119	49	St. Louis	127	62
Houston	135	50	San Francisco	161	42
Los Angeles	140	36	Seattle	106	38
Miami	126	30	Washington, D.C.	179	48

Is there a positive association between daily room rates and car rental rates? Test at the 5 percent level by having MINITAB calculate the rank correlation coefficient.

M▶ 15.93   Use MINITAB to calculate the rank correlation coefficient for the data in Exercise 15.91.

M▶ 15.94   The chapter opener referred to a study conducted by *MONEY* magazine (March, 1990) in which 50 tax professionals were hired to complete a 1040 Federal income tax return for a hypothetical family. Only 2 of the 50 professionals determined the correct tax due. The table on the following page gives the fee charged by each preparer and the amount by which their result differed from the correct tax (all figures are in dollars).

Fee	Error	Fee	Error	Fee	Error	Fee	Error
990	2232	300	53	1015	1484	750	4228
795	2019	450	189	600	1659	2500	4307
750	1665	950	256	1100	1831	276	5064
1950	1400	800	457	422	1852	900	5188
400	1345	1450	494	550	2051	280	5454
960	1111	1425	609	1200	2074	4000	5473
1300	875	650	614	2000	2098	750	5518
850	651	770	737	1500	2198	970	5973
1450	618	750	808	1100	2788	1150	8103
1360	215	975	912	1100	2867	400	8288
1285	0	720	1124	1750	3079	520	9178
750	0	960	1300	1350	3240		
271	54	450	1407	640	3338		

Assess if the fee charged is related to performance by using MINITAB to calculate the rank correlation coefficient between the fee and the size of the error.

Chapter 15 discussed **nonparametric tests** that can be used in place of their parametric counterparts that were considered in previous chapters. Appealing features of nonparametric tests are their less restrictive assumptions and simpler calculations. Generally, they are slightly less powerful than corresponding parametric tests when the underlying assumptions of the latter hold. However, in situations for which the assumptions are not true, a nonparametric test is often more powerful.

The following nonparametric tests of hypotheses were discussed in this chapter.

1. The One-Sample Sign Test for a Population Median $\tilde{\mu}$.

$$H_0: \tilde{\mu} = \tilde{\mu}_0 \quad (\tilde{\mu}_0 \text{ is a constant})$$

PROCEDURE

The test statistic is

$x =$ the number of sample values that exceed $\tilde{\mu}_0$.

> **PROCEDURE**
>
> A large-sample test statistic is
>
> $$z = \frac{x - 0.5n}{0.5\sqrt{n}},$$
>
> where n is the number of sample values retained after discarding those equal to $\tilde{\mu}_0$.

2. The Two-Sample Sign Test: Paired Samples.

$H_0: \tilde{\mu}_D = 0$ ($\tilde{\mu}_D$ is the median of the population of differences)

> **PROCEDURE**
>
> The test statistic is
>
> $x =$ the number of pairs with a positive difference.

> **PROCEDURE**
>
> A large-sample test statistic is
>
> $$z = \frac{x - 0.5n}{0.5\sqrt{n}},$$
>
> where n is the number of pairs with nonzero differences.

3. The Mann-Whitney U-Test: Independent Samples.

$H_0: (\tilde{\mu}_1 - \tilde{\mu}_2) = 0$

> **PROCEDURE**
>
> The test statistic is
>
> $$z = \frac{U_1 - \dfrac{n_1 n_2}{2}}{\sqrt{\dfrac{n_1 n_2(n_1 + n_2 + 1)}{12}}},$$
>
> where
>
> $$U_1 = n_1 n_2 + \frac{n_2(n_2 + 1)}{2} - R_2,$$
>
> R_2 is the rank sum for sample two,
> n_1 and n_2 are the sample sizes.

4. The Kruskal-Wallis H-Test for a Completely Randomized Design.

$$H_0: \tilde{\mu}_1 = \tilde{\mu}_2 = \cdots = \tilde{\mu}_k$$

PROCEDURE

The test statistic is

$$H = \frac{12}{n(n+1)} \Sigma \frac{R_i^2}{n_i} - 3(n+1),$$

where R_i is the rank sum of sample i, and $n = n_1 + \cdots + n_k$.

5. The Friedman F_r-Test for a Randomized Block Design.

$$H_0: \tilde{\mu}_1 = \tilde{\mu}_2 = \cdots = \tilde{\mu}_k$$

PROCEDURE

The test statistic is

$$F_r = \frac{12}{bk(k+1)} \Sigma R_i^2 - 3b(k+1),$$

where R_i is the rank sum of the ith treatment, k is the number of treatments, and b is the number of blocks.

6. The Runs Test for Randomness.

$$H_0: \text{The sequence of sample observations is random.}$$

PROCEDURE

The test statistic is

$$x = \text{the number of runs in the sequence.}$$

PROCEDURE

A large-sample test statistic is

$$z = \frac{x - \mu}{\sigma},$$

where

$$\mu = \frac{2n_1 n_2}{n_1 + n_2} + 1, \qquad \sigma = \sqrt{\frac{2n_1 n_2 (2n_1 n_2 - n_1 - n_2)}{(n_1 + n_2)^2 (n_1 + n_2 - 1)}}$$

n_1 and n_2 denote the number of occurrences of the two types in the sample.

7. Spearman's Test for Rank Correlation.

$H_0: \rho_s = 0$ (there is no association between the variables)

PROCEDURE

The test statistic is

$$r_s = 1 - \frac{6\Sigma d_i^2}{n(n^2 - 1)},$$

where n is the number of pairs, and d_i is the difference between ranks for the two sample values in the ith pair.

Key Words

In reviewing this chapter, you should be able to define, explain, and illustrate each of the following.

nonparametric test *(page 687)*

one-sample sign test *(page 688)*

two-sample sign test *(page 691)*

Mann-Whitney U-test *(page 700)*

rank sum *(page 701)*

Kruskal-Wallis H-test *(page 709)*

Friedman F_r-test *(page 717)*

run *(page 726)*

runs test for randomness *(page 727)*

M▶ MINITAB Commands

READ _ _ *(page 693)*

END *(page 693)*

LET _ *(page 693)*

STEST _ _; *(page 693)*
ALTERNATIVE _.

SET _ *(page 704)*

MANN-WHITNEY _ _; *(page 704)*
ALTERNATIVE _.

KRUSKAL-WALLIS _ _ *(page 712)*

NAME _ _ *(page 720)*

FRIEDMAN _ _ _ *(page 720)*

MEDIAN _ *(page 732)*

RUNS _ _ *(page 732)*

RANK _ _ *(page 740)*

CORRELATION _ _ *(page 740)*

REVIEW EXERCISES

15.95   A movie theater shows its featured film at 7:00 P.M. and 9:30 P.M. Random samples of 14 patrons at each showing were selected and asked their age. The results were the following.

Showing	Patrons' Ages													
7:00 P.M.	57	49	41	56	39	40	37	26	19	68	49	73	28	44
9:30 P.M.	23	20	32	43	21	27	25	52	40	26	34	38	25	21

Use the Mann-Whitney U-test to determine if there is a difference in the median age of patrons for the two showings. Test at the 5 percent significance level.

15.96   During a recent broadcast, a radio talk show received 28 calls. The sexes of the callers, in order of occurrence, are given below.

F M M F F F F M F M F F M M M M M F M F F F F M M M M M

Use Table 7 to determine if there is a lack of randomness with regard to the sex of the callers. Let $\alpha = 0.05$.

15.97   Use the large-sample runs test to solve Exercise 15.96.

15.98   Eleven secretaries participated in a one-day workshop designed to increase their typing speeds. The participants' speeds (words per minute) were measured before and after the workshop. They are given below.

Secretary	Speed Before the Program	Speed After the Program
1	59	64
2	43	51
3	58	55
4	47	59
5	37	48
6	62	62
7	41	45
8	50	52
9	43	53
10	49	52
11	63	65

Use a nonparametric test to determine if there is sufficient evidence at the 0.05 level to conclude that the course tends to increase one's typing speed.

15.99   Determine the p-value for the test in Exercise 15.98.

15.100   A Ferris wheel ride is supposed to last at least 5 minutes. Thirty-six operations of the ride were timed, and the following times (in seconds) were obtained.

283	274	296	301	294	288	302	275	297
291	306	316	285	296	289	295	300	291
305	289	298	287	281	295	291	290	316
275	296	284	303	295	303	290	278	299

Determine if there is sufficient evidence at the 0.01 level to conclude that the median duration of the ride is less than 5 minutes.

15.101 Determine the p-value for the test in Exercise 15.100.

15.102 A company interviewed 10 recent college graduates for an engineering position. Each applicant was assigned a composite score based on a written exam and personal interviews. The scores and college grade point averages of the candidates are given in the following table.

Applicant	Interview Score	GPA
1	47	3.12
2	43	3.59
3	37	2.98
4	46	3.71
5	30	2.76
6	31	3.21
7	44	3.34
8	34	2.95
9	40	3.06
10	35	2.88

Is there sufficient evidence to indicate that the interview score is positively correlated with grade point average? Use a nonparametric test and $\alpha = 0.01$.

15.103 Determine the approximate p-value for the test in Exercise 15.102.

15.104 A completely randomized design produced the following sample results.

Sample 1	Sample 2	Sample 3	Sample 4
65	59	72	56
76	60	75	54
68	62	67	53
63	58	74	51
64	61	71	55
59			58

Use the Kruskal-Wallis H-test to determine if at least two of the sampled populations differ in location. Test with $\alpha = 1$ percent.

15.105 A state conducts a daily lottery for which a 3-digit number is selected. The selected numbers for 30 consecutive drawings are given on the following page (in the order row 1, then row 2).

| 711 | 356 | 189 | 347 | 982 | 597 | 735 | 916 | 069 | 901 | 735 | 914 | 036 | 300 | 918 |
| 131 | 016 | 733 | 868 | 971 | 350 | 071 | 415 | 555 | 783 | 027 | 271 | 244 | 612 | 085 |

Use Table 7 to determine if the data suggest nonrandomness in the numbers drawn. Use $\alpha = 5$ percent (the sample median is 485).

15.106   For Exercise 15.105, use the large-sample runs test and a significance level of 5 percent.

15.107   Dangerous chemicals from industrial wastes linked to cancer and other diseases can enter the food chain through their presence in lake sediments. The amounts of DDT were determined for samples of trout from four lakes and are given below.

Lake	Levels of DDT in parts per million					
One	1.7	1.4	1.9	1.1	2.1	1.8
Two	0.3	0.7	0.5	0.1	1.1	0.9
Three	2.7	1.9	2.0	1.5	2.6	
Four	1.2	3.1	1.9	3.7	2.8	3.5

Do the data provide sufficient evidence at the 0.05 level to conclude that there is a difference in the median DDT levels for the four lakes?

15.108   Determine the approximate p-value for the test in Exercise 15.107.

15.109   A randomized block design was used to determine if at least 2 of 3 treatments differ in their central location. Six blocks were used, and the resulting data are given below. Formulate a suitable set of hypotheses and use a nonparametric method to test at the 5 percent significance level

	Treatment		
Block	1	2	3
1	211.8	210.9	212.8
2	211.1	211.1	213.0
3	211.1	210.1	211.9
4	211.9	210.5	212.9
5	211.4	210.3	212.3
6	212.8	210.7	212.7

15.110   The following data were obtained from a randomized block design that was used to compare the medians of 5 treatments. Use a nonparametric method to test if a significant difference exists at the 1 percent level.

Block	Treatment				
	1	**2**	**3**	**4**	**5**
1	211.8	210.9	212.8	211.9	210.6
2	212.1	211.1	213.0	212.1	211.1
3	211.1	210.1	211.9	211.8	209.8
4	211.9	210.5	212.9	212.0	210.3
5	211.4	210.3	212.3	211.9	210.3
6	211.7	210.7	212.7	212.0	209.3
7	210.9	210.1	211.6	211.7	209.8
8	212.4	211.4	213.5	212.1	211.1
9	208.6	208.4	208.6	208.1	207.9
10	199.6	200.1	200.6	199.5	199.2

15.111 A computer magazine evaluated and rated 4 new color-capable laser prin-ters. Each model was assigned a rating in each of 8 performance categories, and the scores achieved are given in the following table.

Performance Category	Laser Printer			
	I	**II**	**III**	**IV**
1	93	79	97	78
2	79	81	93	83
3	85	82	95	80
4	72	84	82	85
5	91	90	96	88
6	88	78	89	81
7	76	75	93	76
8	68	77	92	78

At the 5 percent level, is there sufficient evidence to conclude that a dif-ference exists in the average ratings for the 4 laser printer models? Use a nonparametric test.

15.112 Find the approximate p-value for the test in Exercise 15.111.

15.113 Refer to Exercise 15.111 and use the Friedman F_r-test to determine if there is a significant difference in the average ratings for models I and II. Let $\alpha = 0.05$.

MINITAB Assignments

M> 15.114 Refer to Exercise 15.100, and use MINITAB to determine if there is suf-ficient evidence at the 0.05 level to conclude that the median duration of the ride is less than 5 minutes.

M> 15.115 Refer to Exercise 15.102, and use MINITAB to calculate the rank corre-lation coefficient between interview score and grade point average.

M> 15.116 Use MINITAB to determine the number of runs above or below the median in Exercise 15.105.

M▶ 15.117 A furniture manufacturer wanted to compare drying times for 4 brands of stains. Each stain was applied to 10 chairs, and the drying times in minutes appear below.

Brand 1	Brand 2	Brand 3	Brand 4
80.6	91.6	90.5	86.7
81.3	83.5	98.5	75.4
82.8	83.4	97.5	79.7
81.5	88.6	99.9	76.5
80.4	96.7	96.9	75.7
79.7	84.8	90.5	84.7
82.3	88.4	96.7	74.5
81.7	89.5	93.8	83.3
80.6	84.4	97.8	84.2
81.5	85.1	96.8	75.3

Is there sufficient evidence to conclude that a difference exists in the median drying times for the 4 brands? Use MINITAB to apply a nonparametric test, and let $\alpha = 0.05$.

M▶ 15.118 Refer to Exercise 15.117, and use MINITAB to apply the Mann-Whitney U-test to determine whether a difference exists in the median drying times for brands 1 and 3. Test at the 0.01 significance level.

M▶ 15.119 A marketing firm conducted a study to determine if there is a difference in the average price of two brands of frozen pizza. Thirty-three food stores were randomly selected, and the prices of the two brands were recorded. They are given below, where the figures are in dollars.

Store	Brand A	Brand B	Store	Brand A	Brand B
1	3.69	3.59	18	3.76	3.70
2	4.19	4.03	19	3.85	3.79
3	3.75	3.75	20	3.64	3.68
4	3.50	3.43	21	3.65	3.57
5	3.65	3.55	22	3.76	3.67
6	3.73	3.69	23	4.09	3.98
7	3.89	3.95	24	3.94	3.90
8	3.93	3.75	25	3.85	3.79
9	3.95	3.88	26	3.69	3.69
10	4.25	4.00	27	3.49	3.79
11	3.41	3.38	28	3.76	3.58
12	3.68	3.60	29	3.84	3.76
13	3.86	3.84	30	3.96	3.87
14	3.83	3.79	31	3.58	3.50
15	3.72	3.75	32	3.74	3.71
16	3.99	3.99	33	3.87	3.76
17	3.85	3.80			

Use MINITAB to perform a nonparametric test to determine if one of the brands tends to be priced higher than the other. Test at the 0.01 significance level.

M▶ 15.120   Use MINITAB to perform the test in Exercise 15.119 by utilizing a second nonparametric procedure.

M▶ 15.121   Use MINITAB to perform the test in Exercise 15.110.

M▶ 15.122   Have MINITAB perform the test in Exercise 15.111.

MULTIPLE REGRESSION MODELS

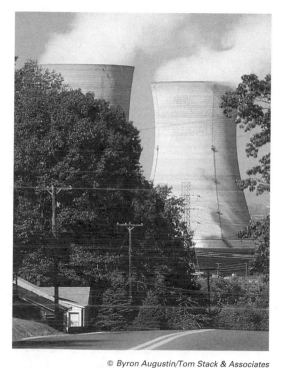

© Byron Augustin/Tom Stack & Associates

THE DEATH BLOW TO NUCLEAR POWER IN THE UNITED STATES?

In a picturesque rural area near Harrisburg, Pennsylvania during the early morning hours of March 28, 1979, a sequence of improbable events began to unfold that would profoundly affect the nation's energy supply for future decades. At the Three Mile Island nuclear power facility, a water valve in the primary cooling system around the nuclear core was accidently closed by a maintenance crew while working on an innocuous problem. An emergency backup system successfully opened a pressure relief valve but later failed to close it, resulting in the evaporation of the nuclear core's primary coolant. These two mishaps were followed by a third unlikely event—the control room operators' failure to notice for several hours the dangerously rising temperature readings being registered by an obscurely placed gauge. The culmination of these mischances was the uncontrolled release of thousands of gallons of radioactive water within the plant's facilities, resulting in the country's worst nuclear accident. The development of nuclear power, with its promises of an inexpensive, clean, and virtually unlimited energy supply, came to an abrupt halt in the United States on that fateful Wednesday morning.

*In the hours following the accident, uncertainty and apprehension about its severity prevailed among local residents. Although the governor of Pennsylvania did not issue a mass evacuation of the surrounding area, the public's reaction produced the largest evacuation in the country's history. Within a 15-mile radius of the plant, an estimated 150,000 people left the area for an average of 5 days. To investigate this unprecedented evacuation, a governmental study was conducted to determine the demographic characteristics of local adult evacuees. A **multiple regression model** was developed that associated evacuation behavior with 24 demographic predictor variables. We will investigate similar models, and revisit the Three Mile Island study to consider some of the demographic variables used in the model.*

LOOKING

A H E A D

Chapter 3 provided a concise introduction to the descriptive aspects of modeling bivariate data with a straight line. Chapter 14 complemented this early coverage with a detailed discussion of inferential methods in relating a dependent (response) variable *y* and a single independent (predictor) variable *x*. Techniques were presented for investigating relationships between two variables such as a woman's systolic blood pressure *y* and her age *x*. Frequently, however, knowledge of a *single* predictor variable *x* will not be adequate for predicting *y*, and *multiple* independent variables must be used to model the relationship for *y*. For example, in addition to age, a woman's blood pressure depends on many factors such as weight, physical condition, level of stress, amount of exercise, and dietary, smoking, and drinking habits. Knowledge of these factors should enable one to predict blood pressures more precisely than is possible by using only information concerning age.

Chapters 3 and 14 considered only **simple regression models.** The models are called **simple** because they involve only one predictor variable *x*. The models discussed in this chapter are referred to as **multiple regression models** because they include two or more predictors. Like the preceding regression chapters, this chapter is primarily concerned with the process of fitting a model to a set of data, evaluating the model's potential usefulness, and utilizing the model for estimation and prediction. Fitting a multiple regression model by hand is a very tedious process that involves complex calculations and requires a knowledge of matrix algebra. Consequently, throughout this chapter we will rely on computer output to aid us in the model building process. Reliance on the computer minimizes hand calculations, emulates how regression is used in practical applications, and allows us to focus on the more important aspects of a multiple regression analysis.

16.1 MULTIPLE REGRESSION MODELS AND ASSUMPTIONS

In Chapter 14, we assumed **probabilistic models** of the form

$$y = \mu_y + \epsilon$$

where ϵ was the **random error component** of the model and μ_y, the **deterministic component,** was written as $\mu_y = \beta_0 + \beta_1 x$. For instance, we considered an ex-

TABLE 16.1			
PLACEMENT SCORES, SAT MATH SCORES, AND STATISTICS GRADES FOR 15 FRESHMEN			
Student	Placement Test Score x_1	SAT Math Score x_2	Numerical Grade y
1	21	610	69
2	17	600	72
3	21	680	94
4	11	550	61
5	15	580	62
6	19	660	80
7	15	620	65
8	23	780	88
9	13	520	54
10	19	590	75
11	16	630	80
12	25	730	93
13	8	490	55
14	14	530	60
15	17	560	64

ample in which the mathematics department of a liberal arts college uses a 25-point placement test to assist in assigning appropriate math courses to incoming freshmen. The department believes that the test is a good predictor of a student's final grade in its introductory statistics course. Based on the appearance of a scatter diagram, we tentatively assumed that the mean course grade μ_y is related to test score x by a straight line. Mathematically, we wrote this as

(16.1) $$\mu_y = \beta_0 + \beta_1 x$$

where this *line of means* is called the **regression line.** One might wonder if course grades y could be predicted more precisely by adding to the model a second independent variable, such as a student's score on the SAT mathematics test. To investigate this, consider Table 16.1 that gives the placement test score x_1, SAT math score x_2, and course grade y for each of the 15 freshmen considered in Chapter 14.

The 15 data points in Table 16.1 can be represented geometrically in 3-dimensional space, where the 3 dimensions correspond to the 3 variables x_1, x_2, and y. A plot of the points appears in Figure 16.1.

To consider if the inclusion of SAT math scores x_2 with placement test scores x_1 enhances the ability to predict course grades y, we will add the term $\beta_2 x_2$ to the model in Equation 16.1. Our tentative model is now

(16.2) $$\mu_y = \beta_0 + \beta_1 x_1 + \beta_2 x_2.$$

It should be noted that Equation 16.2 is a **deterministic model** that relates the *average* course grade of all students with a particular placement score x_1 and SAT

Figure 16.1
Geometrical display of the 15 data points.

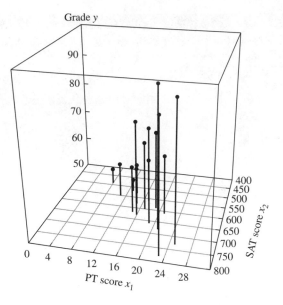

score x_2. To have the model describe an individual's course grade y, we must add a random error component ϵ that represents the deviation of each person's grade from the mean grade of all who have a placement score of x_1 and a SAT score of x_2. The resulting **probabilistic model** is given by Equation 16.3, together with its related assumptions that are needed in the following sections.

KEY CONCEPT

Multiple Regression Model with Two Independent Variables*

(16.3) $$y = \beta_0 + \beta_1 x_1 + \beta_2 x_2 + \epsilon$$

y is the **dependent (response) variable** being modeled,
x_1 and x_2 are **independent (predictor) variables**,
β_0, β_1, and β_2 are **parameters** to be estimated,
ϵ is a random variable called the **random error component.**

Assumptions:

1. For each pair of values of x_1 and x_2, the random error component ϵ has
   a. a normal probability distribution,
   b. a mean of 0,
   c. a constant variance that is denoted by σ^2.

2. For each pair of observations y_i and y_j, the associated random errors ϵ_i and ϵ_j are independent. That is, the error associated with one y-value does not affect the error associated with another y-value.

Note: The mean of y is given by $\mu_y = \beta_0 + \beta_1 x_1 + \beta_2 x_2$ and individual values of y deviate from this with a variance of σ^2.

---

* This model is called a **linear** model because it can be viewed as a linear function of the parameters β_0, β_1, and β_2. Later examples will show that linear models can contain terms that are nonlinear in the predictor variables x_i.

The coefficients β_0, β_1, and β_2 of the assumed model are unknown parameters that must be estimated from the sample data. The estimates are obtained by applying the same procedure as used for the straight line models considered in Chapter 14, namely, the method of least squares.

KEY CONCEPT

The Least Squares Estimated Model

For the assumed multiple regression model

$$y = \beta_0 + \beta_1 x_1 + \beta_2 x_2 + \epsilon$$

the estimated model is the least squares equation

$$\hat{y} = \hat{\beta}_0 + \hat{\beta}_1 x_1 + \hat{\beta}_2 x_2.$$

For n given observations y_1, y_2, . . . , y_n, the least squares equation minimizes the sum of the squared errors $SSE = \Sigma(y_i - \hat{y}_i)^2$.

Formulas exist for determining the least squares equation in a multiple regression model, but they are rarely used in practice because of their complexity and laborious calculations. Instead, computer software packages are employed. To have MINITAB obtain the estimated model for the data in Table 16.1, the values of x_1, x_2, and y are first stored in columns C1, C2, and C3, respectively.

```
READ C1-C3
21 610 69
17 600 72
21 680 94
11 550 61
15 580 62
19 660 80
15 620 65
23 780 88
13 520 54
19 590 75
16 630 80
25 730 93
 8 490 55
14 530 60
17 560 64
END
NAME C1 'X1' C2 'X2' C3 'Y'
```

Menu Commands:
Stat ➤
Regression ➤
Regression
In **Response** *box*
enter **C3**
In **Predictors** *box*
enter **C1 C2**
Select OK

The least squares equation is produced by typing the command **REGRESS** as follows.

```
REGRESS C3 2 C1 C2
```

In the above, the **REGRESS** command is followed by the column of y-values; then the number 2 (the number of predictor variables); and lastly, by the columns containing the values of the predictors. The output produced appears in Exhibit 16.1

Exhibit 16.1   MTB > REGR C3 2 C1 C2

```
The regression equation is
Y = - 9.7 + 0.771 X1 + 0.112 X2

Predictor          Coef        Stdev       t-ratio           p
Constant          -9.73        14.85         -0.66       0.525
X1               0.7708       0.6697          1.15       0.272
X2              0.11195      0.03860          2.90       0.013

s = 5.683        R-sq = 83.9%      R-sq(adj) = 81.3%

Analysis of Variance
SOURCE          DF          SS          MS          F          p
Regression       2       2026.2      1013.1      31.37      0.000
Error           12        387.5        32.3
Total           14       2413.7
```

In Exhibit 16.1, the least squares equation is labeled as the regression equation and is

(16.4) $\hat{y} = -9.7 + 0.771x_1 + 0.112x_2.$

This is the estimate of the assumed model $\mu_y = \beta_0 + \beta_1 x_1 + \beta_2 x_2$ that relates mean course grade μ_y to placement score x_1 and SAT score x_2. The estimate of the regression coefficient β_1 is 0.771, indicating that the average course grade is expected to increase by 0.771 if x_1 is increased by 1 unit and x_2 is held fixed. Similarly, the estimate of β_2, 0.112, implies that the average course grade is expected to increase by 0.112 for each 1-unit increase in SAT score x_2 when placement score x_1 is held constant. The estimate of β_0, the constant term in the model, will only have a practical interpretation when zero is included in the range of the data for both predictors x_1 and x_2. A useful interpretation of the constant term -9.7 in Equation 16.4 does not apply here, because a placement score with $x_1 = 0$ and an SAT score of $x_2 = 0$ are well outside the range of data in Table 16.1.

In Chapter 14, we saw that the simple regression model $\mu_y = \beta_0 + \beta_1 x$ represents geometrically in 2 dimensions a *line of means* with slope β_1. Similarly, the multiple regression model $\mu_y = \beta_0 + \beta_1 x_1 + \beta_2 x_2$ represents in 3-dimensional space a *plane of means,* called the **response surface** of the model. The coefficient of x_1, β_1, is called a **partial slope,** because β_1 measures the change in the mean of y for a 1-unit increase in x_1, when x_2 is held fixed. β_2 is also called a partial slope and has an analogous interpretation. For our example, the response surface of the estimated model in Equation 16.4 is shown in Figure 16.2.

At this point in our investigation of the relationship between final course grades with placement scores and SAT math scores, we have obtained an estimated model. However, before we adopt it and utilize it for estimation and prediction purposes,

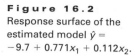

Figure 16.2
Response surface of the estimated model $\hat{y} = -9.7 + 0.771x_1 + 0.112x_2$.

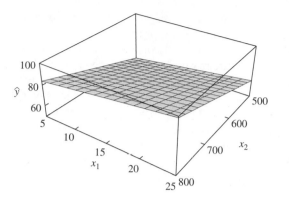

we need to evaluate the model's potential usefulness. In later sections, we will address the general issue of evaluating a model and testing its utility.

As we progress through the sections, we will see that multiple regression models are not limited to only two predictor variables. The techniques of this chapter apply to the following general multiple regression model involving k independent variables $x_1, x_2, \ldots, x_k$.

KEY CONCEPT

Multiple Regression Model with k Independent Variables

(16.5) $\qquad y = \beta_0 + \beta_1 x_1 + \beta_2 x_2 + \beta_3 x_3 + \cdots + \beta_k x_k + \epsilon$

y is the **dependent (response) variable** being modeled,
$x_1, x_2, \ldots, x_k$ are **independent (predictor) variables,**
$\beta_0, \beta_1, \beta_2, \ldots, \beta_k$ are **parameters** to be estimated,
ϵ is a random variable called the **random error component.**

Assumptions:

1. For each combination of values of $x_1, x_2, \ldots, x_k$, the random error component ϵ has
   a. a normal probability distribution,
   b. a mean of 0,
   c. a constant variance that is denoted by σ^2.

2. For each pair of observations y_i and y_j, the associated random errors ϵ_i and ϵ_j are independent. That is, the error associated with one y-value does not affect the error associated with another y-value.

Note:

1. The mean of y is given by $\mu_y = \beta_0 + \beta_1 x_1 + \beta_2 x_2 + \cdots + \beta_k x_k$ and individual values of y deviate from this with a variance of σ^2.

2. β_i (the coefficient of x_i) is called a **partial slope** and measures the change in the mean of y for a 1-unit increase in x_i *while the other predictor variables are held fixed.*

In the summary on page 761, the assumptions concerning the behavior of the random error components are needed for making inferences in following sections. In Section 16.7, we will consider some methods for checking if these regression assumptions are reasonable for a given data set. As for σ^2, the common variance of the random error components, we saw for the straight line models in Chapter 14 that σ^2 is a measure of how individual y-values fluctuate around the assumed model $\mu_y = \beta_0 + \beta_1 x$. Analogously, σ^2 in the multiple regression case measures the variation of y-values about the assumed model $\mu_y = \beta_0 + \beta_1 x_1 + \beta_2 x_2 + \cdots + \beta_k x_k$. As in the simple linear case, our estimate of σ^2 is based on the sum of the squared errors, $SSE = \Sigma(y_i - \hat{y}_i)^2$. In particular, σ^2 is estimated by dividing SSE by its degrees of freedom, df.

PROCEDURE

Estimating the Variance σ^2 of the Random Error Components

For the k-predictor model $y = \beta_0 + \beta_1 x_1 + \beta_2 x_2 + \beta_3 x_3 + \cdots + \beta_k x_k + \epsilon$,

$$\hat{\sigma}^2 = s^2 = \frac{SSE}{df} = \frac{SSE}{n - \text{number of } \beta\text{-parameters}} = \frac{SSE}{n - (k + 1)}$$

Note:

1. σ is called the **standard deviation of the model.**

2. s is the **estimated standard deviation of the model.**

3. s^2 appears as the **mean square error *(MSE)*** in MINITAB's analysis of variance table for regression.

To obtain a better feeling for the practical significance of s, you should remember that σ measures variability of y with respect to the *assumed model* and s measures the variability of the sample data with respect to the *estimated model* (the least squares equation). Consequently, a simple interpretation of s follows from the Empirical rule. We would expect that roughly 95 percent of the sample y-values will differ from their least squares predictions by at most $2s$ units. Few data points would be expected to have an error $(y - \hat{y})$ that is larger than $2s$.

EXAMPLE 16.1   Determine SSE and estimate σ, the standard deviation of the model, for the data pertaining to placement test scores x_1, SAT math scores x_2, and final statistics grades y.

Solution
Copied below is the MINITAB output obtained earlier and shown in Exhibit 16.1.

```
MTB > REGR C3 2 C1 C2

The regression equation is
Y = - 9.7 + 0.771 X1 + 0.112 X2
```

Predictor	Coef	Stdev	t-ratio	p
Constant	-9.73	14.85	-0.66	0.525
X1	0.7708	0.6697	1.15	0.272
X2	0.11195	0.03860	2.90	0.013

s = 5.683   R-sq = 83.9%   R-sq(adj) = 81.3%

Analysis of Variance

SOURCE	DF	SS	MS	F	p
Regression	2	2026.2	1013.1	31.37	0.000
Error	12	387.5	32.3		
Total	14	2413.7			

The sum of squares for error, *SSE*, its degrees of freedom, *df*, and its mean square, *MSE*, are highlighted in the analysis of variance table. Notice that $df = 12$. This follows from the fact that the assumed model $y = \beta_0 + \beta_1 x_1 + \beta_2 x_2 + \epsilon$ has 3 parameters (β_0, β_1, and β_2), and n is 15. Thus, $df = (15 - 3) = 12$.

The estimated variance of the model is

$$s^2 = \frac{SSE}{df} = \frac{387.5}{12} = 32.3.$$

The estimated standard deviation s could be obtained by taking the square root of $s^2 = 32.3$, but this is not necessary because s is given in the MINITAB output. Its value is also highlighted and appears just before the analysis of variance table. We see that $s = 5.683$.

16.2 R^2 and the Global F-Test of a Model's Usefulness

In the previous section, we began investigating how the final course grades y in a statistics course at a particular college are related to the students' placement test scores x_1 and their SAT math scores x_2. We started by assuming tentatively that an individual student's grade y is given by the probabilistic model

(16.6) $y = \beta_0 + \beta_1 x_1 + \beta_2 x_2 + \epsilon$

where the deterministic portion of the model is $\mu_y = \beta_0 + \beta_1 x_1 + \beta_2 x_2$ and denotes the mean course grade of all students with placement score x_1 and SAT score x_2. Next, MINITAB used the sample of 15 data points in Table 16.1 to obtain the least squares estimated model

(16.7) $\hat{y} = -9.7 + 0.771 x_1 + 0.112 x_2.$

Before adopting the assumed model in Equation 16.6 and using its estimated model in Equation 16.7, we need to check how well the model describes the relationship between y and the predictors x_1 and x_2.

The Multiple Coefficient of Determination R^2

In our study of simple linear models in Chapter 14, we used the coefficient of determination r^2 as a measure of how well a model fits a set of data. The same concept is applicable with multiple regression models. However, notation and terminology differ slightly in that R^2 is used and is called the **multiple coefficient of determination.**

In defining r^2 for straight line models, we used $SS(y)$, the **total sum of squares,** to measure the variability in the y-values with respect to their mean $\bar{y}$, since $SS(y)$ equals the sum of the squared deviations of the y-values from $\bar{y}$. SSE, the **error sum of squares,** measured the variability in the y-values with respect to the least squares line, since SSE equals the sum of the squared deviations of the data points from the line. The difference between these quantities, $SS(y) - SSE$, gave the reduction in the total variation in y that was explained by the model. This **explained variation in y** is called the **regression sum of squares.** We divided the regression sum of squares, $SS(y) - SSE$, by the total sum of squares, $SS(y)$, to obtain the ratio r^2 that gave the **proportion of the total variation in y that is explained by the model.** This same ratio of explained variation over total variation is used in the multiple regression case and is defined below.

PROCEDURE

The Multiple Coefficient of Determination R^2

$$(16.8) \qquad R^2 = \frac{\text{Regression } SS}{\text{Total } SS} = \frac{SS(y) - SSE}{SS(y)}$$

Note:

1. R^2 is the proportion of $SS(y)$, the total variation in y, that is explained by the least squares model.

2. $R^2 = 0$ when $SSE = SS(y)$, indicating a complete lack of fit of the least squares model to the data.

3. $R^2 = 1$ when $SSE = 0$, indicating a perfect fit of the least squares model to the data.

EXAMPLE 16.2

Obtain and interpret the value of the multiple coefficient of determination for the data pertaining to placement test scores x_1, SAT math scores x_2, and final statistics grades y.

Solution

The MINITAB output from Exhibit 16.1 is displayed below.

```
MTB > REGR C3 2 C1 C2

The regression equation is
Y = - 9.7 + 0.771 X1 + 0.112 X2
```

Predictor	Coef	Stdev	t-ratio	p
Constant	-9.73	14.85	-0.66	0.525
X1	0.7708	0.6697	1.15	0.272
X2	0.11195	0.03860	2.90	0.013

```
s = 5.683         R-sq = 83.9%      R-sq(adj) = 81.3%
```

Analysis of Variance

SOURCE	DF	SS	MS	F	p
Regression	2	2026.2	1013.1	31.37	0.000
Error	12	387.5	32.3		
Total	14	2413.7			

From the analysis of variance table, we obtain $SSE = 387.5$ and $SS(y) = 2,413.7$. Note that the regression sum of squares, 2,026.2, is equal to $SS(y) - SSE$.

$$R^2 = \frac{\text{Regression } SS}{\text{Total } SS} = \frac{SS(y) - SSE}{SS(y)} = \frac{2,026.2}{2,413.7} = 0.839$$

Since $R^2 = 0.839$, we have that 83.9 percent of the total variation in the y-values (course grades) is accounted for by the least squares equation

$$\hat{y} = -9.7 + 0.771x_1 + 0.112x_2.$$

We performed the above calculations to obtain R^2 in order to illustrate the use of Formula 16.8. Because the value of R^2 is explicitly given in the MINITAB output, the calculations were unnecessary. R^2 appears as R-sq = 83.9% immediately above the analysis of variance table.

The size of the value of R^2 is helpful in intuitively judging how well a particular model fits a sample of data points. We know that R^2 varies between 0 and 1, where these extremes respectively indicate a complete lack of fit and a perfect fit. A value of R^2 close to 1 indicates a better fit than that associated with a smaller value of R^2. Here, however, relying on one's intuition can be misleading, and caution must be exercised in the interpretation of the R^2-value. It can be shown that R^2 will automatically increase as more predictor variables are added to a model.* In fact, R^2 can be forced to equal 1 (a perfect fit achieved) by using a model with n β-parameters, where n is the number of data points. For our course grades example with $n = 15$ points, a value of 1 for R^2 could be obtained by using a model with 14 predictors (15 β-parameters, including the constant term β_0).

NOTE

In developing the multiple regression model

$$y = \beta_0 + \beta_1 x_1 + \beta_2 x_2 + \beta_3 x_3 + \cdots + \beta_k x_k + \epsilon$$

the number of data points used, n, should be substantially more than the number of predictor variables, k. While no generally accepted rule of thumb exists, we suggest that you try to use at least 10 more points than the value of k ($n \geq k + 10$).

* To prevent the coefficient of determination from automatically increasing when a predictor with no explanatory ability is added, statisticians sometimes use a modified version of R^2, called the **adjusted multiple coefficient of determination**. MINITAB denotes this by R-sq(adj) and displays it next to the R^2 value. It is calculated as $1 - [(n - 1)SSE]/[(n - k - 1)SS(y)]$, and it is always less than R^2. However, the two coefficients will have approximately the same value when the number of data points n is large relative to the number of predictors k.

The Global *F*-Test of a Model's Usefulness

After obtaining the least squares equation, we usually check the value of R^2 to determine what proportion of the variation in the *y*-values is explained by the model. But as our preceding comments noted, a value of R^2 that appears quite large does not necessarily indicate that the model contributes useful information for predicting *y* (see Exercise 16.11). Before concluding that the model is useful and deciding to retain it for further evaluation, we need to perform a formal statistical test of the model's potential usefulness. The test described below is based on the *F*-distribution and uses R^2 in determining the value of the test statistic. It is called a global test because it checks the collective contribution of all predictors in the model, as contrasted with later tests that consider only a subset of the variables.

PROCEDURE

The Global *F*-Test of a Model's Utility

Assumed Model: $y = \beta_0 + \beta_1 x_1 + \beta_2 x_2 + \beta_3 x_3 + \cdots + \beta_k x_k + \epsilon$

For testing the hypotheses

H_0: $\beta_1 = \beta_2 = \cdots = \beta_k = 0$ (None of the predictor variables contributes significantly.)

H_a: At least one $\beta_i \neq 0$ (At least one variable makes a significant contribution for predicting *y*.)

The test statistic is

$$F = \frac{\dfrac{R^2}{k}}{\dfrac{1 - R^2}{n - (k + 1)}}.$$

The null hypothesis is rejected when $F > F_\alpha$, where α is the significance level of the test.

Assumptions:
The usual multiple regression assumptions given earlier with Equation 16.5.

Note:

1. In the formula for *F*, R^2 is the multiple coefficient of determination, *n* is the sample size, and *k* is the number of predictor variables in the model.

2. The *F*-distribution has *ndf* = *k*, the number of predictor variables, *ddf* = $n - (k + 1)$ = (the sample size) − (the number of β-parameters).

3. The *F*-value can be obtained from the analysis of variance table in the MINITAB regression output. There it is calculated as

$$F = (\text{Mean Square for Regression})/(\text{Mean Square for Error}).$$

EXAMPLE 16.3   For the course grades example, we assumed the model

$$y = \beta_0 + \beta_1 x_1 + \beta_2 x_2 + \epsilon$$

for the relationship between placement test scores x_1, SAT math scores x_2, and final statistics grades y. Is there sufficient evidence to conclude that the model contributes information for predicting final course grades? Test at the 0.05 significance level.

Solution

Earlier we obtained the estimated model $\hat{y} = -9.7 + 0.771 x_1 + 0.112 x_2$. Reproduced below are the value of R^2 and the ANOVA table from the MINITAB output in Exhibit 16.1.

```
s = 5.683        R-sq = 83.9%     R-sq(adj) = 81.3%
```

```
Analysis of Variance
SOURCE         DF        SS         MS        F        p
Regression      2      2026.2     1013.1    31.37    0.000
Error          12       387.5      32.3
Total          14      2413.7
```

We will use this information to test the utility of the model.

Step 1: Hypotheses.

H_0: $\beta_1 = \beta_2 = 0$ (x_1 and x_2 do not contribute significant information for predicting y.)

H_a: At least one $\beta_i \neq 0$ (At least one of the predictors contributes significantly.)

Step 2: Significance level.

$$\alpha = 0.05$$

Step 3: Calculations.

The value $R^2 = 0.839$ is shaded above in the computer output, and $k = 2$, since there are 2 predictor variables in the model. The number of data points is $n = 15$.

$$F = \frac{\dfrac{R^2}{k}}{\dfrac{1 - R^2}{n - (k + 1)}} = \frac{\dfrac{0.839}{2}}{\dfrac{1 - 0.839}{15 - 3}} = 31.27$$

These calculations are not necessary when the ANOVA table is available, because its *F*-ratio can be shown to be equivalent to the above formula based on R^2. In the ANOVA table we have shaded the *F*-value 31.37. The slight difference between this and our calculated value of 31.27 is due to round-off error in the ANOVA table and in the value of R^2.

Step 4: Rejection Region (or *p*-Value).

The rejection region consists of *F*-values for which $F > F_\alpha = F_{.05} = 3.89$. This value is obtained from part b of Table 6 in Appendix A using $ndf = k = 2$ and $ddf = n - (k + 1) = 15 - 3 = 12$.

Note that since the p-value for F is given in the ANOVA table, we can utilize it instead of obtaining the rejection region. For $F = 31.37$, we have

$$p\text{-value} = 0.000.$$

Step 5: Conclusion.

Because the value of F computed from the data is larger than the table value 3.89, we reject H_0 and accept H_a. This conclusion can also be drawn from the fact that the p-value is smaller than $\alpha = 0.05$. Thus, the data indicate that knowledge of at least one (x_1 and/or x_2) of the independent variables contributes significantly for predicting course grades; that is, the model appears to be useful.

EXAMPLE 16.4

A 1991 salary study of college faculty in a region of the United States produced the data in Table 16.2.* The table gives a person's salary, years of service at his/her institution, an experience rating based on one's record at the time of initial appointment, gender, and present academic rank (Pr: professor; Ao: associate professor; Ai: assistant professor; In: instructor). In the next section, we will investigate a model that utilizes all the variables whose values are displayed in Table 16.2. Presently, we wish to consider the usefulness of the following model that relates salary (y) to years of service (x_1) and experience (x_2) at the time of initial appointment.

$$y = \beta_0 + \beta_1 x_1 + \beta_2 x_2 + \epsilon$$

a. Obtain the estimated model.

b. Determine the value of R^2 and interpret its value.

c. Is there sufficient evidence at the 0.05 significance level to conclude that the model contributes information for predicting salaries?

Solution

The **READ** statement was used to store in columns C1, C2, and C3, respectively, the values of y, x_1, and x_2 from Table 16.2. The following command was then used to generate the ouput that appears in Exhibit 16.2 on page 771.

REGRESS C1 2 C2 C3

a. From Exhibit 16.2, the estimated model is $\hat{y} = 16344 + 1187x_1 + 537x_2$.

b. The multiple coefficient of determination appears in Exhibit 16.2 as R-sq, and its value is $R^2 = 0.870$. Thus, the estimated model is accounting for 87.0 percent of the total variation in the y-values (salaries). The large value of R^2 suggests that the model fits the data quite well, but we will formally check this by performing in part (c) a global test of the model's usefulness.

c. To determine if the model is useful, we need to conduct the following global F-test.

Step 1: Hypotheses.

H_0: $\beta_1 = \beta_2 = 0$ (x_1 and x_2 do not contribute significant information for predicting y.)

H_a: At least one $\beta_i \neq 0$ (At least one of the predictors contributes significantly.)

The author thanks Minitab®, Inc. for permission to use these data.

Menu Commands:
Stat ➤
Regression ➤
Regression
In **Response** *box*
enter **C1**
In **Predictors** *box*
enter **C2 C3**
Select OK

Step 2: Significance level.

$$\alpha = 0.05$$

Step 3: Calculations.

From the ANOVA table in Exhibit 16.2, the test statistic value is $F = 560.69$. This is the easiest way to obtain the value of F, but let's check it by also calculating F with the formula based on the value of R^2. We have $n = 171$ data points, and the number of predictor variables is $k = 2$.

$$F = \frac{\dfrac{R^2}{k}}{\dfrac{1 - R^2}{n - (k + 1)}} = \frac{\dfrac{0.870}{2}}{\dfrac{1 - 0.870}{171 - 3}} = 562.15$$

Theoretically, the two F-values should be the same. The difference is relatively very small and is due to round-off error.

Step 4: Rejection Region (or p-Value).

For the F-value in the ANOVA table in Exhibit 16.2, we have

$$p\text{-value} = 0.000.$$

Note that since the p-value is available, we do not need the rejection region (It consists of $F > F_{.05} \approx 3.00$, based on $ndf = k = 2$ and $ddf = n - (k + 1) = 168$).

Step 5: Conclusion.

Since the p-value is smaller than the significance level α (p-value $= 0.000 < 0.05$), we reject H_0 and accept H_a. The extremely large F-value and small p-value provide strong evidence that knowledge of the predictors as given by the model is useful for predicting faculty salaries in the sampled region. A more thorough investigation of the salary data will be conducted in the following sections.

TABLE 16.2

FACULTY MEMBER, 1991 SALARY, YEARS OF SERVICE, EXPERIENCE RATING AT ENTRANCE, SEX, AND RANK

FM	Slry.	Yrs.	ER	S	Rk.	FM	Slry.	Yrs.	ER	S	Rk.	FM	Slry.	Yrs.	ER	S	Rk.
1	35000	16	1.00	F	Ao	58	52300	22	11.00	F	Pr	115	43000	21	5.25	F	Pr
2	43000	23	4.00	M	Pr	59	47600	24	5.75	M	Pr	116	45600	23	4.00	M	Pr
3	26000	7	8.00	F	Ai	60	47400	22	5.00	M	Pr	117	38700	19	2.00	F	Pr
4	51100	19	4.00	M	Pr	61	43900	22	2.00	F	Pr	118	39500	20	1.00	F	Pr
5	49200	13	19.50	M	Pr	62	45500	23	4.00	F	Pr	119	40600	20	2.75	M	Pr
6	44900	16	3.50	M	Pr	63	47400	24	3.00	M	Pr	120	36400	16	6.83	F	Pr
7	34400	8	5.00	M	Ai	64	51100	22	12.00	M	Pr	121	43400	24	0.00	F	Pr
8	40600	12	5.00	F	Pr	65	42600	17	8.00	F	Pr	122	41000	21	4.00	F	Pr
9	35200	10	7.50	M	Ao	66	48200	25	3.00	M	Pr	123	30300	10	5.50	F	Ao
10	40600	19	3.00	M	Pr	67	46100	24	4.00	M	Pr	124	42400	21	6.00	M	Pr
11	29400	8	6.00	M	Ai	68	43100	22	1.50	M	Pr	125	44800	23	4.00	F	Pr

(Continued on next page)

TABLE 16.2 (Continued)

FACULTY MEMBER, 1991 SALARY, YEARS OF SERVICE, EXPERIENCE RATING AT ENTRANCE, SEX, AND RANK

12	43800	18	4.00	M	Pr	69	30300	10	5.00	M	Ao	126	33100	13	4.57	F	Pr
13	36800	10	6.00	M	Pr	70	42400	19	9.67	M	Pr	127	39300	20	1.00	F	Pr
14	49200	23	5.00	M	Pr	71	50900	25	9.50	M	Pr	128	50900	24	14.50	M	Pr
15	30200	10	4.83	M	Ao	72	28300	8	4.48	F	Ai	129	44700	24	8.00	M	Pr
16	47500	20	4.00	M	Pr	73	47900	24	7.50	M	Pr	130	44000	23	6.00	M	Pr
17	45800	19	5.00	M	Pr	74	51300	25	7.50	F	Pr	131	43900	22	1.00	M	Pr
18	46800	20	5.00	M	Pr	75	50700	24	7.00	M	Pr	132	31200	11	5.50	M	Ao
19	43200	19	4.00	F	Pr	76	42800	22	1.00	M	Pr	133	33100	14	2.50	F	Pr
20	24500	5	0.00	M	In	77	45100	20	4.00	F	Pr	134	32800	11	7.75	F	Pr
21	34100	9	7.50	F	Pr	78	44600	21	5.00	F	Pr	135	45300	24	5.75	M	Pr
22	29600	5	6.00	F	Ao	79	48200	24	7.00	M	Pr	136	39700	21	0.00	F	Pr
23	51000	23	6.33	M	Pr	80	48100	25	10.17	M	Pr	137	54300	26	12.00	M	Pr
24	47600	23	6.00	M	Pr	81	47600	26	5.00	M	Pr	138	47300	26	4.00	M	Pr
25	45900	19	4.00	M	Pr	82	44400	21	5.00	F	Pr	139	38400	17	5.00	M	Pr
26	51500	24	3.00	M	Pr	83	52500	24	11.00	M	Pr	140	40100	19	1.00	F	Pr
27	36300	14	1.00	F	Pr	84	43400	20	2.00	M	Pr	141	42000	21	3.25	M	Pr
28	27600	8	4.50	F	Ao	85	50900	26	10.67	F	Pr	142	25300	6	5.17	F	Ai
29	46800	21	9.67	M	Pr	86	39800	19	4.00	M	Pr	143	24600	6	4.92	M	In
30	47600	22	6.00	M	Pr	87	44600	22	4.00	M	Pr	144	35600	17	1.00	F	Pr
31	43000	19	4.50	F	Pr	88	53800	25	10.00	M	Pr	145	40000	19	3.00	F	Pr
32	28700	6	8.50	F	Ai	89	41300	21	2.50	F	Pr	146	43100	22	3.00	M	Pr
33	43800	20	6.33	M	Pr	90	41500	21	1.00	M	Pr	147	35200	16	3.25	M	Pr
34	48400	22	6.00	M	Pr	91	29400	6	8.50	M	Ai	148	27800	9	2.00	M	Ao
35	47700	24	4.50	M	Pr	92	42600	22	1.00	M	Pr	149	40400	18	8.00	M	Pr
36	53900	24	10.50	F	Pr	93	42600	20	8.25	F	Pr	150	39800	21	2.00	M	Pr
37	39900	16	4.00	M	Pr	94	41100	15	7.50	M	Pr	151	36100	16	1.00	F	Pr
38	47000	24	3.00	M	Pr	95	47500	26	4.50	F	Pr	152	28700	11	6.50	F	Ao
39	27500	9	6.00	M	Ai	96	44100	23	5.00	M	Pr	153	42600	22	6.00	F	Pr
40	25500	4	6.00	F	Ai	97	52600	24	10.00	M	Pr	154	32100	10	10.50	F	Pr
41	42200	20	3.00	M	Pr	98	44300	21	8.00	M	Pr	155	33700	15	3.00	F	Pr
42	46500	24	1.00	M	Pr	99	44000	24	0.00	F	Pr	156	31800	10	15.10	M	Pr
43	41600	15	5.00	M	Pr	100	51000	25	11.00	M	Pr	157	43600	25	4.00	M	Pr
44	41000	19	2.00	M	Pr	101	43100	22	1.25	F	Pr	158	34600	16	9.75	F	Pr
45	46700	21	6.00	M	Pr	102	46700	25	5.00	M	Pr	159	28000	13	1.00	F	Pr
46	41600	20	4.00	M	Pr	103	43500	22	5.50	M	Pr	160	29500	12	8.00	F	Ai
47	45900	22	5.00	M	Pr	104	44400	23	5.00	M	Pr	161	44400	23	8.00	M	Pr
48	43100	20	5.00	M	Pr	105	43900	24	0.00	F	Pr	162	30200	10	9.50	F	Pr
49	37700	19	0.50	M	Pr	106	48300	24	9.00	M	Pr	163	28100	10	4.00	M	Ao
50	35100	8	11.33	M	Pr	107	46300	24	5.00	M	Pr	164	37600	16	11.83	F	Pr
51	49400	24	5.00	M	Pr	108	44500	23	5.00	F	Pr	165	37600	16	9.00	F	Pr
52	47600	22	6.00	M	Pr	109	46300	22	4.00	M	Pr	166	31700	15	5.00	F	Pr
53	41000	19	3.00	F	Pr	110	44400	24	1.00	F	Pr	167	26600	9	3.00	M	Ao
54	40300	15	6.33	M	Pr	111	43700	22	0.00	M	Pr	168	22600	8	6.33	F	Ai
55	42300	18	5.00	F	Pr	112	51600	23	11.00	M	Pr	169	31900	15	2.00	F	Pr
56	43200	19	3.00	M	Pr	113	44100	22	2.00	F	Pr	170	24000	7	1.00	F	Ai
57	46900	23	5.00	M	Pr	114	47200	25	4.00	F	Pr	171	26700	10	9.00	M	Ai

Exhibit 16.2

```
MTB > NAME C1 'Y' C2 'X1' C3 'X2'
MTB > REGRESS C1 2 C2 C3

The regression equation is
Y = 16344 + 1187 X1 + 537 X2
```

Predictor	Coef	Stdev	t-ratio	p
Constant	16344.2	795.5	20.55	0.000
X1	1187.22	36.06	32.93	0.000
X2	537.16	63.92	8.40	0.000

s = 2722   R-sq = 87.0%   R-sq(adj) = 86.8%

Analysis of Variance

SOURCE	DF	SS	MS	F	p
Regression	2	831011E+04	4155054592	560.69	0.000
Error	168	124499E+04	7410632		
Total	170	955510E+04			

For each of the two models considered in Examples 16.3 and 16.4, the results of the global F-test led to the rejection of the null hypothesis and the acceptance of the alternative hypothesis that the model is useful. This tells us that the model contributes information for predicting y, that is, at least one of the predictor terms contributes significant information. We caution, however, that a conclusion of model usefulness does not preclude the likelihood that a better model exists. After all, there are numerous other possible predictor variables that could have been included in the model. Furthermore, it is possible that a variable x used in the model might contribute more information if it appears in a different form such as x^2, e^x, and so forth. These and other considerations in developing a regression model are discussed in the following sections.

NOTE

When the results of the global F-test confirm that a model is useful, do *not* conclude that the model is superior to all others for the purpose of predicting y.

SECTIONS 16.1 AND 16.2 EXERCISES

Multiple Regression Models and Assumptions; R^2 and the Global F-Test of a Model's Usefulness

16.1   The owner of a small business assumed that her monthly sales (y) are related to the monthly traffic flow (x_1) and the number of business hours (x_2) by the model $y = \beta_0 + \beta_1 x_1 + \beta_2 x_2$. Explain why this is a deterministic model and thus, is not appropriate for modeling y.

16.2   How can the model in Exercise 16.1 be changed so that it becomes a probabilistic model? If the owner were interested in modeling the relationship between monthly sales and monthly profit, which of these two variables do you think would be the response variable y?

16.3 The multiple regression model $\mu_y = \beta_0 + \beta_1 x_1 + \beta_2 x_2$ was fit to 15 data points, and the following least squares equation was obtained: $\hat{y} = -2.8 + 47.6x_1 + 9.8x_2$. The multiple coefficient of determination was found to be 0.38.
a. Give the estimates of the partial slopes and interpret their values.
b. Use the estimated model to predict the value of y for $x_1 = 2$ and $x_2 = 10$.
c. Using $\alpha = 0.05$, test if the model contributes information for predicting y.

16.4. A soil engineer investigated the relationship between the number of tons of crop yield (y) and the numbers of applied units of fertilizers A (x_1), B (x_2), and C (x_3). The model $y = \beta_0 + \beta_1 x_1 + \beta_2 x_2 + \beta_3 x_3 + \epsilon$ was fit to 29 data points, and the estimated coefficients were $\hat{\beta}_0 = 94.31$, $\hat{\beta}_1 = 6.65$, $\hat{\beta}_2 = 8.43$, and $\hat{\beta}_3 = -0.80$. Also obtained was $R^2 = 0.71$.
a. Give the estimates of the partial slopes and interpret their values.
b. What value would you predict for y when $x_1 = 4$, $x_2 = 1$, and $x_3 = 2$?
c. Test if the model contributes information for predicting y. Use $\alpha = 0.05$.

16.5 Find the approximate p-value for the global test in part c of Exercise 16.3.

16.6 Obtain the approximate p-value for the global test in part c of Exercise 16.4.

16.7 A multiple regression model was fit to a sample of data points for which $SS(y) = 394$ and $SSE = 85.8$. Determine R^2 and interpret its value.

16.8 For a multiple regression model and 50 data points, $\Sigma y^2 = 2{,}513$, $\Sigma y = 107$, and $SSE = 310$. Determine the multiple coefficient of determination and interpret its value.

16.9 The mean of y is assumed to be related to four predictor variables by the model $\mu_y = \beta_0 + \beta_1 x_1 + \beta_2 x_2 + \beta_3 x_3 + \beta_4 x_4$. This is fit to $n = 30$ data points, and it is found that $R^2 = 0.79$. With $\alpha = 0.05$, test if sufficient evidence exists to conclude that the model contributes information for predicting y.

16.10 The model $y = \beta_0 + \beta_1 x_1 + \beta_2 x_2 + \beta_3 x_3 + \beta_4 x_4 + \beta_5 x_5 + \epsilon$ was fit to 32 data points, and it was determined that $SS(y) = 17{,}984$ and $SSE = 3{,}512$.
a. Calculate R^2, the multiple coefficient of determination.
b. Calculate s, the estimated standard deviation of the model.

c. Do the data provide sufficient evidence to say that the model contributes information for predicting y? Test using a 5 percent significance level.

16.11 In Section 16.2, the reader was cautioned that the size of R^2 can be misleading when the number of predictors in a model is close to the sample size. To illustrate, suppose you only had $n = 15$ data points, but you fit the following model and obtained $R^2 = 0.96$:

$$y = \beta_0 + \beta_1 x_1 + \beta_2 x_2 + \cdots + \beta_{12} x_{12} + \epsilon$$

Test if this large value of R^2 provides sufficient evidence to conclude that the model is useful for predicting y. Use $\alpha = 0.05$.

16.12 MINITAB was used to fit the model $y = \beta_0 + \beta_1 x_1 + \beta_2 x_2 + \epsilon$ to $n = 32$ points, and a portion of the resulting output is given below.
a. What is the estimated model?
b. Estimate the partial slope β_2 and interpret its value.
c. Find the multiple coefficient of determination and interpret its value.
d. What is the estimated standard deviation of the model?
e. Using $\alpha = 0.05$, test the utility of the model.

```
The regression equation is
Y = 0.881 + 0.249 X1 - 0.211 X2

Predictor        Coef        Stdev      t-ratio          p
Constant       0.8811      0.2466         3.57      0.001
X1            0.24882     0.02848         8.74      0.000
X2            -0.2110      0.1004        -2.10      0.044

s = 0.4462        R-sq = 72.5%       R-sq(adj) = 70.6%

Analysis of Variance
SOURCE       DF          SS          MS         F         p
Regression    2     15.1947      7.5973     38.16     0.000
Error        29      5.7741      0.1991
Total        31     20.9688
```

16.13 The rapid rise in health care costs has been accompanied by an increased emphasis on comparison shopping for medical services. Some states now routinely conduct and publicize studies of hospital prices and outcomes of care. To obtain a quick assessment of the relative costs of different hospitals, could one use as a barometer their charges for two common procedures? To investigate this, the table below gives the average costs of a birth and an appendectomy for 15 Pennsylvania hospitals. Also given are the average charges for a hospital stay, after adjusting for differences in the types of patients treated (*The Patriot,* Harrisburg, Pennsylvania, July 21, 1993).

Hospital	Birth (x_1)	Appendectomy (x_2)	Average Adjusted Hospitalization Cost (y)
Carlisle	$2,031	$3,785	$6,222
Chambersburg	$1,948	$2,859	$5,191
Com. H. of Lancaster	$1,146	$2,893	$5,580
Ephrata Community	$1,252	$2,637	$4,833
Gettysburg	$1,853	$4,584	$5,637
Good Samaritan	$1,763	$4,101	$6,827
Hanover General	$1,303	$2,838	$4,461
Harrisburg	$2,118	$4,560	$8,295
Holy Spirit	$1,858	$3,395	$7,023
Lancaster General	$1,056	$2,659	$4,767
Memorial H./York	$2,700	$4,978	$7,711
Penn State University	$1,832	$4,492	$7,460
Polyclinic Med. Ctr.	$2,550	$3,821	$7,806
St. Joseph/Lancaster	$1,919	$3,691	$5,753
York	$1,587	$3,380	$5,601

Using MINITAB, the model $\mu_y = \beta_0 + \beta_1 x_1 + \beta_2 x_2$ was fit to the data, and a portion of the resulting output appears below.

a. Give the estimated model.

b. Estimate the partial slope β_1 and interpret its value.

c. Find the multiple coefficient of determination and interpret its value.

d. Give the estimated standard deviation of the model.

e. Test if the model is useful for estimating average costs for a hospital stay. Use a significance level of 5 percent.

```
The regression equation is
Y = 1548 + 1.10 X1 + 0.738 X2


Predictor       Coef       Stdev     t-ratio          p
Constant      1547.8       949.4        1.63      0.129
X1            1.0999      0.6213        1.77      0.102
X2            0.7379      0.3795        1.94      0.076


s = 737.6        R-sq = 69.1%      R-sq(adj) = 64.0%


Analysis of Variance
SOURCE          DF           SS           MS         F         p
Regression       2     14603772      7301886     13.42     0.001
Error           12      6528652       544054
Total           14     21132424
```

MINITAB Assignments

M▶ 16.14   Use MINITAB to fit the model $\mu_y = \beta_0 + \beta_1 x_1 + \beta_2 x_2$ to the data that are given in Exercise 16.13.

M▶ 16.15   Exercise 14.32 gave the number of wins and the earned run average for each of the 14 American League baseball teams during the 1995 season. The data suggest that knowledge of a team's earned run average is useful for predicting the number of games won. For predicting the number of wins, would it be useful to also include in the model a team's batting average and the number of fielding errors committed? These figures appear below for the American League teams during the 1995 season.

Team	Earned Run Avg. (x_1)	Batting Avg. (x_2)	Fielding Errors (x_3)	Number of Wins (y)
Cleveland	3.84	.291	89	100
Boston	4.40	.280	104	86
Seattle	4.52	.275	96	79
New York	4.55	.276	68	79
California	4.52	.277	84	78
Texas	4.67	.265	90	74
Baltimore	4.32	.262	59	71
Kansas City	4.49	.260	82	70
Chicago	4.85	.280	96	68
Oakland	4.97	.264	98	67
Milwaukee	4.83	.266	89	65
Detroit	5.50	.247	95	60
Toronto	4.90	.260	85	56
Minnesota	5.77	.279	91	56

Use MINITAB to solve the following.
a. Fit the model $y = \beta_0 + \beta_1 x_1 + \beta_2 x_2 + \beta_3 x_3 + \epsilon$ to the data.
b. Find the multiple coefficient of determination and interpret its value in the context of this problem.
c. Test the utility of the model at the 5 percent significance level.

M▶ 16.16   The owner of an expensive gift shop believes that her weekly sales are related to the performance of the stock market for that week and the amount of advertising expenditures during the week. To explore this possibility, she determines the amounts of sales, advertising expenses, and the means of the Dow Jones Industrial Average for 30 randomly selected weeks. These figures are given on the following page, with the sales and advertising figures in units of $1,000.

Week	DJIA (x_1)	Adv. Expenses (x_2)	Sales (y)
1	7015	2.9	58.3
2	7318	3.1	62.9
3	6581	2.5	46.3
4	6623	5.1	48.2
5	6917	2.9	58.2
6	7503	3.4	65.8
7	6223	1.6	36.7
8	7332	3.0	62.3
9	7197	3.2	59.8
10	7513	3.5	71.8
11	7703	3.1	66.9
12	6523	1.9	37.9
13	6332	1.7	30.6
14	7197	3.3	69.8
15	7713	3.8	65.2
16	6679	3.0	52.7
17	6513	2.1	39.3
18	6922	3.4	58.7
19	7146	3.3	60.9
20	6429	2.2	40.5
21	7409	3.9	70.3
22	6315	2.4	39.1
23	6487	2.5	45.9
24	6715	2.8	55.1
25	7543	3.6	70.2
26	7619	3.7	70.3
27	6125	2.1	39.1
28	6987	2.5	45.9
29	7215	2.9	55.1
30	7707	3.4	70.2

Use MINITAB to fit the model $y = \beta_0 + \beta_1 x_1 + \beta_2 x_2 + \epsilon$ to the data.

M▶ 16.17 Refer to Exercise 16.16 and test if the model is useful. Use a significance level of 1 percent.

16.3 USING QUALITATIVE VARIABLES IN A MODEL

Each of the predictor variables used in Examples 16.1 through 16.4 is a quantitative variable, because its possible values are measured numerically. Placement test scores, SAT math scores, years of service, and one's amount of professional experience are all quantitative. In developing a regression model, an experimenter will frequently need to include nonquantitative variables that are categorical or attributive in nature, that is, **qualitative variables.** For instance, in Example 16.7 we will broaden the faculty salaries model by including the qualitative variables gender and academic rank.

A qualitative variable can be included in a model by coding the variable's different levels (values). The coding scheme can be accomplished by using one or more **dummy variables.** A dummy variable is just an **indicator variable** that uses the values 0 and 1 to *indicate* a particular level of the qualitative variable with which it is associated. These concepts are illustrated in the following example.

EXAMPLE 16.5   An entrepreneur owns two pizza establishments that are located on the east and west sides of a large city. The owner wants a model that relates average weekly sales with shop location. Obtain such a model, and interpret the significance of the β-parameters in the model.

Solution

Pizza shop location is a qualitative variable at two levels (east and west). We can insert location in the model by defining a dummy variable x so that its 0-value is associated with one location and its 1-value is associated with the other. Let's assign the value $x = 1$ to the east location and $x = 0$ to the west location.

$$x = \begin{cases} 1 & \text{if east location} \\ 0 & \text{if not} \end{cases}$$

Letting μ_y denote the mean weekly sales, the model that relates average weekly sales to store location is

(16.9) $\mu_y = \beta_0 + \beta_1 x.$

A very important point about Equation 16.9 is that this *one* equation actually describes *two* relationships between sales and location. To see this and understand the meaning of each β-parameter, let's substitute the possible x-values 1 and 0 into the model. This will give us the mean weekly sales for the east and west locations, respectively. In Equation 16.9,

let $x = 1$ to get the mean sales of the east location: $\mu_y = \beta_0 + \beta_1(1) = \beta_0 + \beta_1.$

Let $x = 0$ to get the mean sales of the west location: $\mu_y = \beta_0 + \beta_1(0) = \beta_0.$

Thus, the mean weekly sales for the east location are given by $\mu_y = \beta_0 + \beta_1$, the sum of the unknown parameters β_0 and β_1, and the mean weekly sales for the west location are given by $\mu_y = \beta_0$, the constant term in Equation 16.9.

In summary, the model $\mu_y = \beta_0 + \beta_1 x$, where x is the dummy variable defined above, gives the mean weekly sales for both the east ($x = 1$) and the west ($x = 0$) locations. The value of the unknown parameter β_0 is equal to the mean sales for the west location, and the sum of the parameters β_0 and β_1 equals the mean sales for the east location. Since $(\beta_0 + \beta_1)$ and β_0 give the mean sales for the east and west locations, respectively, we have that

$$\beta_1 = (\beta_0 + \beta_1) - \beta_0 = (\text{east mean sales}) - (\text{west mean sales}).$$

Thus, the value of the parameter β_1 is the difference in mean sales for the east and west locations.

EXAMPLE 16.6 Suppose the entrepreneur in Example 16.5 opens a third pizza shop in the south end of the city. Give a model that relates mean weekly sales with the three shop locations, and interpret the β-parameters.

Solution

Because there are now pizza shops in three areas, the qualitative variable, location, has three levels—east, west, and south. Location can be included in the model by defining two dummy variables (one less than the number of levels). With 3 levels, we will see that only 2 dummy variables x_1 and x_2 are needed, because one level of a qualitative variable always serves as a baseline level (with all dummy variables set at 0). We proceed by arbitrarily picking two locations and assigning the 1-value of x_1 to one location and the 1-value of x_2 to the other. Let's assign $x_1 = 1$ to the east location and $x_2 = 1$ to the south location. Then the west location becomes the baseline location that is given by putting $x_1 = 0$ and $x_2 = 0$.

$$x_1 = \begin{cases} 1 & \text{if east location} \\ 0 & \text{if not} \end{cases} \qquad x_2 = \begin{cases} 1 & \text{if south location} \\ 0 & \text{if not} \end{cases}$$

With μ_y denoting mean weekly sales, the following model relates average weekly sales to store location.

(16.10) $$\mu_y = \beta_0 + \beta_1 x_1 + \beta_2 x_2$$

Note that *one* equation now describes *three* relationships, because the model relates sales to each of the three levels of the qualitative variable location. These three relations are obtained as follows. In Equation 16.10,

let $x_1 = 1$ and $x_2 = 0$ to get the east mean sales: $\mu_y = \beta_0 + \beta_1(1) + \beta_2(0) = \beta_0 + \beta_1$.

Let $x_1 = 0$ and $x_2 = 1$ to get the south mean sales: $\mu_y = \beta_0 + \beta_1(0) + \beta_2(1) = \beta_0 + \beta_2$.

Let $x_1 = 0$ and $x_2 = 0$ to get the west mean sales: $\mu_y = \beta_0 + \beta_1(0) + \beta_2(0) = \beta_0$.

Thus, the model $\mu_y = \beta_0 + \beta_1 x_1 + \beta_2 x_2$ gives the mean weekly sales for the shops located in the east ($x_1 = 1$, $x_2 = 0$), south ($x_1 = 0$, $x_2 = 1$), and west ($x_1 = 0$, $x_2 = 0$). The mean sales are ($\beta_0 + \beta_1$), ($\beta_0 + \beta_2$), and β_0, respectively, for the east, south and west locations.* The interpretation of each β-parameter is given by the following:

$$\beta_0 = \text{(west mean sales)},$$

$$\beta_1 = (\beta_0 + \beta_1) - \beta_0 = \text{(east mean sales)} - \text{(west mean sales)},$$

$$\beta_2 = (\beta_0 + \beta_2) - \beta_0 = \text{(south mean sales)} - \text{(west mean sales)}.$$

In Example 16.5, 1 dummy variable was used for a qualitative variable with 2 levels, and Example 16.6 employed 2 dummy variables for a qualitative variable with 3 levels. As you probably have surmised, 3 dummy variables would be used

---

* Instead of using 2 dummy variables to specify the 3 locations, it is possible to use a model with only 1 variable x such as $\mu_y = \beta_0 + \beta_1 x$ where $x = 1, 2, 3$ for the 3 locations. This, however, would usually not be appropriate, because it implies that the difference in mean weekly sales between locations 2 and 1 is the same as that between locations 3 and 2, namely, the value of β_1.

for 4 levels, and in general, $(k - 1)$ dummy variables would be utilized for a qualitative variable with k levels.

PROCEDURE

Relating the Mean of y to a Qualitative Variable with k Levels

Define $(k - 1)$ dummy variables such as the following:

$$x_1 = \begin{cases} 1 \text{ if level 1} \\ 0 \text{ if not} \end{cases} \qquad x_2 = \begin{cases} 1 \text{ if level 2} \\ 0 \text{ if not} \end{cases} \quad \cdots \quad x_{k-1} = \begin{cases} 1 \text{ if level } (k - 1) \\ 0 \text{ if not} \end{cases}$$

The mean of y is then related to the qualitative variable by

$$\mu_y = \beta_0 + \beta_1 x_1 + \beta_2 x_2 + \cdots + \beta_{k-1} x_{k-1}$$

where the parameters denote the following:

β_0 = (mean of y for level k),
β_1 = (mean of y for level 1) − (mean of y for level k),
β_2 = (mean of y for level 2) − (mean of y for level k),
β_{k-1} = (mean of y for level $k - 1$) − (mean of y for level k).

The next example illustrates how qualitative variables can be added to a model that contains quantitative variables.

EXAMPLE 16.7   In Example 16.4, we considered a study of college faculty salaries. The following model was proposed for relating salary (y) to years of service (x_1) and experience (x_2) at the time of initial appointment.

$$y = \beta_0 + \beta_1 x_1 + \beta_2 x_2 + \epsilon$$

Add appropriate terms to the model so that it also relates y with the qualitative variables gender and academic rank.

Solution
Since gender has 2 levels, we need $(2 - 1) = 1$ dummy variable for its inclusion. We will denote it by x_3 since x_1 and x_2 are already in the model. Define x_3 as

$$x_3 = \begin{cases} 1 & \text{if male} \\ 0 & \text{if not} \end{cases}$$

Because the variable rank has 4 levels (professor, associate professor, assistant professor, and instructor), we define the following $(4 - 1) = 3$ dummy variables.

$$x_4 = \begin{cases} 1 \text{ if prof.} \\ 0 \text{ if not} \end{cases} \qquad x_5 = \begin{cases} 1 \text{ if assoc. prof.} \\ 0 \text{ if not} \end{cases} \qquad x_6 = \begin{cases} 1 \text{ if assist. prof.} \\ 0 \text{ if not} \end{cases}$$

Notice that the instructor level will be the baseline level for rank, since we arbitrarily left it out in defining x_4, x_5, and x_6.

Dummy variables are now sufficiently defined so that we can include gender and rank in the model. The following model relates salary (y) with the quantitative variables years of service (x_1) and experience (x_2), and the qualitative variables gender and academic rank.

$$y = \beta_0 + \beta_1 x_1 + \beta_2 x_2 + \beta_3 x_3 + \beta_4 x_4 + \beta_5 x_5 + \beta_6 x_6 + \epsilon$$

EXAMPLE 16.8   At the beginning of the chapter, we cited a report that investigated the Three Mile Island nuclear accident and the resulting evacuation—the largest in the country's history. The study's primary objective was to determine the demographic characteristics associated with adult evacuees who resided within five miles of the plant. A multiple regression model was utilized that associated evacuation behavior with 24 predictor variables. The dependent variable y was the cumulative number of days evacuated. Most of the independent variables used in the model were qualitative and consisted of the following: gender (2 levels), race (2 levels), marital status (2 levels), presence of a pregnant woman in the house (2 levels), presence of a preschooler in the house (2 levels), presence of a TMI worker in the house (2 levels), type of occupation (12 levels), a history of cancer in the household (2 levels), and a history of thyroid disease in the household (2 levels). The report concluded that the presence of a preschooler in the house and a pregnancy were the two most influential factors affecting the number of evacuation days.

16.4   INTERACTION AND SECOND-ORDER MODELS

Chapter 14 was concerned with relating the mean of y to a single quantitative variable x_1 by use of the model

(16.11) $\mu_y = \beta_0 + \beta_1 x_1.$

In Section 16.1, we extended the model in Equation 16.11 by adding a second quantitative variable x_2. We then assumed the model

(16.12) $\mu_y = \beta_0 + \beta_1 x_1 + \beta_2 x_2.$

The models given by Equations 16.11 and 16.12 are **first-order models,** because each predictor variable appears only as a first-order term (exponent of 1).* A first-order model cannot contain predictor variables that have been mathematically transformed (such as e^x, $\sin(x)$, $\log(x)$, $\sqrt{x}$, x^2, etc.), and there cannot be any products or quotients of predictors (such as $x_1 x_2$, x_1/x_2, $1/x_1$, etc.).

Although first-order models are frequently employed in regression analyses, they involve underlying assumptions that are not always appropriate in modeling data. These are outlined on the following page for the general first-order model with two quantitative predictors.

---

* Recall from algebra that if the exponent of a variable is not written, it is understood as 1. Because the exponent of x is 1, it is thus a first-order term.

> **NOTE**
>
> **Some Restrictions of the First-Order Model**
>
> $$\mu_y = \beta_0 + \beta_1 x_1 + \beta_2 x_2$$
>
> where x_1 and x_2 are quantitative predictor variables.
>
> 1. The graph of its response surface contains no curvature.
>
> 2. The mean of y changes by the *constant amount* β_1 for each 1-unit increase in x_1 while x_2 is held fixed.
>
> 3. The mean of y changes by the *constant amount* β_2 for each 1-unit increase in x_2 while x_1 is held fixed.
>
> 4. x_1 and x_2 act independently in affecting the mean of y. That is, the effect on μ_y from a change in x_1 is the same for all settings of x_2. Similarly, the effect on μ_y from a change in x_2 is the same for all settings of x_1. When x_1 and x_2 act independently in this way, we say that **x_1 and x_2 do not interact.**

The restrictions outlined above are illustrated in the following example.

EXAMPLE 16.9
A manufacturer of hard disk drives for personal computers advertises its products in two trade magazines, A and B. The mean (μ_y) of the company's yearly sales is related to advertising expenditures in A (x_1) and B (x_2) by the following model, where all units are in millions of dollars.

$$\mu_y = 4 + 1.25 x_1 + 1.5 x_2$$

The response surface of the model is a plane that is displayed in Figure 16.3. The response surface contains no curvature because the model is first order.

Since the coefficient (partial slope) of x_1 is 1.25, the model assumes that mean sales will increase by the *constant amount* $1.25 for each additional $1 spent with A (while advertising in B is held fixed at some level). Similarly, because the coefficient of x_2 is 1.5, the model implies that mean sales will increase by the *constant amount* $1.5 for each additional $1 spent with B (while advertising in A is held fixed

Figure 16.3
Response surface of the first-order model $\mu_y = 4 + 1.25 x_1 + 1.5 x_2$.

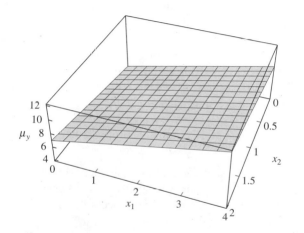

Figure 16.4

Relating mean sales μ_y to A expenditures x_1 when B expenditures are fixed at

1. $x_2 = 2$ (upper line),
   $\mu_y = 7 + 1.25x_1$;
2. $x_2 = 1$ (lower line),
   $\mu_y = 5.5 + 1.25x_1$.

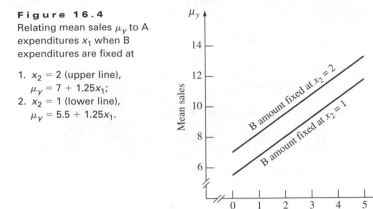

at some level). Moreover, the first-order model assumes that x_1 *and* x_2 *act independently in affecting the mean of y,* that is, x_1 *and* x_2 *do not interact.* Thus, the effect on mean sales from advertising changes in A will not depend on the amount spent with B. For instance, suppose the company decides to spend an amount $x_2 = \$1$ million with magazine B. Then the relationship between mean sales and expenditures with A (x_1) is obtained by putting $x_2 = 1$ in the assumed model $\mu_y = 4 + 1.25x_1 + 1.5x_2$. This gives $\mu_y = 4 + 1.25x_1 + 1.5(1) = 5.5 + 1.25x_1$. Thus, with x_2 fixed at 1, the model relating mean sales to x_1 is

(16.13) $$\mu_y = 5.5 + 1.25x_1.$$

If the company later decides to spend $x_2 = \$2$ million with B, then the relationship between mean sales and expenditures with A (x_1) is obtained by putting $x_2 = 2$ in the assumed model $\mu_y = 4 + 1.25x_1 + 1.5x_2$. This gives $\mu_y = 4 + 1.25x_1 + 1.5(2) = 7 + 1.25x_1$. Thus, with x_2 fixed at 2, the model relating mean sales to x_1 is

(16.14) $$\mu_y = 7 + 1.25x_1.$$

An examination of Equations 16.13 and 16.14 reveals that in both cases the coefficient of x_1 is 1.25, and consequently, mean sales will increase \$1.25 for each additional \$1 spent with A. Thus, mean sales will change at the constant rate of \$1.25 per \$1 change in A, regardless of whether B expenditures are fixed at the value $x_2 = 1$ or $x_2 = 2$. This absence of interaction between x_1 and x_2 in their effect on y is further indicated by looking at the graphs of Equations 16.13 and 16.14 in Figure 16.4. The figure shows the line relating μ_y to x_1 when x_2 is fixed at 1, and also the line that relates μ_y to x_1 when x_2 is fixed at 2. The two lines are parallel with a common slope of 1.25, regardless of whether x_2 is fixed at 1 or 2.

The simplicity of first-order models makes them a popular choice in a researcher's initial attempt to model some physical phenomenon. However, in subsequent stages of the modeling process the investigator often finds the restrictions

Figure 16.5
Response surface for the second-order interaction model $\mu_y = \beta_0 + \beta_1 x_1 + \beta_2 x_2 + \beta_3 x_1 x_2$.

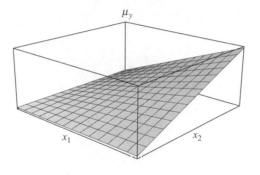

of a first-order model inappropriate, and higher-order models are considered. For instance, the manufacturer of hard disk drives in Example 16.9 may find that magazines A and B have a considerable overlap in readership. Consequently, the size of the increase in mean sales from additional advertising in A (x_1) is expected to lessen as more advertising occurs in B (x_2). Thus, the duplication in readership renders as unrealistic the "no interaction restriction" of a first-order model. It is no longer reasonable to assume that x_1 and x_2 act independently in affecting mean sales, and therefore a model that allows for interaction is needed.

Interaction can be incorporated in the model $\mu_y = \beta_0 + \beta_1 x_1 + \beta_2 x_2$ by adding the **interaction term** $\beta_3 x_1 x_2$. The resulting model becomes

(16.15) $$\mu_y = \beta_0 + \beta_1 x_1 + \beta_2 x_2 + \beta_3 x_1 x_2.$$

The model defined by Equation 16.15 is said to be a **second-order model,** because its highest-order term is $\beta_3 x_1 x_2$, which is of order two.* Its response surface will contain curvature in the shape of a twisted plane (see Figure 16.5).

KEY CONCEPT

Order of a Model

The **order of a model** is equal to the order of the highest-order term in the model.

EXAMPLE 16.10   Consider again the first-order model assumed in Example 16.9, namely,

$$\mu_y = 4 + 1.25 x_1 + 1.5 x_2.$$

Let's examine the effects of changing it to a second-order model by adding the interaction term $-0.2 x_1 x_2$. For illustration purposes, assume the actual model is

(16.16) $$\mu_y = 4 + 1.25 x_1 + 1.5 x_2 - 0.2 x_1 x_2.$$

Figure 16.6 shows the model's response surface, a twisted plane, although its curvature is not prominent for the region displayed in the figure.

If $x_2 = \$1$ million were spent on advertising in magazine B, then the relation between mean sales and expenditures with magazine A (x_1) is obtained by substituting $x_2 = 1$ in the model $\mu_y = 4 + 1.25 x_1 + 1.5 x_2 - 0.2 x_1 x_2$, resulting in

$$\mu_y = 4 + 1.25 x_1 + 1.5(1) - 0.2 x_1(1) = 5.5 + 1.05 x_1.$$

---

* The order of $x_1 x_2$ is equal to $(1 + 1) = 2$, the sum of the exponents. In general, the order of a term is found by adding the exponents of all variables in the term. For instance, the order of the term $20 x_1^5 x_2^8 x_3$ is $(5 + 8 + 1) = 14$.

Figure 16.6
Response surface of the interaction model $\mu_y =$
$4 + 1.25x_1 + 1.5x_2 - 0.2x_1x_2$.

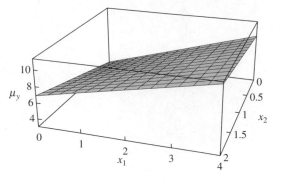

Thus, with x_2 fixed at 1, the model that relates mean sales to x_1 is

(16.17) $\mu_y = 5.5 + 1.05x_1.$

If the company were to spend $x_2 = \$2$ million with B, then the relation between mean sales and expenditures with A (x_1) would be found by setting $x_2 = 2$ in Equation 16.16. This gives $\mu_y = 4 + 1.25x_1 + 1.5(2) - 0.2x_1(2) = 7 + 0.85x_1$. Thus, with x_2 fixed at 2, the model relating mean sales to x_1 is

(16.18) $\mu_y = 7 + 0.85x_1.$

The graphs of Equations 16.17 and 16.18 appear in Figure 16.7. Notice that the lines are not parallel, because the presence of interaction causes the lines to have different slopes of 1.05 and 0.85. With B expenditures fixed at $x_2 = \$1$ million, mean sales would increase only at the rate of $\$1.05$ for each $\$1$ spent with A. The situation is impractical if B expenditures were fixed at $x_2 = \$2$ million. Then mean sales would only increase at the rate of 85 cents for each additional dollar of advertising with magazine A. The existence of interaction in a model complicates its interpretation and produces implications that are sometimes surprising. For this example, advertising expenditures with magazine B should be set at some amount less

Figure 16.7
Relating mean sales μ_y to A expenditures x_1 when B expenditures are fixed at

1. $x_2 = 2$ (upper line),
   $\mu_y = 7 + 0.85x_1$;
2. $x_2 = 1$ (lower line),
   $\mu_y = 5.5 + 1.05x_1$.

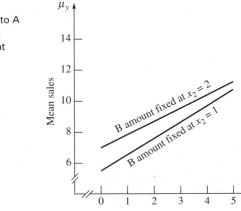

than \$2 million if each dollar spent with A is to yield at least a dollar increase in sales. (Can you show that this amount is $x_2 = \$1.25$ million?)

KEY CONCEPT

Interaction Between Two Predictor Variables
The predictor variables x_1 and x_2 **interact** if the change in μ_y from a 1-unit increase in x_1 is dependent on the level at which x_2 is held fixed. When interaction is present, a term $\beta x_1 x_2$ should be included in the model.

In developing a model, it is not always clear if interaction terms should be included, that is, if their addition to the model would make a significant contribution to predicting the response variable y. Generally, a good procedure to follow is to allow for interaction in the assumed model and then later test the significance of the interaction term(s) in the estimated model. Procedures for doing this are covered in the next section.

The interaction model defined by Equation 16.16 in the previous example is a special case of the following complete second-order model in two quantitative variables.

KEY CONCEPT

Complete Second-Order Model in Two Quantitative Variables x_1, x_2

$$\mu_y = \beta_0 + \beta_1 x_1 + \beta_2 x_2 + \beta_3 x_1 x_2 + \beta_4 x_1^2 + \beta_5 x_2^2$$

or equivalently,

$$y = \beta_0 + \beta_1 x_1 + \beta_2 x_2 + \beta_3 x_1 x_2 + \beta_4 x_1^2 + \beta_5 x_2^2 + \epsilon$$

Second-order models can be defined analogously for more than two predictor variables. The complete second order model in three quantitative variables is given below.

KEY CONCEPT

Complete Second-Order Model in Three Quantitative Variables x_1, x_2, x_3

$$\mu_y = \beta_0 + \beta_1 x_1 + \beta_2 x_2 + \beta_3 x_3 + \beta_4 x_1 x_2 + \beta_5 x_1 x_3$$
$$+ \beta_6 x_2 x_3 + \beta_7 x_1^2 + \beta_8 x_2^2 + \beta_9 x_3^2$$

or equivalently,

$$y = \beta_0 + \beta_1 x_1 + \beta_2 x_2 + \beta_3 x_3 + \beta_4 x_1 x_2 + \beta_5 x_1 x_3$$
$$+ \beta_6 x_2 x_3 + \beta_7 x_1^2 + \beta_8 x_2^2 + \beta_9 x_3^2 + \epsilon$$

We have given the complete second-order models in 2 and in 3 quantitative variables. Second-order models can be defined for any number of quantitative variables, including only one. A second-order model in one variable x is called a **quadratic model.**

Figure 16.8
Two quadratic models.
(a) x^2-coefficient $\beta_2 > 0$.
Response curve is
concave upward:
$\mu_y = 4 - 2x + x^2$.
(b) x^2-coefficient $\beta_2 < 0$.
Response curve is
concave downward:
$\mu_y = -16 + 10x - x^2$.

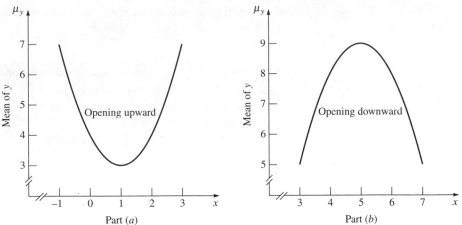

Part (a)   Part (b)

KEY CONCEPT

Second-Order Quadratic Model in One Quantitative Variable x

$$\mu_y = \beta_0 + \beta_1 x + \beta_2 x^2$$

or equivalently,

$$y = \beta_0 + \beta_1 x + \beta_2 x^2 + \epsilon$$

Note:

1. The model's graph (response curve) is a parabola (see Figure 16.8).

2. The sign of β_2, the coefficient of x^2, determines if the parabola opens upward (when $\beta_2 > 0$) or downward (when $\beta_2 < 0$).

Note that although the quadratic model expresses μ_y as a function of only one quantitative variable x, it can be considered as a multple regression model involving two predictor variables. This follows from the fact that the quadratic model can be written as

$$\mu_y = \beta_0 + \beta_1 x_1 + \beta_2 x_2$$

where the first predictor is $x_1 = x$ and the second predictor is $x_2 = x^2$.

EXAMPLE 16.11   A paper mill utilizes a delicate sensing unit in the manufacturing process for its premium writing paper. Each unit requires some daily service and is replaced every 30 days. During its break-in period, a unit needs a relatively large amount of service because of initial adjustments and fine tuning. After this initial period, required service usually decreases to a low level, from which it begins to increase as the unit's replacement time nears. From the daily service records of sensing units, 27 were randomly selected and examined. Table 16.3 shows the number of service minutes

TABLE 16.3

REQUIRED MINUTES OF SERVICE AND NUMBER OF DAYS IN USE FOR $n = 27$ RECORDS

Service y (minutes)	Days in Use x	Service y (minutes)	Days in Use x	Service y (minutes)	Days in Use x
7	14	18	5	2	17
16	25	19	4	4	19
19	8	12	12	19	28
17	8	25	3	12	11
27	2	20	30	18	27
13	23	23	27	25	4
28	6	28	3	6	14
22	9	19	27	9	19
5	17	24	29	10	14

required (y) on the selected date and the number of days the unit had been in operation (x). A scatter diagram of the data appears in Figure 16.9.

a. Formulate a model for relating a unit's daily service minutes to its days in operation.

b. Obtain the least squares estimated model and test its utility at the 0.05 level of significance.

Solution

a. We are told that a unit's daily service requirements are expected to be relatively high for an initial period following installation and also for a period just prior to its replacement. This suggests the following quadratic model, where β_2 is positive.

$$y = \beta_0 + \beta_1 x + \beta_2 x^2 + \epsilon$$

Figure 16.9
Scatter diagram of service requirements (y) versus days in use (x).

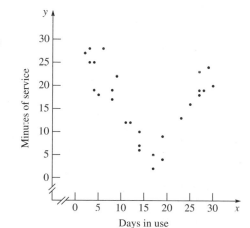

This assumed model appears to be supported by the scatter diagram in Figure 16.9, where the pattern of points suggests the presence of curvature that is quite similar to that in a portion of a parabola.

b. To obtain the estimated model, the **READ** statement was used to store the values of y and x in columns C1 and C2, respectively. The values of x^2 were then generated and stored in column C3 by using the command

```
LET C3 = C2**2
```

Next, the following command was used to generate the regression output that appears in Exhibit 16.3.

```
REGRESS C1 2 C2 C3
```

From Exhibit 16.3, the estimated model is $\hat{y} = 35.9 - 3.28x + 0.0966x^2$. To determine if the model is useful, we will perform the following global F-test.

> *Menu Commands:*
> **Stat** ➤
> **Regression** ➤
> **Regression**
> *In* **Response** *box*
> *enter* **C1**
> *In* **Predictors** *box*
> *enter* **C2 C3**
> *Select* OK

Step 1: Hypotheses.

> H_0: $\beta_1 = \beta_2 = 0$ (x and x^2 do not contribute significant information for predicting y.)

> H_a: At least one $\beta_i \neq 0$ (At least one of the predictors contributes significantly.)

Step 2: Significance level.

$$\alpha = 0.05$$

Step 3: Calculations.

From the ANOVA table in Exhibit 16.3, the test statistic value is $F = 42.71$. This could also be computed using $R^2 = 0.781$, the multiple coefficient of determination in Exhibit 16.3.

Step 4: Rejection Region (or *p*-Value).

For the F-value in the ANOVA table, we have

$$p\text{-value} = 0.000.$$

(Although it's not needed, the rejection region consists of $F > F_{.05} = 3.40$, based on $ndf = k = 2$ and $ddf = n - (k + 1) = 27 - 3 = 24$.)

Step 5: Conclusion.

Since the p-value of 0.000 is smaller than $\alpha = 0.05$, we reject the null hypothesis and accept the alternative. There is considerable evidence that the model is useful for relating a unit's required service time to its number of days in service.

The scatter diagram strongly suggests that the second-order term x^2 is making an important contribution to the model. In the next section, we will show how this belief can be formally tested.

Exhibit 16.3

```
MTB > NAME C1 'Y' C2 'X' C3 'X-SQ'
MTB > REGRESS C1 2 C2 C3

The regression equation is
Y = 35.9 - 3.28 X + 0.0966 X-SQ
```

Predictor	Coef	Stdev	t-ratio	p
Constant	35.859	2.296	15.62	0.000
X	-3.2774	0.3552	-9.23	0.000
X-SQ	0.09663	0.01089	8.88	0.000

s = 3.752   R-sq = 78.1%   R-sq(adj) = 76.2%

Analysis of Variance

SOURCE	DF	SS	MS	F	p
Regression	2	1202.76	601.38	42.71	0.000
Error	24	337.91	14.08		
Total	26	1540.67			

Before concluding this section, it should be emphasized that *only quantitative variables determine the order of a model. A dummy variable that is used to define a qualitative variable has no effect on a model's order.* The next example illustrates this important point.

EXAMPLE 16.12

A company wants to model its employees' contributions to a voluntary savings plan. The following model is proposed:

(16.19)
$$\mu_y = \beta_0 + \beta_1 x_1 + \beta_2 x_2 + \beta_3 x_1 x_2$$

where μ_y is the mean percentage of yearly income contributed to the plan, x_1 is a quantitative variable that denotes years of service, and x_2 is a dummy variable used to distinguish between full-time and part-time workers. It is defined as

$$x_2 = \begin{cases} 1 \text{ if full-time} \\ 0 \text{ if not} \end{cases}$$

© George Semple

The interaction term $\beta_3 x_1 x_2$ may convey the false impression that the model is second order. It is a first-order model, however, because the dummy variable x_2 serves only to indicate the presence or absence of a term. It enables us to use the single Equation 16.19 to describe two relationships between savings and years of service—one for full-time and one for part-time employees. For part-time workers we let $x_2 = 0$, and for full-time employees we set $x_2 = 1$, giving the relationships

$$\mu_y = \beta_0 + \beta_1 x_1 \qquad \text{for part-time employees}$$
$$\mu_y = (\beta_0 + \beta_2) + (\beta_1 + \beta_3)x_1 \qquad \text{for full-time employees}$$

Thus, for each of the two possible values of the dummy variable x_2, the resulting equation is first order in the quantitative variable x_1 and graphs as a straight line.

SECTIONS 16.3 AND 16.4 EXERCISES

Using Qualitative Variables in a Model; Interaction and Second-Order Models

16.18　A chain of food stores sells 3 types of milk: whole, low fat, and skim. The company wants to model the mean μ_y of its daily milk sales for the 3 types sold.

　　a. Is type of milk a qualitative or quantitative variable?

　　b. How many levels does the variable have?

　　c. Give a model that relates mean daily milk sales to the type of milk.

16.19　The manufacturer of a popular pickup truck offers a choice of 3 engine sizes. Its trucks can be purchased with a 2.0, 2.5, or 3.0 liter engine. The manufacturer wants to model the truck's mean miles-per-gallon as a function of engine size. Treat the variable engine size as a categorical variable with 3 levels, and write a model that relates average miles-per-gallon to the size of the engine.

16.20　The chain of food stores referred to in Exercise 16.18 offers the 3 types of milk in a chocolate flavor or plain. Write a model that relates the mean of its daily milk sales as a function of type (whole, low fat, skim) and flavor (chocolate, plain).

16.21　The pickup trucks discussed in Exercise 16.19 can be purchased with either a 5-speed or automatic transmission. Give a model that relates average miles-per-gallon to engine size and type of transmission.

16.22　A flower nursery used the following to model the mean number of roses sold as a function of the season.

$$\mu_y = \beta_0 + \beta_1 x_1 + \beta_2 x_2 + \beta_3 x_3$$

where

$$x_1 = \begin{cases} 1 \text{ if fall} \\ 0 \text{ if not} \end{cases} \quad x_2 = \begin{cases} 1 \text{ if winter} \\ 0 \text{ if not} \end{cases} \quad x_3 = \begin{cases} 1 \text{ if spring} \\ 0 \text{ if not} \end{cases}$$

　　a. Determine the mean number of roses sold for the winter season.

　　b. Determine the mean number of roses sold during the summer.

　　c. Interpret each of the four β-parameters.

16.23　The number of calls to an all-talk radio station greatly depends on the time of day. The station used the following equation to model the mean number of daily calls for the time periods morning, afternoon and evening.

$$\mu_y = \beta_0 + \beta_1 x_1 + \beta_2 x_2$$

where

$$x_1 = \begin{cases} 1 \text{ if morning} \\ 0 \text{ if not} \end{cases} \quad x_2 = \begin{cases} 1 \text{ if evening} \\ 0 \text{ if not} \end{cases}$$

a. What is the mean number of calls received during the afternoon period?

b. What is the mean number of calls received during the morning period?

c. Interpret each of the three β-parameters.

16.24 The following was used to relate the mean of a response variable y to a qualitative variable with five levels.

$$\mu_y = \beta_0 + \beta_1 x_1 + \beta_2 x_2 + \beta_3 x_3 + \beta_4 x_4$$

where

$$x_1 = \begin{cases} 1 \text{ if level 1} \\ 0 \text{ if not} \end{cases} \quad x_2 = \begin{cases} 1 \text{ if level 2} \\ 0 \text{ if not} \end{cases} \quad x_3 = \begin{cases} 1 \text{ if level 3} \\ 0 \text{ if not} \end{cases} \quad x_4 = \begin{cases} 1 \text{ if level 4} \\ 0 \text{ if not} \end{cases}$$

MINITAB was used to fit the model to 45 data points, and it produced the estimated model

$$\hat{y} = 125 + 31.8x_1 - 26.1x_2 - 20.5x_3 + 2.9x_4.$$

a. Estimate the mean of y for the third level of the qualitative variable.

b. Estimate μ_y for level 1 of the variable.

c. What does $\hat{\beta}_0 = 125$ represent?

16.25 For the model in Exercise 16.24, the ANOVA table in the MINITAB output gave the following values for the total sum of squares and error sum of squares.

$$SS(y) = 7,374 \text{ and } SSE = 1,595$$

Test the utility of the model at the 5 percent significance level.

16.26 Consider the first-order model $\mu_y = \beta_0 + \beta_1 x_1 + \beta_2 x_2$ where x_1 and x_2 are quantitative variables.

a. What do we mean when we say that x_1 and x_2 do not interact?

b. Change the model so that it allows for interaction.

c. What is the order of the model in part b?

16.27 Write the complete second-order model in four quantitative predictor variables $x_1, x_2, x_3,$ and x_4.

16.28 A large corporation uses the following to model the number of overtime hours (y) worked by its employees during the year.

$$y = \beta_0 + \beta_1 x_1 + \beta_2 x_2 + \beta_3 x_3 + \beta_4 x_1 x_2 + \beta_5 x_1 x_3 + \beta_6 x_2 x_3 + \beta_7 x_1 x_2 x_3 + \epsilon$$

In the above model, x_1 and x_2 are quantitative variables that denote a worker's hourly rate and age, respectively, and x_3 is a dummy variable that places gender in the model ($x_3 = 1$ if female, $x_3 = 0$ if male).

a. Write the model that relates a man's number of overtime hours to his hourly rate and age.

b. Write the model for women that relates overtime with hourly rate and age.

c. What is the order of the given model?

16.29   The mean of a dependent variable y is assumed to be related to two quantitative variables x_1 and x_2 by the model $\mu_y = 7 + 2x_1 + 5x_2$.

a. Describe the appearance of the response surface if the model were graphed.

b. If x_1 is fixed at 4, by how much does the mean of y change for each 1-unit increase in x_2?

c. If x_1 is fixed at 10, by how much does the mean of y change for each 1-unit increase in x_2?

d. Explain how the model could be changed so that the answers to parts b and c would differ.

16.30   The interaction term $4x_1x_2$ is added to the model in Exercise 16.29, resulting in the new model $\mu_y = 7 + 2x_1 + 5x_2 + 4x_1x_2$.

a. Describe the appearance of the response surface if the model were graphed.

b. If x_1 is fixed at 4, by how much does the mean of y change for each 1-unit increase in x_2?

c. If x_1 is fixed at 10, by how much does the mean of y change for each 1-unit increase in x_2?

16.31   A large apartment complex in the southeast relies completely on electricity for its heating and cooling. The number of kilowatt-hours (kwh) used in a day is related to the average daily temperature. Consumption is greatest for temperature extremes at the low and high ends of the scale, while kwh use is minimal near the middle of the temperature range. Give a model that might be used to relate the number of kwh consumed in a day (y) to the average temperature that day (x).

16.32   A snack food company subjects its peanuts to periodic taste tests. It has found that preference scores for its product are very sensitive to the quantity of salt used in the seasoning. Scores tend to be highest for a moderate amount, and the scores decrease sharply when relatively large or small quantities of salt are used. Write a model that might be suitable for relating mean rating scores (μ_y) to the amount of salt used (x) in the seasoning.

16.33   For the apartment complex referred to in Exercise 16.31, electricity consumption and the average temperature are given below for a random sample of 36 days. The figures for kwh are in units of 1,000 and temperatures are given in degrees Fahrenheit.

Temp. x	Kwh y	Temp. x	Kwh y	Temp. x	Kwh y
82	33	87	31	81	31
43	30	46	27	44	30
73	23	72	23	70	21
35	32	33	33	38	32
29	33	30	31	31	32
60	25	62	23	61	23
75	19	78	18	76	18
61	23	61	21	60	20
49	26	45	26	50	28
57	23	55	27	53	23
94	33	88	34	92	34
77	22	77	24	84	22

MINITAB was used to fit the model $y = \beta_0 + \beta_1 x + \beta_2 x^2 + \epsilon$ to the data, and a portion of the output is given below.

a. Give the estimated model.

b. Test if sufficient evidence exists to conclude that the model provides information for predicting electricity consumption. Use $\alpha = 0.01$.

```
The regression equation is
Y = 67.1 - 1.41 X + 0.0111 X-SQ

Predictor         Coef        Stdev      t-ratio          p
Constant        67.111        5.676        11.82      0.000
X              -1.4098        0.1985        -7.10      0.000
X-SQ          0.011141     0.001623         6.87      0.000

s = 3.258       R-sq = 61.0%      R-sq(adj) = 58.7%

Analysis of Variance
SOURCE          DF          SS           MS          F        p
Regression       2       548.65       274.32      25.84    0.000
Error           33       350.35        10.62
Total           35       899.00
```

MINITAB Assignments

M▶ 16.34 For the data in Exercise 16.33, use MINITAB to perform the following.
a. Construct a scatter diagram of the 36 points.
b. Fit the model $y = \beta_0 + \beta_1 x + \beta_2 x^2 + \epsilon$ to the data.

M▶ 16.35 In Exercise 16.16, the owner of an expensive gift shop thought that her weekly sales were related to the performance of the stock market for that week and the amount of advertising expenditures during the week. She now believes that sales are also affected by whether or not the city serves as host to a convention during the week. To explore this, she records for each of 30 randomly selected weeks her sales, advertising expenses, the mean of the Dow Jones Industrial Average, and if a convention was hosted that week. This information is given below, with the sales and advertising figures in units of $1,000.

Week	DJIA (x_1)	Adv. Expenses (x_2)	Sales (y)	Convention
1	7015	2.9	58.3	no
2	7318	3.1	62.9	yes
3	6581	2.5	46.3	no
4	6623	5.1	48.2	no
5	6917	2.9	58.2	no
6	7503	3.4	65.8	yes
7	6223	1.6	36.7	no
8	7332	3.0	62.3	no
9	7197	3.2	59.8	no
10	7513	3.5	71.8	yes
11	7703	3.1	66.9	yes
12	6523	1.9	37.9	no
13	6332	1.7	30.6	no
14	7197	3.3	69.8	yes
15	7713	3.8	65.2	yes
16	6679	3.0	52.7	no
17	6513	2.1	39.3	no
18	6922	3.4	58.7	no
19	7146	3.3	60.9	no
20	6429	2.2	40.5	yes
21	7409	3.9	70.3	yes
22	6315	2.4	39.1	no
23	6487	2.5	45.9	no
24	6715	2.8	55.1	no
25	7543	3.6	70.2	yes
26	7619	3.7	70.3	yes
27	6125	2.1	39.1	yes
28	6987	2.5	45.9	no
29	7215	2.9	55.1	no
30	7707	3.4	70.2	yes

Let the dummy variable $x_3 = 1$ if the city hosted a convention during the specified week, and then use MINITAB to fit the model

$$y = \beta_0 + \beta_1 x_1 + \beta_2 x_2 + \beta_3 x_3 + \epsilon.$$

M▶ 16.36 Exercise 16.13 was concerned with comparison shopping for medical services. To obtain a quick assessment of the relative costs of different hospitals, could one use as a barometer their charges for two common procedures? To investigate this, the table below gives the average costs of a birth and an appendectomy for 15 Pennsylvania hospitals. Also given are the average charges for a hospital stay, after adjusting for differences in the types of patients treated (*The Patriot,* Harrisburg, Pennsylvania, July 21, 1993).

Hospital	Birth (x_1)	Appendectomy (x_2)	Average Adjusted Hospitalization Cost (y)
Carlisle	$2,031	$3,785	$6,222
Chambersburg	$1,948	$2,859	$5,191
Com. H. of Lancaster	$1,146	$2,893	$5,580
Ephrata Community	$1,252	$2,637	$4,833
Gettysburg	$1,853	$4,584	$5,637
Good Samaritan	$1,763	$4,101	$6,827
Hanover General	$1,303	$2,838	$4,461
Harrisburg	$2,118	$4,560	$8,295
Holy Spirit	$1,858	$3,395	$7,023
Lancaster General	$1,056	$2,659	$4,767
Memorial H./York	$2,700	$4,978	$7,711
Penn State University	$1,832	$4,492	$7,460
Polyclinic Med. Ctr.	$2,550	$3,821	$7,806
St. Joseph/Lancaster	$1,919	$3,691	$5,753
York	$1,587	$3,380	$5,601

Use MINITAB to fit the interaction model $\mu_y = \beta_0 + \beta_1 x_1 + \beta_2 x_2 + \beta_3 x_1 x_2$.

M▶ 16.37 At the 1 percent significance level, test if the model in Exercise 16.35 is useful.

M▶ 16.38 Test if the model in Exercise 16.36 contributes useful information for estimating average costs for a hospital stay. Use $\alpha = 0.05$.

16.5 TESTING PORTIONS OF A MODEL

During the model building process, we usually proceed by (a) formulating an assumed model, (b) obtaining an estimated model, (c) employing the global F-test to check the overall usefulness of the model. If the global test fails to reject the null hypothesis (H_0: the model is not useful), then the assumed model is discarded and the process is begun anew. If, however, the global test provides sufficient evidence that the model is useful, then the researcher will often subject the model to further analysis. Part of this might be an attempt to refine the model by conducting additional tests concerning some of the β-parameters. For instance, in Section 16.2 we considered the assumed model

$$y = \beta_0 + \beta_1 x_1 + \beta_2 x_2 + \epsilon$$

where y, x_1, and x_2 respectively denote final course grade, placement test score, and SAT math score for students enrolled in an introductory statistics course. The least squares estimated model was obtained, and the global F-test concluded that the model was useful. At this point, we may wonder whether the inclusion of x_2 is important. That is, *with x_1 already in the model,* does the addition of x_2 significantly contribute useful information for predicting y? We can address this question by testing the null hypothesis H_0: $\beta_2 = 0$ (x_2 does not contribute significantly). Two applicable tests will be considered in this section. The first test is the simpler, but the second is more general in that it also allows one to test hypotheses involving more than one parameter.

t-Test of a Single Coefficient

For the multiple regression model, the following procedure can be used to test the null hypothesis that a particular coefficient is equal to zero. That is, we can test $H_0: \beta_i = 0$ for any specified parameter in the assumed model

$$y = \beta_0 + \beta_1 x_1 + \beta_2 x_2 + \beta_3 x_3 + \cdots + \beta_k x_k + \epsilon.$$

The test is based on the least squares estimate $\hat{\beta}_i$ of the coefficient β_i. Under the assumptions that we have made concerning the random error components, the sampling distribution of $\hat{\beta}_i$ is normal with a mean of β_i. We will denote its estimated standard deviation by $s_{\hat{\beta}_i}$. By transforming $\hat{\beta}_i$ to standard units, we obtain the test statistic t that appears in Equation 16.20.

PROCEDURE

t-Test of a Single Coefficient
Assumed Model: $y = \beta_0 + \beta_1 x_1 + \beta_2 x_2 + \beta_3 x_3 + \cdots + \beta_k x_k + \epsilon$

For testing the null hypothesis

$H_0: \beta_i = 0$ (In the presence of the other variables, x_i does not contribute significantly.)

the test statistic is

(16.20)
$$t = \frac{\hat{\beta}_i - \beta_i}{s_{\hat{\beta}_i}} = \frac{\hat{\beta}_i}{s_{\hat{\beta}_i}}.$$

The rejection region is given by the following:

For $H_a: \beta_i < 0$, H_0 is rejected for $t < -t_\alpha$;

For $H_a: \beta_i > 0$, H_0 is rejected for $t > t_\alpha$;

For $H_a: \beta_i \neq 0$, H_0 is rejected for $t < -t_{\alpha/2}$ or $t > t_{\alpha/2}$.

Assumptions:
The usual multiple regression assumptions given earlier with Equation 16.5.

Note:

1. The t-distribution has $df = n - (k + 1) = n - $ (number of β's in the model).

2. The value of the t-statistic in Equation 16.20 can be obtained from the coefficient table in the MINITAB regression output.

3. $\hat{\beta}_i$ is the least squares estimate with estimated standard deviation $s_{\hat{\beta}_i}$.

4. The t-statistic in Equation 16.20 can be used to test $H_0: \beta_i = b_0$ (b_0 is a specified constant) by substituting the value of $(\hat{\beta}_i - b_0)$ for $(\hat{\beta}_i - \beta_i)$.

EXAMPLE 16.13 For the course grades example, do the data provide sufficient evidence that the inclusion of the variable x_2 (SAT math score) contributes significantly to predicting course grades? Test at the 0.05 significance level.

Solution

The model assumed earlier was

$$y = \beta_0 + \beta_1 x_1 + \beta_2 x_2 + \epsilon$$

where x_1 is placement test score, x_2 is SAT math score, and y denotes final course grade. In Section 16.1, MINITAB was used to obtain the least squares estimated model, and the resulting output appears in Exhibit 16.4. The global F-test was performed in Example 16.3, and we concluded that the model is useful. We now want to determine if x_2, *in the presence of x_1 in the model,* makes a significant contribution to predicting y. Is the model significantly better than that without x_2 ($y = \beta_0 + \beta_1 x_1 + \epsilon$)? To determine this, we perform the following test.

Step 1: Hypotheses.

H_0: $\beta_2 - 0$ (x_2 does not contribute significant information for predicting y.)

H_a: $\beta_2 \neq 0$ (x_2 does contribute significantly.)

Step 2: Significance level.

$$\alpha = 0.05$$

Step 3: Calculations.

MINITAB gives the value of the test statistic in the coefficient table that appears below the regression equation in Exhibit 16.4. From this we have that

$$t = \frac{\hat{\beta}_2}{s_{\hat{\beta}_2}} - \frac{0.11195}{0.03860} = 2.90$$

Step 4: Rejection Region (or p-Value).

The p-value for $t = 2.90$ is given immediately to its right in the coefficient table.

$$p\text{-value} = 0.013$$

(Although it's unnecessary, the rejection region is obtainable from the t-table using $df = n - (k + 1) = 15 - 3 = 12$. We reject H_0 if $t < -t_{.025} = -2.179$ or $t > t_{.025} = 2.179$.)

Step 5: Conclusion.

Since the p-value of 0.013 is smaller than $\alpha = 0.05$, we reject the null hypothesis and accept the alternative. At the 5 percent level there is sufficient evidence that the presence of x_2 in the model does make a significant contribution, and thus, the SAT math score x_2 should be retained in the model.

Exhibit 16.4   MTB > REGR C3 2 C1 C2

The regression equation is
Y = - 9.7 + 0.771 X1 + 0.112 X2

Predictor	Coef	Stdev	t-ratio	p
Constant	-9.73	14.85	-0.66	0.525
X1	0.7708	0.6697	1.15	0.272
X2	0.11195	0.03860	2.90	0.013

s = 5.683   R-sq = 83.9%   R-sq(adj) = 81.3%

Analysis of Variance

SOURCE	DF	SS	MS	F	p
Regression	2	2026.2	1013.1	31.37	0.000
Error	12	387.5	32.3		
Total	14	2413.7			

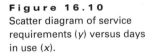

EXAMPLE 16.14   In Example 16.11, we considered a paper mill that utilizes a delicate sensing unit in the manufacturing process for its premium writing paper. Each unit requires daily service, but the number of minutes (y) depends on the number of days (x) that the unit has been in use. Service time is expected to be high during the break-in period and near the end of a unit's useful life, while it should be low near its mid-life period. Because of these expectations, a researcher hypothesized the quadratic model $y = \beta_0 + \beta_1 x + \beta_2 x^2 + \epsilon$, where $\beta_2 > 0$ (so that its response curve, a parabola, will open upward). The model was fit to $n = 27$ data points (see the scatter diagram in Figure 16.10), and a global F-test concluded that the model fits the data well. We now want to test if the x^2 term (the term that gives the model curvature) is really

Figure 16.10
Scatter diagram of service requirements (y) versus days in use (x).

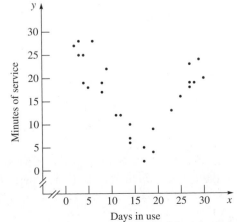

needed. More specifically, is there sufficient evidence at the 0.01 level to conclude that its coefficient $\beta_2 > 0$?

Step 1: Hypotheses.

$$H_0: \beta_2 = 0 \text{ (The response curve does not have curvature.)}$$

$$H_a: \beta_2 > 0 \text{ (The response curve is concave upward.)}$$

Step 2: Significance level.

$$\alpha = 0.01$$

Step 3: Calculations.

The value of the test statistic t is obtained from the MINITAB output that was generated in Example 16.11 and is repeated in Exhibit 16.5. From the coefficient table we have

$$t = \frac{\hat{\beta}_2}{s_{\hat{\beta}_2}} = \frac{0.09663}{0.01089} \approx 8.88.$$

Step 4: Rejection Region (or p-Value).

The p-value next to $t = 8.88$ in Exhibit 16.5 is 0.000. However, the p-values for the t tests given by MINITAB are for two-sided alternative hypotheses ($\neq$ in H_a). To obtain the p-value for our one-sided alternative hypothesis ($>$ in H_a), the given value must be divided by 2 (it might be helpful to review Examples 14 and 15 in Section 10.6).

$$(p\text{-value for } H_a: \beta_2 > 0) = (p\text{-value for } H_a: \beta_2 \neq 0)/2 = 0.000/2 = 0.000$$

(It is not needed, but the rejection region can be obtained from the t-table using $df = n - (k + 1) = 27 - 3 = 24$. We reject H_0 if $t > t_{.01} = 2.492$.)

Step 5: Conclusion.

Since the p-value equals 0.000 and this is smaller than $\alpha = 0.01$, we reject the null hypothesis and accept the alternative. There is sufficient evidence at the 0.01 level to conclude that upward curvature is present in the response curve. The $\beta_2 x^2$ term should be retained in the model.

In addition to testing the null hypothesis that a particular coefficient in a model is zero, a researcher may want to estimate the coefficient with a confidence interval. This is easily accomplished by using the least squares estimate of β_i and its estimated standard deviation that appear in the computer output. The confidence interval formula for β_i is summarized on the following page and illustrated in the next example.

Exhibit 16.5

```
MTB > NAME C1 'Y' C2 'X' C3 'X-SQ'
MTB > REGRESS C1 2 C2 C3

The regression equation is
Y = 35.9 - 3.28 X + 0.0966 X-SQ
```

Predictor	Coef	Stdev	t-ratio	p
Constant	35.859	2.296	15.62	0.000
X	-3.2774	0.3552	-9.23	0.000
X-SQ	0.09663	0.01089	8.88	0.000

```
s = 3.752      R-sq = 78.1%      R-sq(adj) = 76.2%
```

Analysis of Variance

SOURCE	DF	SS	MS	F	p
Regression	2	1202.76	601.38	42.71	0.000
Error	24	337.91	14.08		
Total	26	1540.67			

PROCEDURE

A $(1 - \alpha)$ Confidence Interval for the Coefficient β_i

Assumed Model: $y = \beta_0 + \beta_1 x_1 + \beta_2 x_2 + \beta_3 x_3 + \cdots + \beta_k x_k + \epsilon$

(16.21) $\hat{\beta}_i \pm t_{\alpha/2} s_{\hat{\beta}_i}$

Assumptions:
The usual multiple regression assumptions given earlier with Equation 16.5.

Note:

1. The t-distribution has $df = n - (k + 1) = n - $ (number of β's in the model).

2. $\hat{\beta}_i$ is the least squares estimate with estimated standard deviation $s_{\hat{\beta}_i}$.

EXAMPLE 16.15

Obtain a 95 percent confidence interval for the true coefficient of x^2 for the assumed quadratic model in Example 16.14.

Solution

For 95 percent confidence, $t_{\alpha/2} = t_{.025}$. The degrees of freedom are $df = n - (k + 1) = 27 - 3 = 24$. From the t-table we have that $t_{.025} = 2.064$. From the MINITAB output in Exhibit 16.5, $\hat{\beta}_2 = 0.09663$ and $s_{\hat{\beta}_2} = 0.01089$. A 95 percent confidence interval for β_2 is given by the following.

$$\hat{\beta}_2 \pm t_{\alpha/2} s_{\hat{\beta}_2}$$

$$0.09663 \pm (2.064)(0.01089)$$

$$0.09663 \pm 0.02248$$

Thus, with 95 percent confidence we estimate that the coefficient β_2 of the quadratic term x^2 is between 0.07415 and 0.11911.

We have illustrated how a t-statistic can be used to test the null hypothesis that a particular coefficient is equal to zero. We will now consider another test procedure that is more general in that it is applicable to *one or more* parameters of a regression model.

F-Test of One or More Coefficients

Although this test could be used for a single parameter, it is usually applied to testing the null hypothesis that two or more coefficients are zero. This type of test is useful in situations where a researcher wants to determine if a group of terms should be included in a model. To illustrate, consider again the course grades example with the assumed model $y = \beta_0 + \beta_1 x_1 + \beta_2 x_2 + \epsilon$, where x_1 is placement test score, x_2 is SAT math score, and y denotes final course grade. We might wonder if better predictions of y could be obtained by including second-order terms. Specifically, suppose we are considering for possible adoption the following complete second-order model in x_1 and x_2.

(16.22)   *Complete Model:* $y = \beta_0 + \beta_1 x_1 + \beta_2 x_2 + \beta_3 x_1 x_2 + \beta_4 x_1^2 + \beta_5 x_2^2 + \epsilon$

The terms that we are uncertain about including in the model are the 3 second-order terms. The model without these is called the **reduced model.**

(16.23)   *Reduced Model:* $y = \beta_0 + \beta_1 x_1 + \beta_2 x_2 + \epsilon$

We would expect that the complete model should provide better predictions of y if its additional terms $(\beta_3 x_1 x_2 + \beta_4 x_1^2 + \beta_5 x_2^2)$ can explain a significant amount of the variation in the y-values. This can be measured by examining the difference in the error sum of squares for the reduced and the complete models (recall that the error sum of squares, SSE, is the variation in y that remains unexplained by a model). We will let SSE_C denote the **error sum of squares for the complete model,** and SSE_R will denote the **error sum of squares for the reduced model.** Since the complete model contains additional terms not in the reduced model, SSE_C will be smaller than SSE_R, and the difference $(SSE_R - SSE_C)$ is a measure of the additional variation in the y-values that is explained by the additional terms $\beta_3 x_1 x_2 + \beta_4 x_1^2 + \beta_5 x_2^2$ in the complete model. This reduction in the error sum of squares in going from the reduced to the complete model forms the basis for testing if the additional terms are worth including in the model. The supporting evidence for preferring the complete model is enhanced as the size of the reduction increases, and this change $(SSE_R - SSE_C)$ is an essential part of the test statistic for testing the hypotheses

H_0: $\beta_3 = \beta_4 = \beta_5 = 0$ (The reduced and complete models do not differ significantly.)

H_a: At least one $\beta_i \neq 0$ (The complete model contributes more information for predicting y.)

The test procedure is outlined below, and it is illustrated in Example 16.16 for the course grades data.

PROCEDURE

F-Test of One or More Coefficients

Complete Model: $y = \beta_0 + \beta_1 x_1 + \beta_2 x_2 + \beta_3 x_3 + \cdots + \beta_k x_k + \epsilon$

Reduced Model: The above model with some terms $(\beta_i x_i + \cdots + \beta_j x_j)$ deleted.

For testing the hypotheses

H_0: $\beta_i = \cdots = \beta_j = 0$ (The reduced and complete models do not differ significantly.)

H_a: At least 1 $\beta \neq 0$ (Complete model contributes more information for predicting y.)

The test statistic is

(16.24) $$F = \frac{(SSE_R - SSE_C)/(\text{number of }\beta\text{'s in }H_0)}{MSE_C}.$$

The null hypothesis is rejected when $F > F_\alpha$, where α is the significance level of the test.

Assumptions:
The usual multiple regression assumptions given earlier with Equation 16.5.

Note:

1. In the formula for F, SSE_R and SSE_C are the error sums of squares for the reduced and complete models, respectively. MSE_C is the mean square error for the complete model. These quantities can be obtained from the ANOVA tables in the MINITAB regression output for the reduced and the complete models.

2. The F distribution has $ndf =$ (the number of β's being tested in H_0) and $ddf = n - (k + 1) =$ (sample size) $-$ (number of β's in the complete model).

EXAMPLE 16.16

Do the data for the course grades example provide sufficient evidence that the complete second-order model in Equation 16.22 is a better prediction equation than the first-order model in Equation 16.23? Test at the 5 percent level.

Solution

The complete and the reduced models are

Complete Model: $y = \beta_0 + \beta_1 x_1 + \beta_2 x_2 + \beta_3 x_1 x_2 + \beta_4 x_1^2 + \beta_5 x_2^2 + \epsilon$

Reduced Model: $y = \beta_0 + \beta_1 x_1 + \beta_2 x_2 + \epsilon$

Earlier MINITAB was used to fit the reduced model to the values of x_1, x_2, and y that were stored in columns C1, C2, and C3, respectively. The MINITAB output is reproduced in Exhibit 16.6 on page 804. From the ANOVA table, the error sum of squares for the reduced model is $SSE_R = \mathbf{387.5}$.

To have MINITAB obtain SSE_C, the error sum of squares for the complete model, we first use the **LET** command to create 3 columns that will contain the values of x_1x_2, x_1^2, and x_2^2. Then the **REGRESS** command is used to obtain the estimated complete model.

```
LET C4 = C1*C2    # CREATES COLUMN OF X1*X2 VALUES
LET C5 = C1**2    # CREATES COLUMN OF X1-SQ VALUES
LET C6 = C2**2    # CREATES COLUMN OF X2-SQ VALUES
REGRESS C3 5 C1 C2 C4 C5 C6 # REGRESSES C3 ON 5 PREDICTORS
```

A portion of the resulting output is displayed in Exhibit 16.7. The ANOVA table shows that $SSE_C = 262.38$ and $MSE_C = 29.15$.

With the values of SSE_R, SSE_C, and MSE_C, we now have enough information to test if the second-order terms in the complete model contribute significant information for predicting y.

Step 1: Hypotheses.

H_0: $\beta_3 = \beta_4 = \beta_5 = 0$ (The reduced and complete models do not differ significantly.)

H_a: At least one $\beta_i \neq 0$ (The complete model contributes more information for predicting y.)

Step 2: Significance level.

$$\alpha = 0.05$$

Step 3: Calculations.

$$F = \frac{(SSE_R - SSE_C)/(\text{number of } \beta\text{'s in } H_0)}{MSE_C} = \frac{(387.5 - 262.38)/3}{29.15} = 1.43$$

Step 4: Rejection Region (or p-Value).

Since we had to calculate the F value in step 3, its p-value is not available from the MINITAB output. Consequently, we will determine the rejection region from part b of Table 6 in Appendix A. We find that H_0 is rejected for $F > F_{.05} = 3.86$. This value was obtained using $ndf = $ (the number of β's being tested in H_0) = 3 and $ddf = n - $ (number of β's in the complete model) $= 15 - 6 = 9$.

Step 5: Conclusion.

Since $F = 1.43$ is not in the rejection region, we do not reject the null hypothesis. At the 0.05 level, there is insufficient evidence to say that the second-order terms in the complete model contribute useful information for predicting y. Thus, we cannot conclude that the complete model with the second-order terms is a better prediction equation than the reduced model.

Recall that in Example 16.3 we performed a global F-test and concluded that the first-order model (reduced model) is useful. Consequently, we would use its estimated model for predicting y. Exhibit 16.6 shows this to be

$$\hat{y} = -9.7 + 0.771x_1 + 0.112x_2.$$

■■■ **Exhibit 16.6**

```
MTB > REGR C3 2 C1 C2
The regression equation is
Y = - 9.7 + 0.771 X1 + 0.112 X2

Predictor        Coef        Stdev      t-ratio         p
Constant        -9.73        14.85       -0.66       0.525
X1             0.7708       0.6697        1.15       0.272
X2            0.11195      0.03860        2.90       0.013
s = 5.683       R-sq = 83.9%     R-sq(adj) = 81.3%

Analysis of Variance
SOURCE        DF          SS          MS         F         p
Regression     2       2026.2      1013.1     31.37     0.000
Error         12        387.5        32.3
Total         14       2413.7
```

■■■ **Exhibit 16.7**

```
MTB > LET C4 = C1*C2
MTB > LET C5 = C1**2
MTB > LET C6 = C2**2
MTB > NAME C1 'X1' C2 'X2' C3 'Y' C4 'X1X2' C5 'X1-SQ' C6 'X2-SQ'
MTB > REGRESS C3 5 C1 C2 C4 C5 C6
The regression equation is
Y = -253 -17.2X1 +1.40X2 +0.0468X1X2 -0.309X1-SQ -0.00169X2-SQ

Predictor        Coef        Stdev      t-ratio         p
Constant       -252.9       125.9        -2.01       0.076
X1             -17.18       10.73        -1.60       0.144
X2             1.3994      0.6487         2.16       0.059
X1X2          0.04676     0.03438         1.36       0.207
X1-SQ         -0.3091      0.3197        -0.97       0.359
X2-SQ      -0.0016942    0.0009617       -1.76       0.112
s = 5.399       R-sq = 89.1%     R-sq(adj) = 83.1%

Analysis of Variance
SOURCE        DF          SS          MS         F         p
Regression     5      2151.35      430.27     14.76     0.000
Error          9       262.38       29.15
Total         14      2413.73
```

EXAMPLE 16.17   In Example 16.4, we discussed a salary study of college faculty in a particular region of the United States. In Example 16.7, the following model was proposed to relate salary (y) to years of service (x_1), experience (x_2) at the time of initial appointment, and the qualitative variables gender and academic rank.

$$y = \beta_0 + \beta_1 x_1 + \beta_2 x_2 + \beta_3 x_3 + \beta_4 x_4 + \beta_5 x_5 + \beta_6 x_6 + \epsilon$$

The variables x_3, x_4, x_5 and x_6 are dummy variables used to place gender and rank in the model. They are defined as follows.

$$x_3 = \begin{cases} 1 \text{ if male} \\ 0 \text{ if not} \end{cases} \qquad x_4 = \begin{cases} 1 \text{ if prof.} \\ 0 \text{ if not} \end{cases} \qquad x_5 = \begin{cases} 1 \text{ if assoc.} \\ 0 \text{ if not} \end{cases} \qquad x_6 = \begin{cases} 1 \text{ if assist.} \\ 0 \text{ if not} \end{cases}$$

The data given earlier in Table 16.2 were coded for the four dummy variables as indicated above, and a portion of the data is shown in Table 16.4.

a. Obtain the estimated model, and test at the 0.01 level if sufficient evidence exists to conclude that the model is useful.

b. Determine if the inclusion of academic rank in the model contributes significant information for predicting salaries. Use $\alpha = 0.05$.

Solution

a. To obtain the estimated model, the values of y and the predictors x_1 through x_6 were stored in columns C7 and C1 through C6, respectively. The following command was then used to generate the MINITAB output in Exhibit 16.8.

```
REGRESS C7 6 C1-C6   # REGRESSES Y IN C7 ON 6 VARS. IN C1-C6
```

Menu Commands:
Stat ➤
Regression ➤
Regression
In **Response** box
enter **C7**
In **Predictors** box
enter **C1-C6**
Select OK

TABLE 16.4

FACULTY MEMBER, 1991 SALARY, YEARS OF SERVICE, EXPERIENCE RATING AT ENTRANCE, SEX, AND RANK

Fac. M.	y	x_1	x_2	x_3	x_4	x_5	x_6
1	35000	16	1.00	0	0	1	0
2	43000	23	4.00	1	1	0	0
3	26000	7	8.00	0	0	0	1
4	51100	19	4.00	1	1	0	0
5	49200	13	19.50	1	1	0	0
6	44900	16	3.50	1	1	0	0
7	34400	8	5.00	1	0	0	1
8	40600	12	5.00	0	1	0	0
9	35200	10	7.50	1	0	1	0
10	40600	19	3.00	1	1	0	0
..	..	..	..	..	..	..	..
..	..	..	..	..	..	..	..
..	..	..	..	..	..	..	..
165	37600	16	9.00	0	1	0	0
166	31700	15	5.00	0	1	0	0
167	26600	9	3.00	1	0	1	0
168	22600	8	6.33	0	0	0	1
169	31900	15	2.00	0	1	0	0
170	24000	7	1.00	0	0	0	1
171	26700	10	9.00	1	0	0	1

▉▉▉ **Exhibit 16.8**

```
MTB > NAME C7 'Y' C1 'X1' C2 'X2' C3 'X3'
MTB > NAME C4 'X4' C5 'X5' C6 'X6'
MTB > REGRESS C7 6 C1-C6

The regression equation is
Y = 16193 + 1023 X1 + 491 X2 + 1525 X3 + 2943 X4 + 563 X5 - 179 X6
```

Predictor	Coef	Stdev	t-ratio	p
Constant	16193	1893	8.55	0.000
X1	1022.67	54.82	18.66	0.000
X2	490.92	62.80	7.82	0.000
X3	1524.8	434.8	3.51	0.001
X4	2943	2063	1.43	0.156
X5	563	2019	0.28	0.781
X6	-179	2018	-0.09	0.930

```
s = 2594       R-sq = 88.5%      R-sq(adj) = 88.0%
```

Analysis of Variance

SOURCE	DF	SS	MS	F	p
Regression	6	845197E+04	1408661376	209.42	0.000
Error	164	110313E+04	6726381		
Total	170	955510E+04			

The least squares estimated model from Exhibit 16.8 is

$$\hat{y} = 16193 + 1023x_1 + 491x_2 + 1525x_3 + 2943x_4 + 563x_5 - 179x_6.$$

To test the usefulness of the model, we'll perform the following global F-test.

Step 1: Hypotheses.

H_0: $\beta_1 = \beta_2 = \cdots = \beta_6 = 0$ (The predictors do not contribute significant information.)

H_a: At least one $\beta_i \neq 0$ (At least one of the predictors contributes significantly.)

Step 2: Significance level.

$$\alpha = 0.01$$

Step 3: Calculations.

From the ANOVA table in Exhibit 16.8, the test statistic value is $F = 209.42$.

Step 4: Rejection Region (or p-Value).

The p-value, next to 209.42 in the ANOVA table, is 0.000.

Step 5: Conclusion.

Since the p-value is less than $\alpha = 0.01$, we reject H_0 and accept H_a. There is very strong evidence that the model is useful for predicting salaries.

b. We want to test if the qualitative variable academic rank contributes useful information to the model for the purpose of predicting salaries. Rank enters the model through the dummy variables x_4, x_5, and x_6. These variables differentiate the ranks. If the β-coefficients are 0 for all terms involving them ($\beta_4 x_4$, $\beta_5 x_5$, $\beta_6 x_6$), then rank is not a factor in the model. Thus, we need to test the null hypothesis

$$H_0: \beta_4 = \beta_5 = \beta_6 = 0.$$

The complete and the reduced models are

Complete Model: $y = \beta_0 + \beta_1 x_1 + \beta_2 x_2 + \beta_3 x_3 + \beta_4 x_4 + \beta_5 x_5 + \beta_6 x_6 + \epsilon$

Reduced Model: $y = \beta_0 + \beta_1 x_1 + \beta_2 x_2 + \beta_3 x_3 + \epsilon$

> **Menu Commands:**
> **Stat ➤**
> **Regression ➤**
> **Regression**
> *In* **Response** *box*
> *enter* **C7**
> *In* **Predictors** *box*
> *enter* **C1-C3**
> *Select* OK

For the complete model, we can obtain SSE_C and MSE_C from the ANOVA table in Exhibit 16.8. Their values are $SSE_C = 110313 \cdot 10^4$ and $MSE_C = 6726381$. We now need to have MINITAB find SSE_R for the reduced model. This can be obtained from the following command.

```
REGRESS C7 3 C1-C3   # REGRESSES C7 ON 3 VARS. IN C1,C2,C3
```

The resulting output is given in Exhibit 16.9. From the ANOVA table, we have that $SSE_R = 117851 \cdot 10^4$. We can now perform the following test for a portion of the model.

Step 1: Hypotheses.

$H_0: \beta_4 = \beta_5 = \beta_6 = 0$ (The reduced and complete models do not differ significantly.)

$H_a:$ At least one $\beta_i \neq 0$ (The complete model contributes more information for predicting y.)

Step 2: Significance level.

$$\alpha = 0.05$$

Step 3: Calculations.

$$F = \frac{(SSE_R - SSE_C)/(\text{number of } \beta\text{'s in } H_0)}{MSE_C} = \frac{(117851 \cdot 10^4 - 110313 \cdot 10^4)/3}{6726381} = 3.74.$$

Step 4: Rejection Region (or p-Value).

H_0 is rejected for $F > F_{.05} \approx 2.60$. This value is from part b of Table 6 using

$ndf = $ (the number of β's tested in H_0) $= 3$

$ddf = n - $ (the number of β's in the complete model) $= 171 - 7 = 164$.

Step 5: Conclusion.

Since $F = 3.74$ is in the rejection region, we reject H_0 and accept H_a. There is sufficient evidence at the 0.05 level to conclude that the inclusion of academic rank in the model does contribute useful information for predicting salaries. Thus, the complete model is preferred to the reduced model, and we would use the estimated model obtained in part a.

Exhibit 16.9 MTB > REGRESS C7 3 C1-C3

The regression equation is
Y = 16179 + 1160 X1 + 507 X2 + 1332 X3

Predictor	Coef	Stdev	t-ratio	p
Constant	16178.9	778.2	20.79	0.000
X1	1160.47	36.25	32.01	0.000
X2	507.22	63.14	8.03	0.000
X3	1331.5	433.8	3.07	0.003

s = 2656   R-sq = 87.7%   R-sq(adj) = 87.4%

Analysis of Variance

SOURCE	DF	SS	MS	F	p
Regression	3	837659E+04	2792196608	395.67	0.000
Error	167	117851E+04	7056916		
Total	170	955510E+04			

16.6 USING THE MODEL TO ESTIMATE AND PREDICT

In developing a multiple regression model, a researcher may be guided by several objectives. Interest may focus mainly on understanding the behavior of the phenomenon under study. Sometimes the major purpose is to find a set of independent variables that strongly influence the response variable y. Often, however, the investigator ultimately plans to use the model for estimation and prediction purposes. To illustrate the methodology for this, consider again the assumed model

$$y = \beta_0 + \beta_1 x_1 + \beta_2 x_2 + \epsilon.$$

where y, x_1, and x_2 respectively denote final course grade, placement test score, and SAT math score for students enrolled in an introductory statistics course. MINITAB was used to obtain the following least squares estimated model.

(16.25) $\hat{y} = -9.7 + 0.771x_1 + 0.112x_2$

The multiple coefficient of determination was found to be $R^2 = 0.839$, suggesting a good fit of the least squares equation to the data. To confirm this, we next conducted a global F-test and concluded that the model is useful. At this stage, we might want to use the estimated model to make an inference or a prediction about final course grades. As we saw with the straight line models in Chapter 14, the least squares equation could be used to

a. **estimate the mean of y, μ_y,** for specified values of the predictor variables;

b. **predict a single value of y** for specified values of the predictor variables.

For instance, we might want to estimate the mean course grade for all students who achieve a score of 17 on the math placement test and have SAT math scores of 600.

Or suppose we have a student with scores of 17 and 600, and we would like to predict the student's final course grade. In each case, the least squares estimated model is used by substituting in it the values $x_1 = 17$ and $x_2 = 600$, and then evaluating $\hat{y}$. This gives the value

$$\begin{aligned}
\hat{y} &= -9.7 + 0.771x_1 + 0.112x_2 \\
&= -9.7 + 0.771(17) + 0.112(600) \\
&= 70.6
\end{aligned}$$

The same value of 70.6 is used to estimate the mean grade and also predict the student's final course grade. As we saw in Chapter 14, however, the two situations differ in regard to their error bound. The mean grade can be estimated more precisely. A 95 percent confidence interval for it will have a smaller width than that of a 95 percent prediction interval for the individual's grade. The following example illustrates this and shows how MINITAB can be used to obtain these intervals.

EXAMPLE 16.18

Obtain a 95 percent confidence interval to estimate the mean course grade for all students who achieve a score of 17 on the math placement test and have a 600 SAT math score. Also, predict with 95 percent probability the course grade of an individual with these scores.

Menu Commands:
Stat ➤
Regression ➤
Regression
In **Response** *box*
enter **C3**
In **Predictors** *box*
enter **C1 C2**
Select Options
In **Pred. intervals**
for new obs. *box*
enter **17 600**
In **Confidence level**
box enter **95**
Select OK
Select OK

Solution

In an earlier example, we had MINITAB store the values of y, x_1, and x_2 in columns C3, C1, and C2, respectively. By using the subcommand **PREDICT** with the **REGRESS** command, MINITAB will construct a 95 percent confidence interval for μ_y and a 95 percent prediction interval for y. To obtain these for the values $x_1 = 17$ and $x_2 = 600$, type the following.

```
REGRESS C3 2 C1 C2;
PREDICT 17 600.   # 1ST ENTER PT SCORE 17, THEN SAT SCORE 600
```

In addition to the desired intervals, MINITAB will also produce the usual output for the **REGRESS** command. As many as 10 **PREDICT** subcommands can be used with the same **REGRESS** command. Furthermore, the values $x_1 = 17$ and $x_2 = 600$ could be replaced by 2 columns of values for x_1 and x_2.

The output generated from the above commands is shown in Exhibit 16.10. The confidence and predictions intervals are shaded and are the following.

95% confidence interval for μ_y when $x_1 = 17$, $x_2 = 600$: $67.25 \le \mu_y \le 73.85$

95% prediction interval for y when $x_1 = 17$, $x_2 = 600$: $57.73 \le y \le 83.36$

Thus, with 95 percent confidence we estimate that the mean course grade is between 67.25 and 73.85 for all students who score 17 and 600 on the placement and SAT math tests. Also, with probability 0.95, we predict that an individual with these scores will have a course grade between 57.73 and 83.36.

Note that the confidence interval is more precise in that its width is only 6.60, compared to a width of 25.63 for the prediction interval. Because the prediction interval is so wide, it would have little, if any, practical value.

Exhibit 16.10

```
MTB > NAME C1 'X1' C2 'X2' C3 'Y'
MTB > REGRESS C3 2 C1 C2;
SUBC> PREDICT 17 600.   # PLACEMENT TEST 17, SAT 600

The regression equation is
Y = - 9.7 + 0.771 X1 + 0.112 X2
```

Predictor	Coef	Stdev	t-ratio	p
Constant	-9.73	14.85	-0.66	0.525
X1	0.7708	0.6697	1.15	0.272
X2	0.11195	0.03860	2.90	0.013

s = 5.683   R-sq = 83.9%   R-sq(adj) = 81.3%

Analysis of Variance

SOURCE	DF	SS	MS	F	p
Regression	2	2026.2	1013.1	31.37	0.000
Error	12	387.5	32.3		
Total	14	2413.7			

Fit	Stdev.Fit	95% C.I.	95% P.I.
70.55	1.51	(67.25, 73.85)	(57.73, 83.36)

For straight line models, we warned that one should be very cautious in using the least squares equation to estimate and predict at values that lie outside the extremes of the x-values for which the model was developed. The assumed model might not be applicable outside this range, and the error bound could be unsuitably large. Similar risks exist when extrapolation is used with multiple regression models. With a simple linear model, it is clear when a value of x falls outside the range of the sample points. The situation is considerably more complex for multiple regression models, and the occurrence of extrapolation is sometimes not easily recognized. To illustrate, consider Figure 16.11 that shows the 15 points (x_1, x_2) for the data

Figure 16.11
Extrapolation occurs at the point $x_1 = 9$ and $x_2 = 760$.

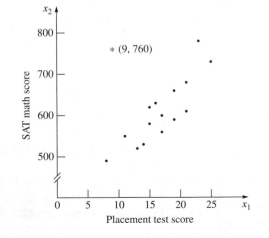

used to obtain the estimated model in Equation 16.25. Also shown is the point (9, 760) that is clearly distant from the 15 points used to fit the model, even though separately $x_1 = 9$ and $x_2 = 760$ each lies within the range of sample values for x_1 and x_2.

EXAMPLE 16.19

For the 171 college faculty salaries in Example 16.17, obtain the predicted salary for each faculty member, and compare it with the actual salary received.

Solution

The predictions, $\hat{y}$, and the amounts by which they differ from the actual y-values, $(y - \hat{y})$, can be obtained as part of MINITAB's output from the **REGRESS** command. Recall that MINITAB allows for three levels of output with this command. The maximum amount displays the values of $\hat{y}$ and $(y - \hat{y})$, and it is obtained by typing the following *before* using **REGRESS**.

BRIEF 3

MINITAB refers to the predictions, $\hat{y}$, as the **fits,** and these are displayed in a column headed by "Fit." The difference between a y-value and its fit, $(y - \hat{y})$, is called the **residual,** and they are also given as part of the output. In the following section, we will examine how the fits and the residuals are useful in the model building process. Their values for the 171 faculty are displayed in Table 16.5.

TABLE 16.5

FACULTY MEMBER, ACTUAL SALARY, PREDICTED SALARY, RESIDUAL

FM	y	$\hat{y}$	$(y - \hat{y})$	FM	y	$\hat{y}$	$(y - \hat{y})$	FM	y	$\hat{y}$	$(y - \hat{y})$
1	35000	33609	1391	58	52300	47035	5265	115	43000	43189	-189
2	43000	46146	-3146	59	47600	48027	-427	116	45600	46146	-546
3	26000	27100	-1100	60	47400	45614	1786	117	38700	39548	-848
4	51100	42055	9045	61	43900	42616	1284	118	39500	40080	-580
5	49200	43528	5672	62	45500	44621	879	119	40600	42464	-1864
6	44900	38741	6159	63	47400	46677	723	120	36400	38851	-2451
7	34400	28175	6225	64	51100	49050	2050	121	43400	43680	-280
8	40600	33862	6738	65	42600	40449	2151	122	41000	42576	1576
9	35200	32189	3011	66	48200	47700	500	123	30300	29682	618
10	40600	41564	-964	67	46100	47168	-1068	124	42400	45082	-2682
11	29400	28666	734	68	43100	43896	-796	125	44800	44621	179
12	43800	41032	2768	69	30300	30962	-662	126	33100	34674	-1574
13	36800	33833	2967	70	42400	44838	-2438	127	39300	40080	-780
14	49200	46637	2563	71	50900	50891	9	128	50900	52323	-1423
15	30200	30878	-678	72	28300	26395	1905	129	44700	49132	-4432

(Continued on next page)

TABLE 8.3 *(Continued)*

FACULTY MEMBER, ACTUAL SALARY, PREDICTED SALARY, RESIDUAL

16	47500	43078	4422	73	47900	48887	−987	130	44000	47127	−3127
17	45800	42546	3254	74	51300	48384	2916	131	43900	43650	250
18	46800	43569	3231	75	50700	48641	2059	132	31200	32230	−1030
19	43200	40530	2670	76	42800	43650	−850	133	33100	34680	−1580
20	24500	22831	1669	77	45100	41553	3547	134	32800	34190	−1390
21	34100	32022	2078	78	44600	43066	1534	135	45300	48027	−2727
22	29600	24814	4786	79	48200	48641	−441	136	39700	40612	−912
23	51000	47289	3711	80	48100	51220	−3120	137	54300	53141	1159
24	47600	47127	473	81	47600	49705	−2105	138	47300	49214	−1914
25	45900	42055	3845	82	44400	43066	1334	139	38400	40500	−2100
26	51500	46677	4823	83	52500	50605	1895	140	40100	39057	1043
27	36300	33944	2356	84	43400	42096	1304	141	42000	43732	−1732
28	27600	27146	454	85	50900	50963	−63	142	25300	24688	612
29	46800	46884	−84	86	39800	42055	−2255	143	24600	26269	−1669
30	47600	46105	1495	87	44600	45123	−523	144	35600	37012	−1412
31	43000	40776	2224	88	53800	51136	2664	145	40000	40039	−39
32	28700	26323	2377	89	41300	41839	−539	146	43100	44632	−1532
33	43800	44221	−421	90	41500	42627	−1127	147	35200	38619	−3419
34	48400	46105	2295	91	29400	27848	1552	148	27800	28466	−666
35	47700	47414	286	92	42600	43650	−1050	149	40400	42996	−2596
36	53900	48835	5065	93	42600	43639	−1039	150	39800	43118	−3318
37	39900	38987	913	94	41100	39682	1418	151	36100	35989	111
38	47000	46677	323	95	47500	47934	−434	152	28700	31196	−2496
39	27500	29689	−2189	96	44100	46637	−2537	153	42600	44580	−1980
40	25500	23051	2449	97	52600	50114	2486	154	32100	34517	−2417
41	42200	42587	−387	98	44300	46064	−1764	155	33700	35949	−2249
42	46500	45696	804	99	44000	43680	320	156	31800	38300	−6500
43	41600	38455	3145	100	51000	51627	−627	157	43600	48191	−4591
44	41000	41073	−73	101	43100	42248	852	158	34600	40285	−5685
45	46700	45082	1618	102	46700	48682	−1982	159	28000	32921	−4921
46	41600	43078	−1478	103	43500	45859	−2359	160	29500	32214	−2714
47	45900	45614	286	104	44400	46637	−2237	161	44400	48109	−3709
48	43100	43569	−469	105	43900	43680	220	162	30200	34026	−3826
49	37700	40337	−2637	106	48300	49623	−1323	163	28100	30471	−2371
50	35100	34404	696	107	46300	47659	−1359	164	37600	41306	−3706
51	49400	47659	1741	108	44500	45112	−612	165	37600	39917	−2317
52	47600	46105	1495	109	46300	45123	1177	166	31700	36930	−5230
53	41000	40039	961	110	44400	44171	229	167	26600	28957	−2357
54	40300	39108	1192	111	43700	43159	541	168	22600	27303	−4703
55	42300	39998	2302	112	51600	49582	2018	169	31900	35458	−3558
56	43200	41564	1636	113	44100	42616	1484	170	24000	23664	336
57	46900	46637	263	114	47200	46666	534	171	26700	32184	−5484

SECTIONS 16.5 AND 16.6 EXERCISES

Testing Portions of a Model; Using the Model to Estimate and Predict

16.39 The multiple regression model $\mu_y = \beta_0 + \beta_1 x_1 + \beta_2 x_2$ was fit to 25 data points, and the following least squares equation was obtained: $\hat{y} = -2.8 + 47.6 x_1 + 9.8 x_2$. The estimated standard deviations of $\hat{\beta}_1$ and $\hat{\beta}_2$ are 17.4 and 2.5, respectively. Using $\alpha = 0.01$, test the null hypothesis $H_0: \beta_1 = 0$ against $H_a: \beta_1 > 0$.

16.40 A soil engineer investigated the relationship between the number of tons of crop yield (y) and the numbers of applied units of fertilizers A (x_1), B (x_2), and C (x_3). The model $\mu_y = \beta_0 + \beta_1 x_1 + \beta_2 x_2 + \beta_3 x_3$ was fit to 29 points, and the estimated model is $\hat{y} = 94.31 + 6.65 x_1 + 8.43 x_2 - 0.80 x_3$. The estimated standard deviations of $\hat{\beta}_1$, $\hat{\beta}_2$, and $\hat{\beta}_3$ are 2.15, 1.79, and 0.38, respectively. With $\alpha = 0.05$, is there sufficient evidence that the fertilizer A term should be retained in the model?

16.41 Find the approximate p-value for the test in Exercise 16.39.

16.42 Obtain the approximate p-value for Exercise 16.40.

16.43 Construct a 95 percent confidence interval to estimate the coefficient of x_2 in Exercise 16.39.

16.44 For the model in Exercise 16.40, estimate β_1 with a 90 percent confidence interval.

16.45 MINITAB was used to fit the model $y = \beta_0 + \beta_1 x_1 + \beta_2 x_2 + \epsilon$ to $n = 32$ points, and a portion of the resulting output is given below.
a. Using $\alpha = 0.05$, test the utility of the model.
b. Is there sufficient evidence to conclude that the predictor variable x_2 contributes significantly to predicting y? Use $\alpha = 0.05$.
c. Construct a 99 percent confidence interval to estimate β_1.

```
The regression equation is
Y = 0.881 + 0.249 X1 - 0.211 X2

Predictor        Coef        Stdev       t-ratio          p
Constant        0.8811      0.2466          3.57      0.001
X1              0.24882     0.02848         8.74      0.000
X2             -0.2110      0.1004         -2.10      0.044

s = 0.4462      R-sq = 72.5%      R-sq(adj) = 70.6%
Analysis of Variance
SOURCE          DF           SS           MS           F          p
Regression       2      15.1947       7.5973       38.16      0.000
Error           29       5.7741       0.1991
Total           31      20.9688
```

16.46 The following complete second-order model in 2 quantitative variables x_1 and x_2 was fit to $n = 46$ data points. The resulting error sum of squares was $SSE = 598.6$, and the multiple coefficient of determination was $R^2 = 0.792$.

$$y = \beta_0 + \beta_1 x_1 + \beta_2 x_2 + \beta_3 x_1^2 + \beta_4 x_2^2 + \beta_5 x_1 x_2 + \epsilon$$

a. With $\alpha = 0.05$, test if the model contributes information for predicting y.
b. Give the null and alternative hypotheses for determining if some of the second-order terms contribute information for predicting y.
c. What hypotheses would you test to see if the interaction term contributes significantly?

16.47 For the second-order model in Exercise 16.46, the least squares estimate of the interaction coefficient was 5.73, and the estimated standard deviation of $\hat{\beta}_5$ was 3.12. Conduct the test of hypotheses referred to in part c. Use $\alpha = 0.05$.

16.48 With reference to Exercise 16.46, the following model without second-order terms was fit to the data, and the resulting error sum of squares was $SSE = 799.6$. Use $\alpha = 0.05$ to conduct the hypotheses test referred to in part b.

$$y = \beta_0 + \beta_1 x_1 + \beta_2 x_2 + \epsilon$$

16.49 A company with plants in the northern and southern parts of its state manufactures colored writing paper. Each plant produces paper in the colors white, beige, yellow, and rust. To compare paper quality for different colors and plant locations, the company proposed the following model, where y is an index measure of relative quality.

$$\mu_y = \beta_0 + \beta_1 x_1 + \beta_2 x_2 + \beta_3 x_3 + \beta_4 x_4$$

where

$$x_1 = \begin{cases} 1 \text{ if white} \\ 0 \text{ if not} \end{cases} \quad x_2 = \begin{cases} 1 \text{ if beige} \\ 0 \text{ if not} \end{cases} \quad x_3 = \begin{cases} 1 \text{ if rust} \\ 0 \text{ if not} \end{cases} \quad x_4 = \begin{cases} 1 \text{ if north plant} \\ 0 \text{ if not} \end{cases}$$

a. Give the null and alternative hypotheses for determining if a difference exists in the mean quality for the four colors.
b. What hypotheses would you test to determine if there is a difference in the mean quality of paper produced at the two plants?

16.50 The company in Exercise 16.49 decided to include in the assumed model interaction terms between the qualitative variables plant location and paper color. The following model was assumed, where x_1, x_2, x_3, and x_4 are defined in Exercise 16.49.

$$\mu_y = \beta_0 + \beta_1 x_1 + \beta_2 x_2 + \beta_3 x_3 + \beta_4 x_4 + \beta_5 x_1 x_4 + \beta_6 x_2 x_4 + \beta_7 x_3 x_4$$

a. Write the hypotheses for determining if there is a difference in the mean paper quality for the four colors.

b. Write the hypotheses for determining if a difference exists in the mean paper quality for the two plants.

16.51   The following model given in Exercise 16.49 was fit to 125 data points, and the resulting error sum of squares was 1,599.97. The variables are as defined earlier.

$$\mu_y = \beta_0 + \beta_1 x_1 + \beta_2 x_2 + \beta_3 x_3 + \beta_4 x_4$$

The reduced model

$$\mu_y = \beta_0 + \beta_1 x_1 + \beta_2 x_2 + \beta_3 x_3$$

was next fit to the same data, and the corresponding error sum of squares was 1,793.88. Is there sufficient evidence to conclude that there is a difference in mean paper quality for the two plants? Test at the 5 percent level.

16.52   The following model is used by a large corporation to model the number of overtime hours (y) worked by its employees during the year.

$$y = \beta_0 + \beta_1 x_1 + \beta_2 x_2 + \beta_3 x_3 + \beta_4 x_1 x_2 + \beta_5 x_1 x_3 + \beta_6 x_2 x_3 + \beta_7 x_1 x_2 x_3 + \epsilon$$

In the above model, x_1 and x_2 are quantitative variables that denote a worker's hourly rate and age, respectively, and x_3 is a dummy variable that places gender in the model ($x_3 = 1$ if female, $x_3 = 0$ if male).
a. What hypotheses would one test to determine if there is a difference for men and women in the relation between overtime hours with hourly rate and age?
b. For the hypotheses test in part a, the complete and reduced models were fit to $n = 35$ data points, and the resulting error sums of squares for the two models were $SSE_C = 912.35$ and $SSE_R = 1,145.82$. Perform the test using $\alpha = 0.01$.

16.53   Exercise 16.15 gave the number of wins (y), earned run average (x_1), batting average (x_2), and number of fielding errors (x_3) for each of the 14 American League baseball teams during the 1995 season. The following model was fit to the data, and the resulting ANOVA table is shown below.

$$y = \beta_0 + \beta_1 x_1 + \beta_2 x_2 + \beta_3 x_3 + \epsilon$$

Analysis of Variance

Source	DF	SS	MS	F	P
Regression	3	1595.72	531.91	19.47	0.000
Error	10	273.20	27.32		
Total	13	1868.93			

Next, the reduced model $y = \beta_0 + \beta_1 x_1 + \epsilon$ was fit and its ANOVA table is below.

```
Analysis of Variance
Source        DF         SS          MS         F        P
Regression     1       1338.1      1338.1     30.25    0.000
Error         12        530.9        44.2
Total         13       1868.9
```

Do the data provide sufficient evidence to conclude that the variables batting average and number of fielding errors should be retained in the model? Test at the 5 percent level of significance.

MINITAB Assignments

M▶ 16.54 The following table gives the number of kilowatt-hours used daily by a large apartment complex in the southeast for a random sample of 36 days. Also given is the average temperature for each of the sampled days. Use MINITAB to fit the model $y = \beta_0 + \beta_1 x + \beta_2 x^2 + \epsilon$, and then test if the x^2 term contributes significantly to the prediction of y. Use $\alpha = 0.05$.

Temp. x	Kwh y	Temp. x	Kwh y	Temp. x	Kwh y
82	33	87	31	81	31
43	30	46	27	44	30
73	23	72	23	70	21
35	32	33	33	38	32
29	33	30	31	31	32
60	25	62	23	61	23
75	19	78	18	76	18
61	23	61	21	60	20
49	26	45	26	50	28
57	23	55	27	53	23
94	33	88	34	92	34
77	22	77	24	84	22

M▶ 16.55 For the data in Exercise 16.54 and the model $y = \beta_0 + \beta_1 x + \beta_2 x^2 + \epsilon$, use the **BRIEF** and **REGRESS** commands to obtain the predicted kwh usage $\hat{y}$ and the residual $(y - \hat{y})$ for each of the sampled days.

M▶ 16.56 For the data in Exercise 16.54 and the model $y = \beta_0 + \beta_1 x + \beta_2 x^2 + \epsilon$, obtain a 95 percent confidence interval to estimate the mean kwh usage for all days when the average temperature is
a. 40 degrees.
b. 70 degrees.

M▶ 16.57 In Exercise 16.35, the owner of an expensive gift shop thought that her weekly sales were related to the stock market performance and advertising expenditures for the week, and whether or not her city hosted a convention during the week. The following data were given for 30 randomly selected weeks. Sales and advertising figures are given in units of $1,000.

Week	DJIA (x_1)	Adv. Expenses (x_2)	Sales (y)	Convention
1	7015	2.9	58.3	no
2	7318	3.1	62.9	yes
3	6581	2.5	46.3	no
4	6623	5.1	48.2	no
5	6917	2.9	58.2	no
6	7503	3.4	65.8	yes
7	6223	1.6	36.7	no
8	7332	3.0	62.3	no
9	7197	3.2	59.8	no
10	7513	3.5	71.8	yes
11	7703	3.1	66.9	yes
12	6523	1.9	37.9	no
13	6332	1.7	30.6	no
14	7197	3.3	69.8	yes
15	7713	3.8	65.2	yes
16	6679	3.0	52.7	no
17	6513	2.1	39.3	no
18	6922	3.4	58.7	no
19	7146	3.3	60.9	no
20	6429	2.2	40.5	yes
21	7409	3.9	70.3	yes
22	6315	2.4	39.1	no
23	6487	2.5	45.9	no
24	6715	2.8	55.1	no
25	7543	3.6	70.2	yes
26	7619	3.7	70.3	yes
27	6125	2.1	39.1	yes
28	6987	2.5	45.9	no
29	7215	2.9	55.1	no
30	7707	3.4	70.2	yes

Let the dummy variable $x_3 = 1$ if the city hosted a convention in a specified week.

a. Use MINITAB to fit the complete model $y = \beta_0 + \beta_1 x_1 + \beta_2 x_2 + \beta_3 x_3 + \epsilon$.

b. Use MINITAB to fit the reduced model $y = \beta_0 + \beta_1 x_1 + \epsilon$.

c. Is there sufficient evidence at the 5 percent level to conclude that the complete model contributes more information for predicting weekly sales?

M▶ 16.58 For the complete model in Exercise 16.57, obtain a 95 percent prediction interval for sales in a week with a convention and $x_1 = 7500$ and $x_2 = 3.5$ thousand.

16.7 ANALYZING RESIDUALS TO EVALUATE THE MODEL

The previous sections have discussed several steps involved in the development of a model. Before adopting it and actually using it in practice, the model builder still needs to address the following three potential problems. Has the functional form of the model been specified incorrectly? Does the estimated model generate some predictions that grossly differ from the actual data? Are there serious violations in the underlying assumptions for the random error components? Considerable information concerning answers to these questions is contained within the residuals that result from applying the estimated model to the data.

KEY CONCEPT

Residuals and Fits
A predicted value $\hat{y}$ obtained from the estimated model is called a **fit**, and the difference between the observed value y and its fit is the **residual**.

$$\text{Residual} = (\text{observed } y - \text{fit}) = (y - \hat{y})$$

In recent years, with the wide availability of statistical computer software, extensive analyses of residuals have become common practice. This section provides an introduction to some of the more frequently used techniques. The interested reader is encouraged to consult the references for a more in-depth discussion of these and other methods.

Misspecification of the Model

An analysis of the residuals often begins by constructing for each predictor variable x_i a plot of the residuals against the corresponding values of x_i. Such a graph is useful in revealing if the order of a model has been under-specified in the predictor x_i. We illustrate this in the next example.

EXAMPLE 16.20

In Example 16.11, the second-order model $y = \beta_0 + \beta_1 x + \beta_2 x^2 + \epsilon$ was fit to the paper mill data shown in Table 16.6. Fit the first-order model $y = \beta_0 + \beta_1 x + \epsilon$ to the data and examine the resulting residuals.

Solution
The values of y and x were stored in columns C1 and C2, respectively, and then the following commands were applied.

```
BRIEF 3
REGRESS C1 1 C2   # Y VALUES IN C1, X VALUES IN C2
```

The resulting output includes the residuals shown in Table 16.7 (they can also be calculated from the estimated model $\hat{y} = 19.604 - 0.2033x$ that is given as part of the output). A plot of the 27 residuals against their corresponding x-values appears in Figure 16.12. Ideally, the residuals should vary randomly around zero (indicated

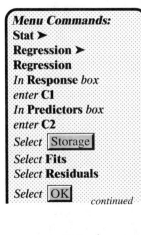

Menu Commands:
Stat ➤
Regression ➤
Regression
In **Response** *box*
enter **C1**
In **Predictors** *box*
enter **C2**
Select Storage
Select **Fits**
Select **Residuals**
Select OK
continued

TABLE 16.6

REQUIRED MINUTES OF SERVICE AND NUMBER OF DAYS IN USE
FOR $n = 27$ RECORDS

Service y (minutes)	Days in Use x	Service y (minutes)	Days in Use x	Service y (minutes)	Days in Use x
7	14	18	5	2	17
16	25	19	4	4	19
19	8	12	12	19	28
17	8	25	3	12	11
27	2	20	30	18	27
13	23	23	27	25	4
28	6	28	3	6	14
22	9	19	27	9	19
5	17	24	29	10	14

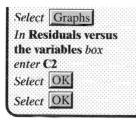

Select Graphs
In **Residuals versus the variables** *box enter* **C2**
Select OK
Select OK

by a horizontal line) with no detectable pattern or trend. Such is not the case in Figure 16.12, where the residual plot deviates from zero in a systematic way. Small and large values of x tend to have positive residuals, while the residuals are negative for intermediate x-values. This systematic pattern in the residual plot frequently indicates the need for a higher-order x term, such as x^2. Figure 16.13 shows the residual plot for the second-order model $y = \beta_0 + \beta_1 x + \beta_2 x^2 + \epsilon$ that was used in Example 16.11. Its appearance indicates that the addition of the x^2 term to the model has greatly reduced the systematic pattern inherent in the residual plot for the first-order model.

TABLE 16.7

RESIDUALS FOR THE ESTIMATED MODEL $\hat{y} = 19.604 - 0.2033x$ AND THE DATA IN TABLE 16.6

x	y	Fit $\hat{y}$	Residual $(y - \hat{y})$	x	y	Fit $\hat{y}$	Residual $(y - \hat{y})$	x	y	Fit $\hat{y}$	Residual $(y - \hat{y})$
14	7	16.76	−9.76	5	18	18.59	−0.59	17	2	16.15	−14.15
25	16	14.52	1.48	4	19	18.79	0.21	19	4	15.74	−11.74
8	19	17.98	1.02	12	12	17.17	−5.17	28	19	13.91	5.09
8	17	17.98	−0.98	3	25	18.99	6.01	11	12	17.37	−5.37
2	27	19.20	7.80	30	20	13.51	6.49	27	18	14.12	3.88
23	13	14.93	−1.93	27	23	14.12	8.88	4	25	18.79	6.21
6	28	18.38	9.62	3	28	18.99	9.01	14	6	16.76	−10.76
9	22	17.78	4.22	27	19	14.12	4.88	19	9	15.74	−6.74
17	5	16.15	−11.15	29	24	13.71	10.29	14	10	16.76	−6.76

Figure 16.12
Plot of residuals against x indicates curvature is needed in the model.

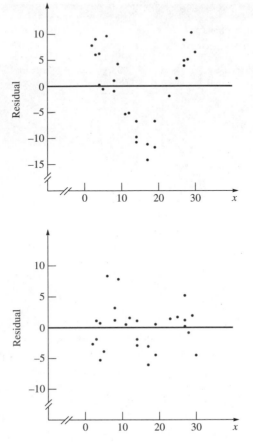

Figure 16.13
Plot of residuals versus x after adding an x^2-term to the assumed model.

Checking for Outliers

An **outlier** is a data point whose residual is extremely large. While there is not universal agreement on its definition, the following provides a reasonable guide as to what constitutes an outlier in a data set.

KEY CONCEPT

A data point will be considered an **outlier** if its standardized residual is greater than 3 in absolute value.

Detecting outliers is facilitated by working with residuals in their standardized form. Since the mean of the residuals is zero, a standardized residual is obtained by dividing the residual by its standard deviation. Although they are difficult to calculate, standardized residuals are readily available with many statistical software packages, including MINITAB.

In developing a model, it is very important to check for outliers, because they often identify potential problems that require additional investigation. Sometimes an outlier occurs because of a measurement or recording error. An outlier may result because of a malfunctioning in the process being observed. It might represent an element that was not intended for inclusion in the sampled population, such as the price of a duplex in a survey of single-unit homes. In instances such as these, outliers should be corrected when possible or discarded. Since outliers do not follow the general trend of the remaining data, they can profoundly influence the choice of the estimated model. However, one should be extremely cautious in discarding an outlier, and it should only be done if ample justification can be found. Rather than representing an abnormality, an outlier may reflect an improperly chosen model or serious violations in its underlying assumptions.

One procedure for locating outliers is to plot the standardized residuals against the fits $\hat{y}$. The next example illustrates how MINITAB can be used for this purpose.

EXAMPLE 16.21

For the 171 faculty salaries in Example 16.17, plot the standardized residuals against the fits in order to detect the presence of outliers.

Solution

Earlier the y-values were stored in column C7, and the values of the predictor variables x_1 through x_6 were placed in columns C1 through C6. Then the following command was used to obtain the estimated model.

```
REGRESS C7 6 C1-C6
```

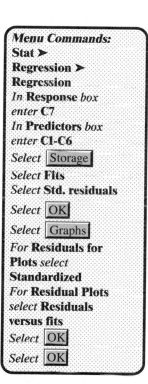

Menu Commands:
Stat ➤
Regression ➤
Regression
In **Response** *box*
enter **C7**
In **Predictors** *box*
enter **C1-C6**
Select Storage
Select **Fits**
Select **Std. residuals**
Select OK
Select Graphs
For **Residuals for Plots** *select*
Standardized
For **Residual Plots** *select* **Residuals versus fits**
Select OK
Select OK

If two storage columns are included at the end of the above command, MINITAB will place in them the standardized residuals and the fits. The following instructs MINITAB to regress the y-values in C7 on the 6 predictors in columns C1 through C6, and then store the standardized residuals in C8 and the fits in C9.

```
REGRESS C7 6 C1-C6 C8 C9
```

The **PLOT** command can now be used to graph the standardized residuals in column C8 against the fits in C9.

```
GSTD    # SWITCHES TO STANDARD GRAPHICS MODE
PLOT C8 C9   # PLOTS STD. RESIDUALS IN C8 VS. FITS IN C9
```

The resulting standardized residual plot appears in Exhibit 16.11 (multiple points at a location are denoted by an integer). We have superimposed 2 horizontal lines at -3 and $+3$ to show the boundaries of the standardized residuals for locating outliers. Only 1 point lies above the $+3$ upper-boundary line (standardized residual equals 3.51, fit equals \$42,055), and it should be examined in detail. No point lies below the -3 lower-boundary line. However, it is also prudent to perform at least a cursory check of any points that just barely fall within the horizontal strip defined by the ± 3 boundary lines. Two such points occur in Exhibit 16.11, with one nearly touching the upper-boundary line and the other falling close to the lower-boundary line.

■ **Exhibit 16.11**

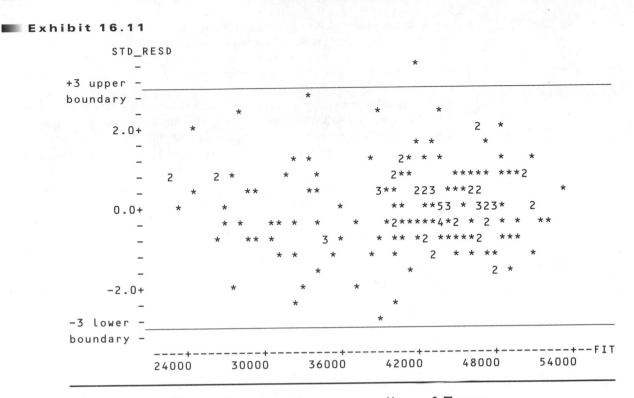

Checking for Nonnormality of Error Components

When the multiple regression model was defined in Section 16.1, certain assumptions were made concerning the behavior of the random error components ϵ. One assumption was that the random error components are normally distributed for each combination of predictor values. As is true with many statistical procedures, moderate departures from normality do not pose major problems, especially when the sample size n is large. On the other hand, gross departures from the normality assumption can seriously affect the validity of hypotheses tests and the precision of predictions and inferences.

A simple check for nonnormality of the random error components can be made by displaying the standardized residuals (or residuals) in some graphical form such as a histogram or a stem-and-leaf display. This is illustrated in the following example.

EXAMPLE 16.22   Construct a stem-and-leaf display of the standardized residuals for the 171 faculty salaries, and note if the normality assumption appears to be unreasonable.

Solution

In Example 16.21, we had MINITAB store the standardized residuals in column C8. The following command will generate the stem-and-leaf display shown in Exhibit 16.12.

    STEM C8

Menu Commands:
Graph ➤
Character Graphs ➤
Stem-and-Leaf
In **Variables** *box enter*
C8
Select OK

If the stem-and-leaf display in Exhibit 16.12 were rotated counterclockwise 90 degrees, its appearance would be mound shaped and resemble a normal distribution. There does not appear to be any gross departures from normality.

■ Exhibit 16.12

```
MTB > STEM C8    # STD. RESIDUALS ARE STORED IN C8

Stem-and-Leaf of STD_RESD    N  = 171
Leaf Unit = 0.10

       1     -2 6
       4     -2 220
       9     -1 98775
      23     -1 44332222100000
      54     -0 999999999888888777766666555555
    (34)     -0 4444443333333222222222111111100000
      83      0 0000001111111112222222333334444
      52      0 5555555555666667778888888999999
      22      1 000112222344
      10      1 789
       7      2 0034
       3      2 56
       1      3
       1      3 5
```

An additional check for nonnormality in the random error components can be made by constructing a **normal probability plot** for the standardized residuals (see Section 9.4 for a discussion of normal probability plots). The procedure consists of first determining the so-called *normal scores* of the standardized residuals. These are then plotted against the corresponding standardized residuals. If the normality assumption holds, then the plotted points should lie approximately on a straight line. The next example illustrates this procedure.

EXAMPLE 16.23

Perform a normal probability plot of the standardized residuals for the 171 faculty salaries.

Solution

The standardized residuals were stored earlier in column C8. In the following, the command **NSCORES** calculates the normal scores for the values in C8 and places the results in column C10. The **PLOT** command graphs the normal scores against the corresponding standardized residuals. The **CORRELATION** command is then applied to the points. The resulting MINITAB output is given in Exhibit 16.13.

```
NSCORES C8 C10 # PUTS IN C10 NORMAL SCORES OF STD RESIDS(C8)
GSTD
PLOT C10 C8
CORRELATION C8 C10
```

Menu Commands:
(First type command
NSCORES C8 C10)
Graph ➤
Plot
In **Graph variables**
Y-box enter **C10**
In **Graph variables**
X-box enter **C8**
Select OK
Stat ➤
Basic Statistics ➤
Correlation
In **Variables** *box*
enter **C8 C10**
Select OK

The normal probability plot appears quite straight, and this is supported by the large correlation coefficient of 0.994. As indicated in Example 16.22, the normality assumption for the random error components appears to be tenable.

■ **Exhibit 16.13**

```
MTB > NAME C10 'N_SCORE'
MTB > NSCORES C8 C10 # PUTS IN C10 NORMAL SCORES OF STD RESIDS(C8)
MTB > PLOT C10 C8
 N_SCORE
     -                                                          *
     -                                                    *
     -                                               *  2*
  2.0+                                          22 * 2*
     -                                        535
     -                                      748*
     -                                   *+662
  0.0+                              5769
     -                           54+5
     -                         5663
     -                       346
     -                 ***32
 -2.0+             **2
     -            *
     -         *                                      STD_RESD
     -    ------+---------+---------+---------+---------+---------+
          -2.4      -1.2      0.0       1.2       2.4       3.6

MTB > CORR C8 C10

Correlation of STD_RESD and N_SCORE = 0.994
```

When the viability of the normality assumption is in doubt, the situation can sometimes be remedied by first performing a transformation on the response variable y (such as $1/y$, $\log y$, $\sqrt{y}$, and so forth) and then regressing the transformed variable on the predictor variables. Another approach is to use an estimation procedure other than the ordinary least squares that we have employed. Robust estimation methods exist that are less sensitive to departures from normality. Rectifying the existence of nonnormality is not a simple issue, and it is further complicated by the fact that the suggestion of its presence may be caused by other factors such as a poorly chosen model or a violation in one of the other underlying assumptions. The interested reader is advised to consult the references on regression for addressing this problem.

Figure 16.14
Funnel effect from
nonconstant variance.

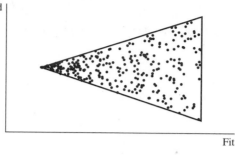

Checking for Nonconstant Error Variances

Another assumption of the multiple regression model is that the random error components have a common variance for each combination of predictor values. In reality, this is not always the case, and frequently the variance tends to increase as the fits become larger. In these instances a plot of the standardized residuals against the fits tends to resemble a portion of a *funnel* as in Figure 16.14. The use of this type of plot to detect nonconstant variance (**heteroscedasticity**) is illustrated in the following example.

EXAMPLE 16.24   For the 171 faculty salaries, examine the plot of standardized residuals against fits to detect the possible presence of unequal variances in the random error components.

Solution
The desired plot was generated earlier in Example 16.21, and it is repeated on p. 826 in Exhibit 16.14. The variation in the standardized residuals produces a slight funnel effect for initial fits at the low end of the salary range, but the pattern is not sustained. The constant variance assumption does not appear to be seriously violated.

The remedies for the detection of nonconstant error variances are similar to those described for the existence of nonnormality. The investigator might attempt a transformation on the response variable to stabilize the variance, or abandon the method of ordinary least squares and use a robust procedure to obtain the estimated model. Frequently the method of **weighted least squares** is applied when the problem of heteroscedasticity is encountered. Details on this and other advanced procedures can be found in the regression references.

Checking for Dependent Errors

In addition to the assumptions of normality and constant variance, we also assume for the multiple regression model that the random error components are independent. This assumption is often violated with **time-series** data where the sample observa-

■ Exhibit 16.14

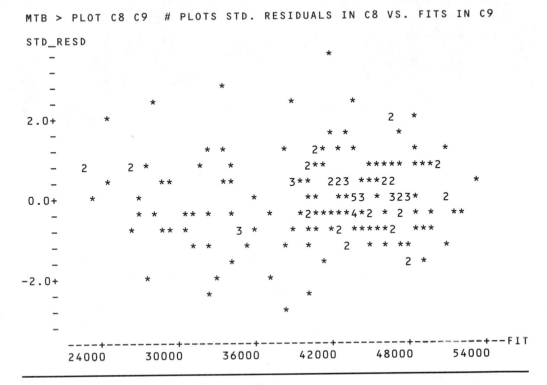

```
MTB > PLOT C8 C9   # PLOTS STD. RESIDUALS IN C8 VS. FITS IN C9

STD_RESD
     -                                          *
     -
     -                            *
     -              *                    *          *
 2.0+        *                                  2   *
     -                                     *  *      *
     -           *  *        *    2*  * *       *   *
     -    2     2  *       *    *      2**    ***** ***2
     -     *      **      **     3**  223 ***22              *
 0.0+   *      *              *     **   **53 * 323*   2
     -       * *    ** *   *      *  *2*****4*2 * 2  * *   **
     -      *   ** *      3  *     * ** *2 *****2  ***
     -           * *     *     *  *    2 * * **      *
     -              *          *         2 *
-2.0+      *           *      *
     -              *       *
     -                  *
     -
       ----+---------+---------+---------+---------+---------+--FIT
         24000     30000     36000     42000     48000     54000
```

TABLE 16.8					

STANDARDIZED RESIDUALS AND ASSOCIATED TIME PERIODS FOR A MULTIPLE REGRESSION FIT

Std. Residual	Time Period	Std. Residual	Time Period	Histogram of STD_RESD	
−0.58	9	0.21	11	Midpoint	Count
−0.99	4	−0.62	5	−2.0	1 *
0.03	12	−1.95	2	−1.5	0
−1.25	1	1.65	19	−1.0	3 ***
0.39	15	−0.27	8	−0.5	4 ****
−0.11	6	0.27	13	0.0	4 ****
0.36	14	2.17	18	0.5	4 ****
−1.16	3	0.49	16	1.0	2 **
−0.10	10	1.08	20	1.5	1 *
−0.58	7	0.96	17	2.0	1 *

tions are collected in a time sequence such as hourly, daily, weekly, and so forth. With this type of data, one should check if the error terms are correlated over time. A simple check is to plot the standardized residuals (or residuals) against the time periods and observe if some type of nonrandom pattern exists.

EXAMPLE 16.25   A multiple regression model was fit to 20 data points, and the resulting standardized residuals are displayed in Table 16.8. Also shown is the time period in which the corresponding data point was observed. The histogram of the standardized residuals has approximately a reassuring mound shape, but it belies the fact that serious doubts exist about the assumption of independent errors. The problem surfaces when the standardized residuals are plotted against the time periods. The plot is shown in Exhibit 16.15. There is a lack of randomness in the points, with a long run of negative residuals followed by a long run of positive residuals. As time progresses, the standardized residuals tend to increase in approximately a linear pattern.

■ **Exhibit 16.15**

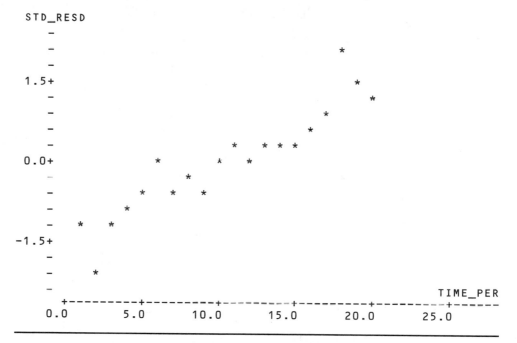

When a plot such as that in the last example reveals a time-related effect, a time series model is called for, where the time period is used as a predictor variable.

Several techniques for analyzing residuals have been presented in this section. A brief summary of these is given on the following page.

PROCEDURE

Suggested Procedure for Analyzing Residuals

Check for

1. **a misspecification of the model.** For each predictor x_i, plot the residuals (or standardized residuals) against the corresponding values of x_i. If curvature is exhibited, try a higher-order term like x_i^2 in the model.

2. **outliers.** Plot the standardized residuals against the fits $\hat{y}$, and investigate data points with standardized residuals greater than 3 in magnitude. Try to correct measurement or recording errors, and discard any point that should not have been included in the sample.

3. **nonconstant error variances.** Inspect the plot of the standardized residuals against the fits, and observe if the standardized residuals tend to fluctuate in a nonrandom pattern.

4. **nonnormality of error components.** Construct a histogram or a stem-and-leaf display of the standardized residuals and see if the distribution lacks the general shape of a normal curve. Another possibility is to construct a normal probability plot for the standardized residuals.

5. **dependent errors.** For time series data, plot the standardized residuals against the time periods in which the data were observed. Look for some type of nonrandom pattern in the standardized residuals. This might consist of an unusually large (or small) number of runs in positive and negative residuals.

Note:

Remedial measures for the above conditions usually require advanced techniques that are beyond the scope of an introductory statistics course. Details on appropriate methodology can be found in the references on regression in Appendix C.

16.8   OTHER CONSIDERATIONS

Before concluding our discussion of multiple regression, we will briefly consider some other important issues in the development of a multiple regression model.

Multicollinearity

Before a multiple regression model can be constructed, one must select an appropriate response variable y and an initial set of predictor variables that will be considered for inclusion in the model. Frequently a statistician is presented with a vaguely worded problem that requires considerable probing of the client to identify appropriate variables for the data collection stage. Although it might be tempting to use a shotgun approach and collect observations on a multitude of variables, this would be an expensive and time-consuming strategy. Furthermore, the careless use of many predictors produces an overly complicated model that can pose serious problems in estimating the model's parameters. A likely source of these problems is the existence of **multicollinearity** among the predictor variables. Multicollinearity refers to some of the predictors being correlated among themselves. While correlations between the independent variables usually exist to some extent, it is a serious problem when the correlations are large and strong linear relationships exist among some predictors. When two variables are highly correlated, they are redundant in that they duplicate information that contributes to the prediction of y. In addition to the inefficiency of this redundancy, it can produce an estimated model with unstable coefficients that are difficult to interpret and that have large standard deviations. Moreover, t-tests concerning the coefficients of the highly correlated variables often yield conclusions that appear to be in contradiction with the global F-test. This is illustrated in the following example.

EXAMPLE 16.26   The registrar at a liberal arts college wanted to model the scores (y) of graduating seniors on a national examination taken during the spring semester of their senior year. Among several predictor variables considered were a student's grade-point average (x_1) at the end of the junior year and the student's percentile class standing (x_2) at the time of the examination. Exhibit 16.16 shows a portion of the MINITAB output when y was regressed on x_1 and x_2 for a sample of 72 students.

Notice that for x_1, the estimated coefficient is 73.75 with a very large estimated standard deviation of 96.51. The situation is much worse for x_2. The value of its estimated coefficient is -0.094, and the estimated standard deviation of the coefficient is 1.728, more than 18 times as large! Also, since the partial slope of x_2 is negative, interpreting its value of 0.094 is confusing and meaningless. Its negative sign contradicts our expectation that test score y will increase as percentile ranking x_2 increases and grade-point average x_1 is held fixed. The contradiction arises because x_1 and x_2 are highly correlated ($r = 0.988$), and the resulting presence of multicollinearity prevents x_1 from being held fixed while x_2 is allowed to increase. Another indication that multicollinearity may be a problem comes from comparing the results of individual t-tests for β_1 and β_2 with the global F-test for the overall model. From the ANOVA table, $F = 10.61$, and its p-value of 0.000 indicates that at least one of the two variables in the model $\mu_y = \beta_0 + \beta_1 x_1 + \beta_2 x_2$ contributes information for predicting y. However, from the coefficient table we see that the t-tests of H_0: $\beta_1 = 0$ and H_0: $\beta_2 = 0$ are both nonsignificant with p-values of 0.447 and 0.957, respectively. Although it may appear otherwise, the results of the t-tests are not contradicting the conclusion from the F-test. The test of H_0: $\beta_1 = 0$ indicates that grade-point average x_1 has a nonsignificant predictive value *if class standing* x_2 *is already in the model.* Similarly, the conclusion from testing H_0: $\beta_2 = 0$ reveals that x_2 does not contribute a significant amount of information for predicting y *once* x_1 *is present in the model.* The results of the t-tests are consistent with the fact that x_1 and x_2 are highly correlated with each other, and thus their predictive capabilities pertaining to y greatly overlap. With the serious multicollinearity problem in this study, only one of the predictors x_1 and x_2 should be used in the final model.

Exhibit 16.16

```
MTB > REGR C1 2 C2-C3

The regression equation is
Y = 256 + 73.8 X1 - 0.09 X2

Predictor       Coef        Stdev      t-ratio          p
Constant       256.4        198.8        1.29      0.201
X1             73.75        96.51        0.76      0.447
X2            -0.094        1.728       -0.05      0.957

s = 60.55        R-sq = 23.5%      R-sq(adj) = 21.3%

Analysis of Variance

SOURCE          DF          SS          MS         F        p
Regression       2        77842       38921     10.61    0.000
Error           69       253016        3667
Total           71       330859

MTB > CORR C2 C3

Correlation of X1 and X2 = 0.988
```

The previous example illustrates primary reasons why you should try to avoid encounters with multicollinearity. At the beginning of a regression study, before the data collection stage, give careful thought to what predictor variables might be strongly related to y. Try to use, however, predictors that are uncorrelated or only weakly related to each other. You want to avoid predictor variables with a large amount of duplication in explanatory information. Considerable help in this endeavor often comes from talking with people who are knowledgeable in the field to which the study pertains.

For situations in which multicollinearity must be dealt with, one should consider replacing the method of least squares with one designed to deal with this problem. One such method is **ridge regression,** and it falls under the general category of **biased estimation techniques.** Information on these can be found in the advanced regression references in Appendix C. Another alternative for coping with multicollinearity is to delete predictor variables that are highly redundant, retaining only a subset of predictors for which multicollinearity is not a problem. **Stepwise regression** is a method that can be used to select such a subset of predictor variables.

Stepwise Regression

Stepwise regression is a technique used to methodically screen a large number of independent variables for the purpose of selecting a useful subset of predictors. The procedure requires an enormous number of calculations, and consequently, requires the use of a computer. Several variations of the method exist, and they do not always produce the same subset of variables. We will briefly outline two frequently used methods.

The **backward elimination** method starts with the model that includes all k variables under consideration. At step 1, the t-ratios for the estimates of the β-parameters are compared, and the variable with the t-ratio closest to 0 is considered for elimination. If its p-value exceeds some predetermined criterion (often $\alpha = 0.10$ is used), then that variable is discarded and the procedure progresses to step 2. At step 2 the model with the remaining $(k - 1)$ variables is fit, and the t-ratios for the estimates of the β-parameters are compared. The variable with the t-ratio closest to 0 is considered for elimination, and it is discarded as long as its p-value exceeds the criterion α. The procedure then progresses to step 3 where the model is fit to the remaining $(k - 2)$ variables, and a comparison of t-ratios for the estimates of the β-parameters is again made as in the previous steps. The process continues to eliminate the least significant variable at each step until a step is reached at which no t-ratio has a p-value that exceeds the predetermined criterion α. The variables that remain at this step constitute the subset of predictor variables selected.

Some computer programs use a backward elimination method with a criterion for elimination that differs from that described here, but the basic procedure is the same. For instance, instead of using a p-value criterion, a program may discard a variable if the size of its associated t-value is less than a certain criterion. Equivalently, it may eliminate a variable if the square of its t-value, which is an F-value, is smaller than some value (MINITAB uses this criterion). Stepwise programs usually have options for other variations in the procedure.

Another type of stepwise regression is the **forward selection** method. Instead of starting with the model that includes all k variables under consideration, it begins by examining models with only one variable. At step 1, a one-predictor model of the form

$$\mu_y = \beta_0 + \beta_1 x_i$$

is fit for each of the k variables. The variable whose estimated coefficient has the largest t-ratio (in absolute value) is selected, and it is retained if its p-value is less than or equal to some predetermined criterion ($\alpha = 0.10$ and $\alpha = 0.05$ are common choices). For convenience, we'll let x_1 denote the selected variable at step 1. At step 2, a two-variable model of the form

$$\mu_y = \beta_0 + \beta_1 x_1 + \beta_2 x_i$$

is fit for each of the $(k - 1)$ unselected variables in step 1. That is, every possible two-variable model that includes x_1 with one of the other $(k - 1)$ variables is fit at step 2. The t-ratio is computed for each of the other $(k - 1)$ variables, and the one with the largest t-ratio is the selected variable at step 2, as long as its p-value is less than or equal to the criterion α. We'll let x_2 denote the variable selected at step 2. At step 3, a three-variable model of the form

$$\mu_y = \beta_0 + \beta_1 x_1 + \beta_2 x_2 + \beta_3 x_i$$

is fit for each of the $(k - 2)$ unselected variables in step 2. Note that these are all the possible three-variable models that include x_1 and x_2 with one of the remaining $(k - 2)$ variables. The t-ratios are computed for each of the other $(k - 2)$ variables, and the one with the largest t-ratio is the selected variable at step 3, provided that its p-value is less than or equal to the α criterion. The process continues to add the

■ **Exhibit 16.17**

```
MTB > STEPWISE C11 C1-C10;   # Y IN C11, X's IN C1 TO C10
SUBC> FENTER = 4. # ENTRY CRITERION OF F>4 (T>2 IN ABS.VALUE)
    STEPWISE REGRESSION OF Y  ON  10 PREDICTORS, WITH N =    90
```

STEP	1	2	3	4
CONSTANT	140.90	89.25	48.70	61.33
X7	0.733	0.561	0.492	0.494
T-RATIO	8.54	5.60	4.62	4.53
X3		0.246	0.249	0.247
T-RATIO		3.44	3.02	2.87
X1			23	21
T-RATIO			2.91	2.21
X5				1.1
T-RATIO				2.08
S	51.3	48.7	47.8	47.5
R-SQ	51.77	59.28	62.30	63.71

Menu Commands:
Stat ➤
Regression ➤
Stepwise
In **Response** box
enter **C11**
In **Predictors** box
enter **C1-C10**
Select Options
In **F to enter** box
enter **4**
Select OK
Select OK

most significant variable at each step until a step is reached at which no t-ratio has a p-value that is less than or equal to the α criterion. The variables chosen up to this step make up the final subset of predictors selected by the method.

In the forward selection procedure, as with the backward elimination method, statistical software vary in the criterion used for variable selection at each step. Many contain options that permit the user considerable flexibility in the way the procedure is implemented, such as allowing the removal of a variable entered earlier if at a later step it no longer makes a significant contribution. Another common option is to force the program to add a variable at each step, never deleting one that was included at an earlier step, and to stop only when all variables have been included. The user then considers the order in which the variables entered and their corresponding effects on R^2 to select an appropriate subset of predictors.

Exhibit 16.17 is an example of MINITAB output obtained by applying the **STEPWISE** command to a set of y-values in column C11 and 10 independent variables with values in columns C1 through C10. The forward selection method was used, and a new variable entered at each step, as long as the absolute value of its t-ratio exceeded 2.* The first variable entered at step 1 was x_7 because its estimated coefficient had the largest t-value (8.54). At step 2 the second variable entered was x_3. With x_7 already in the model, x_3 had the largest t-ratio (3.44) among the 9 variables that did not enter at step 1. At step 3 the predictor x_1 joined x_7 and x_3 in the model, and at step 4 the variable x_5 came in. The procedure ended at that point because with x_7, x_3, x_1, and x_5 in the model, none of the remaining 6 predictors had a t-ratio with an absolute value exceeding the entry criterion of 2 (F-ratio of

* $|t| > 2$ is equivalent to $F > 2^2$, and this criterion is specified in MINITAB by using the subcommand **FENTER 4** after the **STEPWISE** main command.

$2^2 = 4$). The coefficients are highlighted for the estimated model with the 4 selected predictors. The estimated model is

$$\hat{y} = 61.33 + 0.494x_7 + 0.247x_3 + 21x_1 + 1.1x_5.$$

MINITAB also gives the multiple coefficient of determination, $R^2 = 63.71$ percent, and the estimated standard deviation of the model, $s = 47.5$.

Validating the Model

As we complete this chapter, you may feel somewhat overwhelmed by the many intricacies involved in the model building process. The feeling is justified, since the ability to consistently construct useful models of various phenomena is probably more of an art than a science. Even if one meticulously follows all the aforementioned considerations during the various developmental phases, the model may fail miserably when it is implemented with newly collected data. The fact that the model works well with the data that was used to construct it is no assurance that it will perform satisfactorily in the future with a different data set. In light of this, a skilled model builder will attempt to validate a model before it is actually put into practice. One way to allow for this is the use of **data splitting,** in which the original data is partitioned into two groups, one for fitting the model and the other for validating it. There are no hard and fast rules for partitioning the original sample, but for time series data the most recent observations are often used for the validation sample. Its size should be at least 10 percent but not more than 50 percent of the original sample. When data splitting is not practical, other validation techniques, such as the PRESS statistic, are available and can be found in the regression references in Appendix C.

SECTIONS 16.7 AND 16.8 EXERCISES

Analyzing Residuals to Evaluate the Model;
Other Considerations

16.59   Explain how residuals can be used to check if a higher-order term should be added to a model.

16.60   What is an outlier?
     a. Explain how outliers can be detected in a data set.
     b. Why is it important to check for outliers?
     c. Should outliers always be discarded?

16.61   The table below gives the number of kilowatt-hours used daily (y) by a large apartment complex in the southeast for 36 randomly selected days. Also given is the average temperature (x) for each sampled day.

Temp. x	Kwh y	Temp. x	Kwh y	Temp. x	Kwh y
82	33	87	31	81	31
43	30	46	27	44	30
73	23	72	23	70	21
35	32	33	33	38	32
29	33	30	31	31	32
60	25	62	23	61	23
75	19	78	18	76	18
61	23	61	21	60	20
49	26	45	26	50	28
57	23	55	27	53	23
94	33	88	34	92	34
77	22	77	24	84	22

The model $\mu_y = \beta_0 + \beta_1 x$ was fit to the data, and a plot of the residuals against the values of x is shown below. Does the plot indicate that the model has been misspecified? If your answer is yes, explain how the model should be altered.

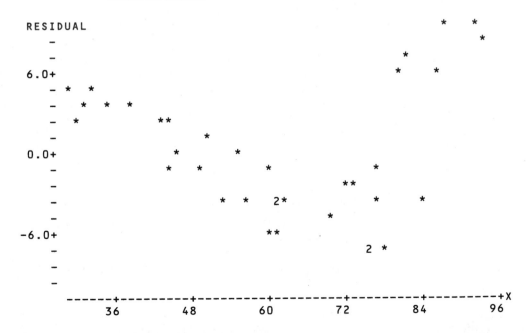

16.62   The model $y = \beta_0 + \beta_1 x_1 + \beta_2 x_2 + \epsilon$ was fit to the 10 data points shown on the following page, and the estimated model obtained was $\hat{y} = 2.416 + 1.0717x_1 - 0.3842x_2$.

a. Use the estimated model to calculate the residual for each data point.

b. Plot the residuals against the values of the second predictor x_2.

c. What change in the assumed model is suggested by the residual plot in part b?

Point	x_1	x_2	y
1	13	5	13.5
2	10	1	10.4
3	18	4	21.0
4	20	4	22.0
5	11	2	13.5
6	10	3	14.5
7	12	2	15.0
8	17	5	17.2
9	15	2	17.5
10	17	3	20.9

16.63　The figure below is a plot of the standardized residuals against the corresponding fits that were obtained in a multiple regression study. Identify any outliers.

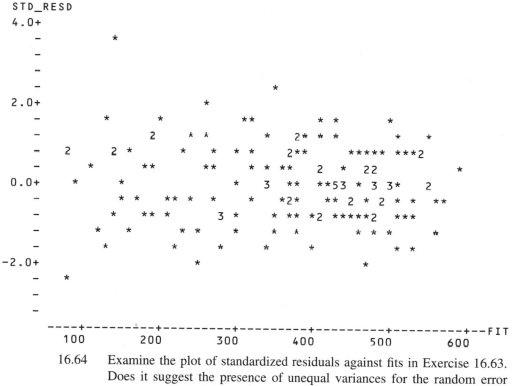

16.64　Examine the plot of standardized residuals against fits in Exercise 16.63. Does it suggest the presence of unequal variances for the random error components? Explain.

16.65　What is multicollinearity, and what are some problems that it can pose with a regression model?

16.66　What is stepwise regression, and why is it useful in the development of a multiple regression model?

16.67　Analyze the following plot of standardized residuals versus fits for the possible existence of outliers.

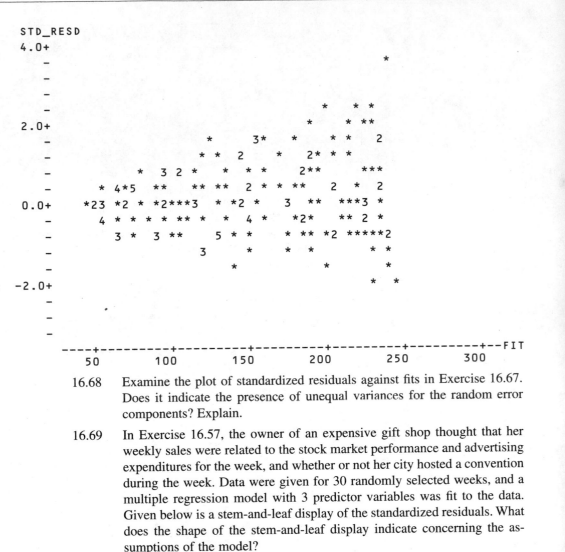

16.68 Examine the plot of standardized residuals against fits in Exercise 16.67. Does it indicate the presence of unequal variances for the random error components? Explain.

16.69 In Exercise 16.57, the owner of an expensive gift shop thought that her weekly sales were related to the stock market performance and advertising expenditures for the week, and whether or not her city hosted a convention during the week. Data were given for 30 randomly selected weeks, and a multiple regression model with 3 predictor variables was fit to the data. Given below is a stem-and-leaf display of the standardized residuals. What does the shape of the stem-and-leaf display indicate concerning the assumptions of the model?

```
Stem-and-leaf of STD_RESD  N = 30
Leaf Unit = 0.10

    1    -3  3
    1    -2
    1    -2
    4    -1  976
    4    -1
    9    -0  98877
   14    -0  42110
   (5)    0  12233
   11     0  55568899
    3     1  24
    1     1  9
```

16.70   In Exercise 16.61, the model $\mu_y = \beta_0 + \beta_1 x$ was fit to the 36 temperatures (x) and daily kilowatt-hours (y). The standardized residuals for the points are shown below. Construct a histogram of the standardized residuals. Does the shape of the histogram indicate that the normality assumption might be violated?

Tmp. x	Kwh y	Sd.R.	Tmp. x	Kwh y	Sd.R.	Tmp. x	Kwh y	Sd.R.
82	33	1.60	87	31	1.27	81	31	1.18
43	30	0.49	46	27	−0.09	44	30	0.50
73	23	−0.57	72	23	−0.58	70	21	−1.01
35	32	0.81	33	33	1.00	38	32	0.84
29	33	0.96	30	31	0.54	31	32	0.77
60	25	−0.32	62	23	−0.70	61	23	−0.71
75	19	−1.36	78	18	−1.53	76	18	−1.55
61	23	−0.71	61	21	−1.12	60	20	−1.33
49	26	−0.26	45	26	−0.31	50	28	0.16
57	23	−0.76	55	27	0.02	53	23	−0.82
94	33	1.80	88	34	1.91	92	34	1.98
77	22	−0.72	77	24	−0.32	84	22	−0.64

16.71   In monitoring the acidity of the lake from which its water supply is drawn, a municipality measured several variables that affect acidity. A multiple regression model was fit to data collected over a 24-month period, and the standardized residuals are displayed below. Check for dependency in the model's random error components by plotting the standardized residuals against the corresponding time periods.

Month	Std. Residual	Month	Std. Residual
1	0.13	13	−0.16
2	0.79	14	−0.16
3	2.13	15	−1.47
4	0.60	16	−1.61
5	−1.61	17	0.61
6	0.01	18	−0.01
7	0.12	19	2.04
8	0.86	20	−1.85
9	0.91	21	−0.18
10	0.35	22	−1.06
11	0.23	23	−0.44
12	0.14	24	−0.38

16.72   Emotional stress can overstimulate the sympathetic nervous system and cause the heart to become electrically unstable. Investigating this phenomenon, a researcher proposed the following model for describing the relationship between a response variable y and six independent variables x_1 through x_6.

$$y = \beta_0 + \beta_1 x_1 + \beta_2 x_2 + \beta_3 x_3 + \beta_4 x_4 + \beta_5 x_5 + \beta_6 x_6 + \epsilon$$

MINITAB was used to produce the following correlation table that shows the correlation coefficient for each pair of variables in the model.

	Y	X1	X2	X3	X4	X5
X1	0.586					
X2	0.666	0.443				
X3	0.549	0.658	0.551			
X4	0.700	0.608	0.906	0.849		
X5	0.456	0.242	0.448	0.334	0.445	
X6	0.427	0.266	0.463	0.335	0.458	0.988

a. Give the correlation between x_4 and x_2; x_4 and x_3; x_6 and x_5.
b. What advice would you offer the researcher concerning the proposed model?

16.73 A dependent variable y was regressed on six predictor variables, and a portion of the MINITAB output is given below. How do the results suggest that multicollinearity is a problem in this analysis?

```
The regression equation is
Y=-119 + 0.270X1 + 3.74X2 - 0.02X3 + 0.18X4 + 112X5 - 1.67X6
```

Predictor	Coef	Stdev	t-ratio	p
Constant	-119.1	199.3	-0.60	0.552
X1	0.27047	0.09137	2.96	0.004
X2	3.744	9.587	0.39	0.697
X3	-0.020	7.761	-0.00	0.998
X4	0.179	1.239	0.14	0.885
X5	111.76	77.83	1.44	0.156
X6	-1.674	1.406	-1.19	0.238

s = 47.68   R-sq = 60.0%   R-sq(adj) = 56.2%

Analysis of Variance

SOURCE	DF	SS	MS	F	p
Regression	6	214837	35806	15.75	0.000
Error	63	143233	2274		
Total	69	358070			

16.74 A stepwise regression analysis using the forward selection method is conducted on 5 predictor variables x_1 through x_5 and $n = 104$ points. At step 1, a one-predictor model of the form

$$y = \beta_0 + \beta_1 x_i + \epsilon$$

is fit for each of the 5 variables. The table below gives for each x_i the estimated coefficient and its estimated standard deviation.

Predictor	$\hat{\beta}_i$	$s_{\hat{\beta}_i}$
x_1	-2.38	0.97
x_2	11.21	4.78
x_3	-1.07	0.21
x_4	0.99	0.35
x_5	17.45	4.92

Determine which variable would be selected for possible inclusion at step 1, and explain your answer.

16.75 Ten predictor variables x_1 through x_{10} were considered for use in a multiple regression model. A stepwise regression analysis using the forward selection method was employed to select a subset of the predictors. The number of data points was $n = 31$, and $\alpha = 0.05$ was used as the criterion for adding a variable at a step. The resulting MINITAB output is given below.

```
      STEP        1       2       3       4
   CONSTANT    24.87   40.32   51.73   62.34

   X2           2.81    2.73    2.49    2.41
   T-RATIO      9.37    5.81    4.13    3.06

   X10                 10.71    9.21    9.26
   T-RATIO              6.93    4.01    3.25

   X6                           4.95    4.72
   T-RATIO                      2.69    1.57

   X5                                   0.85
   T-RATIO                              2.19

   S           26.9    25.3    24.3    24.2
   R-SQ        65.87   73.29   78.30   79.34
```

a. List the variables that were selected for inclusion in the model.
b. Determine the approximate p-value of the t-ratio for the variable that entered at step 4.
c. Find the increase at step 4 in the multiple coefficient of determination.
d. Write the estimated model at step 4.
e. Test the usefulness of the model produced at step. 4. Use $\alpha = 0.05$.

MINITAB Assignments

M▶ 16.76 In Exercise 16.62, the model $y = \beta_0 + \beta_1 x_1 + \beta_2 x_2 + \epsilon$ was fit to the data on the following page. Use MINITAB to perform the following.
a. Obtain the estimated model and the standardized residuals.
b. Plot the standardized residuals against the values of x_2.
c. What change in the assumed model is suggested by the plot in part b?

Point	x_1	x_2	y
1	13	5	13.5
2	10	1	10.4
3	18	4	21.0
4	20	4	22.0
5	11	2	13.5
6	10	3	14.5
7	12	2	15.0
8	17	5	17.2
9	15	2	17.5
10	17	3	20.9

M▶ 16.77 Refer to the data in Exercise 16.76, and use MINITAB to do the following.
   a. Fit the model $y = \beta_0 + \beta_1 x_1 + \beta_2 x_2 + \beta_3 x_2^2 + \epsilon$ and obtain the standardized residuals.
   b. Plot the standardized residuals against the values of x_2.
   c. Does the plot in part b indicate that a change is needed in the assumed model?

M▶ 16.78 In Exercise 16.61, the model $\mu_y = \beta_0 + \beta_1 x$ was fit to the following data.

Temp. x	Kwh y	Temp. x	Kwh y	Temp. x	Kwh y
82	33	87	31	81	31
43	30	46	27	44	30
73	23	72	23	70	21
35	32	33	33	38	32
29	33	30	31	31	32
60	25	62	23	61	23
75	19	78	18	76	18
61	23	61	21	60	20
49	26	45	26	50	28
57	23	55	27	53	23
94	33	88	34	92	34
77	22	77	24	84	22

Use MINITAB to do the following.
   a. Obtain the estimated model and the standardized residuals.
   b. Plot the standardized residuals against the values of x.
   c. Does the plot in part b indicate that the model has been misspecified? If yes, how should the model be altered?

M▶ 16.79 Use MINITAB to construct a normal probability plot of the standardized residuals that were obtained in Exercise 16.78. Does it appear that the normality assumption might be violated?

M> 16.80 Use MINITAB to fit the model $\mu_y = \beta_0 + \beta_1 x + \beta_2 x^2$ to the data in Exercise 16.78, and obtain the standardized residuals and the fits. Have MINITAB plot the standardized residuals against the fits to detect outliers and to check the assumption of constant error variance.

M> 16.81 The table below gives the earned run average, batting average, number of fielding errors, number of runs scored, number of runs given up, and the number of wins for each American League baseball team during the 1995 season. Store the data in columns C1 through C6, and use the command **STEPWISE C6 C1-C5** to have MINITAB select a suitable set of predictors. Give the estimated model and the value of R^2.

Team	Earned Run Avg. (x_1)	Batting Avg. (x_2)	Fielding Errors (x_3)	Runs Scored (x_4)	Opponents' Runs (x_5)	Wins (y)
Cleveland	3.84	.291	89	840	607	100
Boston	4.40	.280	104	791	698	86
Seattle	4.52	.275	96	794	708	79
New York	4.55	.276	68	749	688	79
California	4.52	.277	84	801	697	78
Texas	4.67	.265	90	691	720	74
Baltimore	4.32	.262	59	704	640	71
Kansas City	4.49	.260	82	631	691	70
Chicago	4.85	.280	96	755	758	68
Oakland	4.97	.264	98	730	761	67
Milwaukee	4.83	.266	89	740	747	65
Detroit	5.50	.247	95	654	844	60
Toronto	4.90	.260	85	642	777	56
Minnesota	5.77	.279	91	703	889	56

LOOKING BACK

This chapter is concerned with using a **multiple regression model** to describe the relationship between a dependent variable y and k independent variables. Specifically, we have assumed that the variables are related by the model

$$y = \beta_0 + \beta_1 x_1 + \beta_2 x_2 + \beta_3 x_3 + \cdots + \beta_k x_k + \epsilon$$

where ϵ is a random error component that is assumed to be normally distributed with 0 mean and variance σ^2 for each combination of predictor values. Additionally, errors for each pair of observations y_i and y_j are assumed to be independent. The β-parameters in the assumed model are estimated from the sample data by using the method of least squares. The values of the estimates are obtained from computer output. The common variance σ^2 of the random error components is estimated by

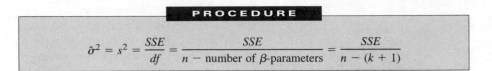

$$\hat{\sigma}^2 = s^2 = \frac{SSE}{df} = \frac{SSE}{n - \text{number of } \beta\text{-parameters}} = \frac{SSE}{n - (k + 1)}$$

In the above, s is called the **estimated standard deviation of the model,** and s^2 is given as the mean square error (MSE) in the regression's ANOVA table.

The **multiple coefficient of determination,** R^2, is useful in assessing how well the least squares model fits the data, since it is the proportion of the total variation in y that is explained by the estimated model.

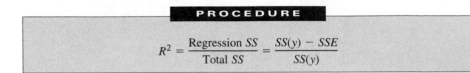

$$R^2 = \frac{\text{Regression } SS}{\text{Total } SS} = \frac{SS(y) - SSE}{SS(y)}$$

A formal test of the model's usefulness is provided by the **global F-test.** To test $H_0: \beta_1 = \beta_2 = \cdots = \beta_k = 0$, the test statistic is

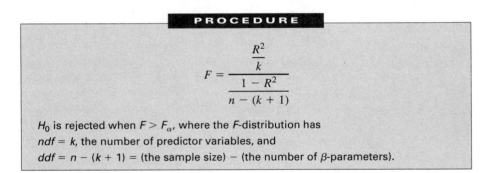

$$F = \frac{\dfrac{R^2}{k}}{\dfrac{1 - R^2}{n - (k + 1)}}$$

H_0 is rejected when $F > F_\alpha$, where the F-distribution has
$ndf = k$, the number of predictor variables, and
$ddf = n - (k + 1) = $ (the sample size) $-$ (the number of β-parameters).

A **t-test** can be used **to check the usefulness of a single term** in the model. To test $H_0: \beta_i = 0$, the test statistic is

$$t = \frac{\hat{\beta}_i}{s_{\hat{\beta}_i}}$$

$\hat{\beta}_i$ is the least squares estimate with estimated standard deviation $s_{\hat{\beta}_i}$. The t-statistic has $df = n - (k + 1) = n - $ (the number of β's in the model). A $(1 - \alpha)$ confidence interval for β_i is $\hat{\beta}_i \pm t_{\alpha/2} s_{\hat{\beta}_i}$.

The following *F*-test can be used **to test the usefulness of a group of terms** in the model.

PROCEDURE

F-Test of One or More Coefficients:

Complete Model: $y = \beta_0 + \beta_1 x_1 + \beta_2 x_2 + \beta_3 x_3 + \cdots + \beta_k x_k + \epsilon$

Reduced Model: The above model with some terms $(\beta_i x_i + \cdots + \beta_j x_j)$ deleted.

For testing the hypotheses

H_0: $\beta_i = \cdots = \beta_j = 0$ (The reduced and complete models do not differ significantly.)

H_a: At least 1 $\beta \neq 0$ (Complete model contributes more information for predicting y.)

the test statistic is

$$F = \frac{(SSE_\text{R} - SSE_\text{C})/(\text{number of } \beta\text{'s in } H_0)}{MSE_\text{C}}$$

The null hypothesis is rejected when $F > F_\alpha$, where α is the significance level of the test. The numerator and denominator degrees of freedom are

$ndf = $ (the number of β's being tested in H_0),

$ddf = n - (k + 1) = $ (sample size) $-$ (number of β's in the complete model).

Before a model is adopted and used in practice, a **residual analysis** should be conducted to evaluate the appropriateness of the model and to check for the presence of outliers and violations in the model's underlying assumptions. The following is a suggested procedure for analyzing residuals.

PROCEDURE

Check for

1. **a misspecification of the model.** For each predictor x_i, plot the residuals (or standardized residuals) against the corresponding values of x_i. If curvature is exhibited, try a higher-order term like x_i^2 in the model.

2. **outliers.** Plot the standardized residuals against the fits $\hat{y}$, and investigate data points with standardized residuals greater than 3 in magnitude. Try to correct measurement or recording errors, and discard any point that should not have been included in the sample.

3. **nonconstant error variances.** Inspect the plot of the standardized residuals against the fits, and observe if the standardized residuals tend to fluctuate in a nonrandom pattern.

4. **nonnormality of error components.** Construct a histogram or a stem-and-leaf display of the standardized residuals and see if the distribution lacks the general shape of a normal curve. Another possibility is to construct a normal probability plot for the standardized residuals.

5. **dependent errors.** For time series data, plot the standardized residuals against the time periods in which the data were observed. Look for some type of nonrandom pattern in the standardized residuals. This might consist of an usually large (or small) number of runs in positive and negative residuals.

Key Words

In reviewing this chapter, you should be able to define, explain, and illustrate each of the following.

probabilistic model *(page 758)*

deterministic model *(page 757)*

random error component *(page 758)*

least squares equation *(page 759)*

response surface *(page 760)*

partial slope *(page 760)*

multiple regression model *(page 761)*

standard deviation of the model (σ) *(page 762)*

estimated standard deviation of the model (s) *(page 762)*

mean square error *(page 762)*

multiple coefficient of determination (R^2) *(page 764)*

global F-test *(page 766)*

regression sum of squares *(page 764)*

qualitative variable *(page 776)*

dummy (indicator) variable *(page 777)*

interaction *(page 781)*

first-order model *(page 780)*

second-order model *(page 783)*

order of a model *(page 783)*

complete second-order model *(page 785)*

quadratic model *(page 785)*

t-test of a single coefficient *(page 796)*

F-test of one or more coefficients
    (page 801)

complete model *(page 801)*

reduced model *(page 801)*

estimating the mean of *y* *(page 808)*

predicting an individual *y* *(page 808)*

residual *(page 818)*

residual analysis *(page 818)*

fit *(page 818)*

misspecification of model *(page 818)*

outlier *(page 820)*

normal probability plot *(page 823)*

funnel effect *(page 825)*

heteroscedasticity *(page 825)*

multicollinearity *(page 829)*

stepwise regression *(page 830)*

backward elimination method
    (page 831)

forward selection method *(page 831)*

validating the model *(page 833)*

data splitting *(page 833)*

M▶ MINITAB Commands

READ _ _ *(page 759)*

END *(page 759)*

NAME _ _ *(page 759)*

REGRESS _ _ _ _ *(page 759)*

LET *(page 788)*

REGRESS _ _ _ _; *(page 809)*
PREDICT _ _.

BRIEF _ *(page 811)*

GSTD *(page 821)*

PLOT _ _ *(page 821)*

STEM _ *(page 822)*

NSCORES _ _ *(page 823)*

CORRELATION _ _ *(page 823)*

STEPWISE _ _ _; *(page 832)*
FENTER _.

REVIEW EXERCISES

16.82   The multiple regression model $y = \beta_0 + \beta_1 x_1 + \beta_2 x_2 + \beta_3 x_1 x_2 + \epsilon$
    was fit to 30 data points, where x_1 and x_2 are quantitative predictor vari-
    ables. A portion of the MINITAB output is shown on the following page.
    a. Give the estimated model.
    b. Determine the order of the model.
    c. Estimate the mean of *y* when $x_1 = 2$ and $x_2 = 10$.
    d. Give the estimated standard deviation of the model.
    e. Find the multiple coefficient of determination and interpret its value.

```
The regression equation is
Y = - 5.43 + 3.64 X1 + 0.733 X2 - 0.228 X1X2

Predictor        Coef         Stdev       t-ratio          p
Constant       -5.433         2.122        -2.56       0.017
X1              3.6422        0.7863         4.63       0.000
X2              0.7333        0.3902         1.88       0.071
X1X2           -0.2280        0.1394        -1.64       0.114

s = 0.2914       R-sq = 93.4%       R-sq(adj) = 92.6%

Analysis of Variance
SOURCE         DF          SS           MS          F          p
Regression      3        31.256       10.419     122.66     0.000
Error          26         2.208        0.085
Total          29        33.465
```

16.83 Test if the model in Exercise 16.82 contributes information for predicting y. Use a significance level of 5 percent.

16.84 Refer to Exercise 16.82, and obtain a 95 percent confidence interval for β_1.

16.85 For the model in Exercise 16.82, test if the interaction term contributes information for predicting y. Use $\alpha = 0.05$.

16.86 A package delivery company offers 3 types of service: priority overnight, standard overnight, and 2-day economy. The company wants to model the mean number of daily deliveries, μ_y, for the types of service.
a. Type of service is what kind of variable?
b. How many levels does the variable have?
c. Write a model that relates the mean number of daily deliveries to the type of service.

16.87 A popular snowmobile is available in 5 models. The choices are M500, M550, M700, M950, and M1000. The company uses the following to model the mean number of yearly sales for the different models.

$$\mu_y = \beta_0 + \beta_1 x_1 + \beta_2 x_2 + \beta_3 x_3 + \beta_4 x_4$$

where

$$x_1 = \begin{cases} 1 \text{ if M500} \\ 0 \text{ if not} \end{cases} \quad x_2 = \begin{cases} 1 \text{ if M550} \\ 0 \text{ if not} \end{cases} \quad x_3 = \begin{cases} 1 \text{ if M700} \\ 0 \text{ if not} \end{cases} \quad x_4 = \begin{cases} 1 \text{ if M950} \\ 0 \text{ if not} \end{cases}$$

a. Find the mean number of yearly sales for the M950 model.
b. Find the mean number of yearly sales for the M1000 model.
c. Interpret each of the five β-parameters in the model.

16.88 Give the assumptions concerning the random error components for a multiple regression model.

16.89   The regression model $\mu_y = \beta_0 + \beta_1 x_1 + \beta_2 x_2 + \beta_3 x_3$ was fit to 24 points, and the following least squares equation was obtained: $\hat{y} = 5.8 + 2.1x_1 - 4.8x_2 + 10.0x_3$. The resulting multiple coefficient of determination was $R^2 = 0.68$.
   a. Estimate the partial slopes and interpret their values.
   b. Estimate the mean of y when $x_1 = 4$, $x_2 = 2$, and $x_3 = 5$.
   c. Using $\alpha = 0.05$, test if the model contributes information for predicting y.

16.90   Obtain the approximate p-value for the global test in Exercise 16.89.

16.91   For the model in Exercise 16.89, the estimated standard deviation of $\hat{\beta}_3$ is 4.71. Is there sufficient evidence that the variable x_3 contributes information for predicting y? Use $\alpha = 0.05$.

16.92   Use the information in Exercises 16.89 and 16.91 to obtain a 95 percent confidence interval for the coefficient β_3.

16.93   A researcher fit the multiple regression model $y = \beta_0 + \beta_1 x_1 + \beta_2 x_2 + \beta_3 x_1 x_2 + \epsilon$ to n data points and then wanted an interval to estimate the mean of y for $x_1 = 10$ and $x_2 = 20$. Should a confidence interval or prediction interval be used? Explain.

16.94   The model $\mu_y = \beta_0 + \beta_1 x_1 + \beta_2 x_2$ was fit to $n = 14$ data points, and $SS(y) = 1,594$ and $SSE = 504$. Determine R^2 and interpret its value.

16.95   Test the utility of the model in Exercise 16.94. Use a 1 percent significance level.

16.96   The following equation is used by an indoor amusement facility to model the amount (y) that a person spends at its facility during a visit.

$$y = \beta_0 + \beta_1 x_1 + \beta_2 x_2 + \beta_3 x_3 + \beta_4 x_1 x_2 + \beta_5 x_1 x_3 + \beta_6 x_2 x_3 + \beta_7 x_1 x_2 x_3 + \epsilon$$

   In the above model, x_1 is the number of minutes at the facility, and x_2 and x_3 are dummy variables that indicate one's sex ($x_2 = 1$ if male, $x_2 = 0$ if female) and adult status ($x_3 = 1$ if an adult, $x_3 = 0$ if a child).
   a. What is the order of the model?
   b. Write the model that relates an adult's expenditures with the person's gender and time spent at the facility.
   c. Write the model that relates a child's expenditures with the person's gender and time spent at the facility.
   d. What hypotheses would one test to determine if there is a difference for adults and children in the relationship between expenditures with gender and time at the facility?

16.97   For the hypotheses in part d of Exercise 16.96, the complete and reduced models were fit to $n = 32$ points, and the resulting error sums of squares for the 2 models are $SSE_C = 721.3$ and $SSE_R = 811.2$. Conduct the test using $\alpha = 0.05$.

16.98 A manufacturer with a national distribution program has determined that its yearly profit (y) is related to its number (x) of salespersons. Profits tend to be highest when the sales force is of moderate size, and profits decline sharply when the size of the force is relatively large or small. Write a model that might be used to relate yearly profit to the number of salespersons employed by the company.

16.99 *Multiple Choice:* For the model $\mu_y = \beta_0 + \beta_1 x + \beta_2 x^2$, the response curve has a maximum point (high point) and opens downward when
a. $\beta_0 < 0$ b. $\beta_0 > 0$ c. $\beta_1 < 0$ d. $\beta_1 > 0$ e. $\beta_2 < 0$ f. $\beta_2 > 0$

16.100 The average prices (in dollars) per ounce of gold, platinum, and silver for the years 1975 through 1990 are given below (*Minerals Yearbook*, U.S. Bureau of Mines).

Year	Gold (x_1)	Platinum (x_2)	Silver (y)
75	161	164	4.42
76	125	162	4.35
77	148	162	4.62
78	194	237	5.40
79	308	352	11.09
80	613	439	20.63
81	460	475	10.52
82	376	475	7.95
83	424	475	11.44
84	361	475	8.14
85	318	475	6.14
86	368	519	5.47
87	448	600	7.01
88	438	600	6.53
89	383	600	5.50
90	387	600	4.82

The model $y = \beta_0 + \beta_1 x_1 + \beta_2 x_2 + \beta_3 x_1 x_2 + \epsilon$ was fit to the data, and a portion of the MINITAB output is displayed on the following page. Use the results to solve the following.
a. Give the estimated model.
b. Is there sufficient evidence to say that the model is useful for predicting silver prices? Test at the 5 percent significance level.
c. Predict with 95 percent probability the price of an ounce of silver for a year in which the average cost of gold and platinum are $325 and $475, respectively.

```
REGRESS 'Y' 3 'X1' 'X2' 'X1X2';
PREDICT 325 475 154375.
The regression equation is
Y = - 3.45 + 0.0743 X1 - 0.00740 X2 -0.000069 X1X2

Predictor          Coef        Stdev      t-ratio          p
Constant         -3.449        2.120        -1.63      0.130
X1              0.074284     0.009099         8.16      0.000
X2             -0.007398     0.007077        -1.05      0.316
X1X2         -0.00006918   0.00002304        -3.00      0.011

s = 1.011         R-sq = 95.3%        R-sq(adj) = 94.1%

Analysis of Variance
SOURCE         DF            SS           MS         F         p
Regression      3        247.703       82.568     80.75    0.000
Error          12         12.270        1.022
Total          15        259.973

   Fit   Stdev.Fit          95% C.I.           95% P.I.
 6.499       0.466     (  5.483,   7.515)   (  4.072,   8.926)
```

16.101 For the model in Exercise 16.100, test if the interaction term contributes significant information for predicting silver prices. Use $\alpha = 0.05$.

16.102 After fitting the complete model in Exercise 16.100, the following reduced model was fit to the data, and the error sum of squares was $SSF_R = 127.93$.

$$y = \beta_0 + \beta_1 x_1 + \epsilon$$

Is there sufficient evidence that the complete model with the variable platinum included contributes more information than the reduced model without it? Test at the 5 percent level.

16.103 The following model is used to relate a dependent variable y to two quantitative variables x_1 and x_2.

$$y = \beta_0 + \beta_1 x_1 + \beta_2 x_2 + \beta_3 x_1 x_2 + \epsilon$$

a. What is the order of the model?
b. Explain what it means to say that the variables x_1 and x_2 interact.
c. How could the model be altered so that it would not allow for interaction?

16.104 The model $y = \beta_0 + \beta_1 x + \epsilon$ was fit to $n = 40$ points. Examine the following plot of the residuals against the x-values, and determine if the model has been misspecified. If it has, explain how the model should be changed.

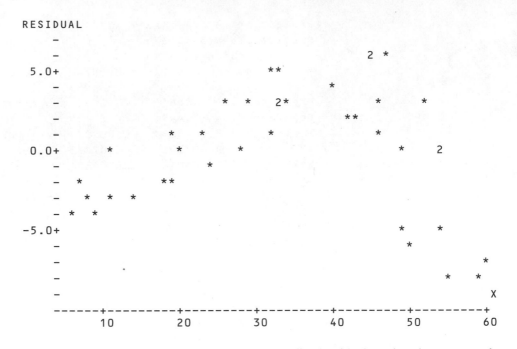

16.105 The following is a plot of the standardized residuals against the corresponding fits that were obtained from fitting a multiple regression model to 410 data points. Identify any outliers.

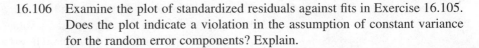

16.106 Examine the plot of standardized residuals against fits in Exercise 16.105. Does the plot indicate a violation in the assumption of constant variance for the random error components? Explain.

16.107   Displayed below is a histogram of the standardized residuals obtained from fitting a multiple regression model to $n = 549$ points. What does its shape indicate concerning the assumptions of the model?

```
Histogram of STD_RESD    N = 549
Each * represents 5 obs.

Midpoint    Count
    -3.0        3   *
    -2.5        6   **
    -2.0       22   *****
    -1.5       33   *******
    -1.0       62   ************
    -0.5       93   ******************
     0.0      110   **********************
     0.5       91   ******************
     1.0       64   *************
     1.5       42   *********
     2.0       14   ***
     2.5        8   **
     3.0        1   *
```

16.108   What are some indications that a problem of multicollinearity might exist in a multiple regression analysis?

16.109   A stepwise regression analysis with the forward selection method was used to select a subset of predictors from the variables x_1 through x_8. The criterion for adding a variable at a step was $\alpha = 0.05$, and $n = 30$ data points were used. Use the MINITAB output below to solve the following.
a. What variable entered the model at step 3?
b. Determine the approximate p-value of the t-ratio for the variable that entered at step 3.
c. Write the estimated model produced at step 3.
d. Test the utility of the model that was produced at step 3. Use $\alpha = 0.05$.

```
STEPWISE REGRESSION OF Y ON 8 PREDICTORS, WITH N = 30

         STEP        1        2        3
CONSTANT         22077    18116    19765

X7                1194     1226      924
T-RATIO          10.26    12.70     7.08

X2                         657      534
T-RATIO                   3.74     3.37

X5                                 5233
T-RATIO                            3.04

S                 3849     3180     2783
R-SQ             78.99    86.17    89.80
```

16.110 The model $\mu_y = \beta_0 + \beta_1 x_1 + \beta_2 x_2$ was fit to 36 points, and a global F-test was highly significant, indicating that the model was useful. However, two t-tests of the null hypotheses $H_0: \beta_1 = 0$ and $H_0: \beta_2 = 0$ were both nonsignificant. Explain how this could occur.

MINITAB Assignments

M▶ 16.111 The average prices (in dollars) per ounce of gold, platinum, and silver for the years 1975 through 1990 were given in Exercise 16.100 and are repeated below. Use MINITAB to fit the model $y = \beta_0 + \beta_1 x_1 + \beta_2 x_2 + \beta_3 x_1 x_2 + \epsilon$ to the data and solve the following.
a. Give the multiple coefficient of determination, and explain its value in the context of this problem.
b. Test the utility of the model at the 5 percent significance level.
c. Does the interaction term contribute information for predicting silver prices? Test with $\alpha = 0.05$.

Year	Gold (x_1)	Platinum (x_2)	Silver (y)
75	161	164	4.42
76	125	162	4.35
77	148	162	4.62
78	194	237	5.40
79	308	352	11.09
80	613	439	20.63
81	460	475	10.52
82	376	475	7.95
83	424	475	11.44
84	361	475	8.14
85	318	475	6.14
86	368	519	5.47
87	448	600	7.01
88	438	600	6.53
89	383	600	5.50
90	387	600	4.82

M▶ 16.112 For the model and data in Exercise 16.111, use MINITAB to obtain a 95 percent prediction interval for the price of an ounce of silver in a year in which the average cost of gold and platinum are \$320 and \$480, respectively.

M▶ 16.113 Use MINITAB to fit the reduced model $y = \beta_0 + \beta_1 x_1 + \epsilon$ to the data in Exercise 16.111. Is there sufficient evidence that the complete model in Exercise 16.111 contributes more information than the reduced model? Use $\alpha = 0.05$.

M▶ 16.114 In Exercise 16.35, the owner of an expensive gift shop thought that her weekly sales were related to the stock market performance and advertising expenditures for the week, and whether or not her city hosted a convention during the week. The following data were given for 30 randomly selected weeks. Sales and advertising figures are given in units of $1,000.

Week	DJIA (x_1)	Adv. Expenses (x_2)	Sales (y)	Convention
1	7015	2.9	58.3	no
2	7318	3.1	62.9	yes
3	6581	2.5	46.3	no
4	6623	5.1	48.2	no
5	6917	2.9	58.2	no
6	7503	3.4	65.8	yes
7	6223	1.6	36.7	no
8	7332	3.0	62.3	no
9	7197	3.2	59.8	no
10	7513	3.5	71.8	yes
11	7703	3.1	66.9	yes
12	6523	1.9	37.9	no
13	6332	1.7	30.6	no
14	7197	3.3	69.8	yes
15	7713	3.8	65.2	yes
16	6679	3.0	52.7	no
17	6513	2.1	39.3	no
18	6922	3.4	58.7	no
19	7146	3.3	60.9	no
20	6429	2.2	40.5	yes
21	7409	3.9	70.3	yes
22	6315	2.4	39.1	no
23	6487	2.5	45.9	no
24	6715	2.8	55.1	no
25	7543	3.6	70.2	yes
26	7619	3.7	70.3	yes
27	6125	2.1	39.1	yes
28	6987	2.5	45.9	no
29	7215	2.9	55.1	no
30	7707	3.4	70.2	yes

Let the dummy variable $x_3 = 1$ if the city hosted a convention in a specified week. Use MINITAB to fit the model $y = \beta_0 + \beta_1 x_1 + \beta_2 x_2 + \beta_3 x_3 + \epsilon$ and complete the following.

a. Obtain the estimated model, the standardized residuals, and the fits.
b. Plot the standardized residuals against the values of x_1.
c. Does the plot in part b indicate that a higher-order term in x_1 is needed?

M▶ 16.115 Use MINITAB to construct a histogram of the standardized residuals obtained in Exercise 16.114. Does its shape suggest that the normality assumption of the random error components might be violated?

M➤ 16.116   Using MINITAB, construct a normal probability plot of the standardized residuals that were obtained in Exercise 16.114. Does it appear that the normality assumption might be violated?

M➤ 16.117   Have MINITAB plot the standardized residuals against the fits that were obtained in Exercise 16.114. Examine the plot for the presence of outliers and for a violation of the assumption of constant error variance.

M➤ 16.118   The following table shows the earned run average, batting average, fielding percentage, number of runs scored, number of runs given up, and the number of wins for each National League baseball team during the 1992 season.

Team	Earned Run Avg. (x_1)	Batting Avg. (x_2)	Fielding % (x_3)	Runs Scored (x_4)	Opponent's Runs (x_5)	Wins (y)
Atlanta	3.14	.254	.982	682	569	98
Pittsburgh	3.35	.255	.984	693	595	96
Cincinnati	3.46	.260	.984	660	609	90
Montreal	3.25	.252	.980	648	581	87
St. Louis	3.38	.262	.985	631	604	83
San Diego	3.56	.255	.981	617	636	82
Houston	3.72	.246	.981	608	668	81
Chicago	3.39	.254	.982	593	624	78
New York	3.66	.235	.981	599	653	72
San Francisco	3.61	.244	.982	574	647	72
Philadelphia	4.11	.253	.978	686	717	70
Los Angeles	3.41	.248	.972	548	636	63

a. Store the data in columns C1 through C6, and have MINITAB select a suitable set of predictor variables by applying the command

```
STEPWISE C6 C1-C5
```

b. Give the estimated model for the variables selected by MINITAB in part a.

c. What percentage of the total variation in the number of wins is explained by the model in part b?

APPENDIX A

TABLE 1   BINOMIAL PROBABILITIES

The Tabulated Values Are the Cumulative Probabilities $P(x \leq k)$ for $k = 0$ through 1.

k	p = .05	p = .10	p = .20	p = .30	p = .40	p = .50	p = .60	p = .70	p = .80	p = .90	p = .95
0	0.950	0.900	0.800	0.700	0.600	0.500	0.400	0.300	0.200	0.100	0.050
1	1.000	1.000	1.000	1.000	1.000	1.000	1.000	1.000	1.000	1.000	1.000

The Tabulated Values Are the Cumulative Probabilities $P(x \leq k)$ for $k = 0$ through 2.

k	p = .05	p = .10	p = .20	p = .30	p = .40	p = .50	p = .60	p = .70	p = .80	p = .90	p = .95
0	0.902	0.810	0.640	0.490	0.360	0.250	0.160	0.090	0.040	0.010	0.003
1	0.998	0.990	0.960	0.910	0.840	0.750	0.640	0.510	0.360	0.190	0.098
2	1.000	1.000	1.000	1.000	1.000	1.000	1.000	1.000	1.000	1.000	1.000

The Tabulated Values Are the Cumulative Probabilities $P(x \leq k)$ for $k = 0$ through 3.

k	p = .05	p = .10	p = .20	p = .30	p = .40	p = .50	p = .60	p = .70	p = .80	p = .90	p = .95
0	0.857	0.729	0.512	0.343	0.216	0.125	0.064	0.027	0.008	0.001	0.000
1	0.993	0.972	0.896	0.784	0.648	0.500	0.352	0.216	0.104	0.028	0.007
2	1.000	0.999	0.992	0.973	0.936	0.875	0.784	0.657	0.488	0.271	0.143
3	1.000	1.000	1.000	1.000	1.000	1.000	1.000	1.000	1.000	1.000	1.000

The Tabulated Values Are the Cumulative Probabilities $P(x \leq k)$ for $k = 0$ through 4.

k	p = .05	p = .10	p = .20	p = .30	p = .40	p = .50	p = .60	p = .70	p = .80	p = .90	p = .95
0	0.815	0.656	0.410	0.240	0.130	0.062	0.026	0.008	0.002	0.000	0.000
1	0.986	0.948	0.819	0.652	0.475	0.312	0.179	0.084	0.027	0.004	0.000
2	1.000	0.996	0.973	0.916	0.821	0.687	0.525	0.348	0.181	0.052	0.014
3	1.000	1.000	0.998	0.992	0.974	0.937	0.870	0.760	0.590	0.344	0.185
4	1.000	1.000	1.000	1.000	1.000	1.000	1.000	1.000	1.000	1.000	1.000

TABLE 1   BINOMIAL PROBABILITIES (continued)

n = 5   **n = 5**

The Tabulated Values Are the Cumulative Probabilities $P(x \le k)$ for $k = 0$ through 5.

k	p = .05	p = .10	p = .20	p = .30	p = .40	p = .50	p = .60	p = .70	p = .80	p = .90	p = .95
0	0.774	0.590	0.328	0.168	0.078	0.031	0.010	0.002	0.000	0.000	0.000
1	0.977	0.919	0.737	0.528	0.337	0.187	0.087	0.031	0.007	0.000	0.000
2	0.999	0.991	0.942	0.837	0.683	0.500	0.317	0.163	0.058	0.009	0.001
3	1.000	1.000	0.993	0.969	0.913	0.812	0.663	0.472	0.263	0.081	0.023
4	1.000	1.000	1.000	0.998	0.990	0.969	0.922	0.832	0.672	0.410	0.226
5	1.000	1.000	1.000	1.000	1.000	1.000	1.000	1.000	1.000	1.000	1.000

n = 6   **n = 6**

The Tabulated Values Are the Cumulative Probabilities $P(x \le k)$ for $k = 0$ through 6.

k	p = .05	p = .10	p = .20	p = .30	p = .40	p = .50	p = .60	p = .70	p = .80	p = .90	p = .95
0	0.735	0.531	0.262	0.118	0.047	0.016	0.004	0.001	0.000	0.000	0.000
1	0.967	0.886	0.655	0.420	0.233	0.109	0.041	0.011	0.002	0.000	0.000
2	0.998	0.984	0.901	0.744	0.544	0.344	0.179	0.070	0.017	0.001	0.000
3	1.000	0.999	0.983	0.930	0.821	0.656	0.456	0.256	0.099	0.016	0.002
4	1.000	1.000	0.998	0.989	0.959	0.891	0.767	0.580	0.345	0.114	0.033
5	1.000	1.000	1.000	0.999	0.996	0.984	0.953	0.882	0.738	0.469	0.265
6	1.000	1.000	1.000	1.000	1.000	1.000	1.000	1.000	1.000	1.000	1.000

n = 7   **n = 7**

The Tabulated Values Are the Cumulative Probabilities $P(x \le k)$ for $k = 0$ through 7.

k	p = .05	p = .10	p = .20	p = .30	p = .40	p = .50	p = .60	p = .70	p = .80	p = .90	p = .95
0	0.698	0.478	0.210	0.082	0.028	0.008	0.002	0.000	0.000	0.000	0.000
1	0.956	0.850	0.577	0.329	0.159	0.062	0.019	0.004	0.000	0.000	0.000
2	0.996	0.974	0.852	0.647	0.420	0.227	0.096	0.029	0.005	0.000	0.000
3	1.000	0.997	0.967	0.874	0.710	0.500	0.290	0.126	0.033	0.003	0.000
4	1.000	1.000	0.995	0.971	0.904	0.773	0.580	0.353	0.148	0.026	0.004
5	1.000	1.000	1.000	0.996	0.981	0.937	0.841	0.671	0.423	0.150	0.044
6	1.000	1.000	1.000	1.000	0.998	0.992	0.972	0.918	0.790	0.522	0.302
7	1.000	1.000	1.000	1.000	1.000	1.000	1.000	1.000	1.000	1.000	1.000

TABLE 1 BINOMIAL PROBABILITIES (continued)

n = 8

The Tabulated Values Are the Cumulative Probabilities $P(x \le k)$ for $k = 0$ through 8.

k	p = .05	p = .10	p = .20	p = .30	p = .40	p = .50	p = .60	p = .70	p = .80	p = .90	p = .95
0	0.663	0.430	0.168	0.058	0.017	0.004	0.001	0.000	0.000	0.000	0.000
1	0.943	0.813	0.503	0.255	0.106	0.035	0.009	0.001	0.000	0.000	0.000
2	0.994	0.962	0.797	0.552	0.315	0.145	0.050	0.011	0.001	0.000	0.000
3	1.000	0.995	0.944	0.806	0.594	0.363	0.174	0.058	0.010	0.000	0.000
4	1.000	1.000	0.990	0.942	0.826	0.637	0.406	0.194	0.056	0.005	0.000
5	1.000	1.000	0.999	0.989	0.950	0.855	0.685	0.448	0.203	0.038	0.006
6	1.000	1.000	1.000	0.999	0.991	0.965	0.894	0.745	0.497	0.187	0.057
7	1.000	1.000	1.000	1.000	0.999	0.996	0.983	0.942	0.832	0.570	0.337
8	1.000	1.000	1.000	1.000	1.000	1.000	1.000	1.000	1.000	1.000	1.000

n = 9

The Tabulated Values Are the Cumulative Probabilities $P(x \le k)$ for $k = 0$ through 9.

k	p = .05	p = .10	p = .20	p = .30	p = .40	p = .50	p = .60	p = .70	p = .80	p = .90	p = .95
0	0.630	0.387	0.134	0.040	0.010	0.002	0.000	0.000	0.000	0.000	0.000
1	0.929	0.775	0.436	0.196	0.071	0.020	0.004	0.000	0.000	0.000	0.000
2	0.992	0.947	0.738	0.463	0.232	0.090	0.025	0.004	0.000	0.000	0.000
3	0.999	0.992	0.914	0.730	0.483	0.254	0.099	0.025	0.003	0.000	0.000
4	1.000	0.999	0.980	0.901	0.733	0.500	0.267	0.099	0.020	0.001	0.000
5	1.000	1.000	0.997	0.975	0.901	0.746	0.517	0.270	0.086	0.008	0.001
6	1.000	1.000	1.000	0.996	0.975	0.910	0.768	0.537	0.262	0.053	0.008
7	1.000	1.000	1.000	1.000	0.996	0.980	0.929	0.804	0.564	0.225	0.071
8	1.000	1.000	1.000	1.000	1.000	0.998	0.990	0.960	0.866	0.613	0.370
9	1.000	1.000	1.000	1.000	1.000	1.000	1.000	1.000	1.000	1.000	1.000

n = 10

The Tabulated Values Are the Cumulative Probabilities $P(x \le k)$ for $k = 0$ through 10

k	p = .05	p = .10	p = .20	p = .30	p = .40	p = .50	p = .60	p = .70	p = .80	p = .90	p = .95
0	0.599	0.349	0.107	0.028	0.006	0.001	0.000	0.000	0.000	0.000	0.000
1	0.914	0.736	0.376	0.149	0.046	0.011	0.002	0.000	0.000	0.000	0.000
2	0.988	0.930	0.678	0.383	0.167	0.055	0.012	0.002	0.000	0.000	0.000
3	0.999	0.987	0.879	0.650	0.382	0.172	0.055	0.011	0.001	0.000	0.000
4	1.000	0.998	0.967	0.850	0.633	0.377	0.166	0.047	0.006	0.000	0.000
5	1.000	1.000	0.994	0.953	0.834	0.623	0.367	0.150	0.033	0.002	0.000
6	1.000	1.000	0.999	0.989	0.945	0.828	0.618	0.350	0.121	0.013	0.001
7	1.000	1.000	1.000	0.998	0.988	0.945	0.833	0.617	0.322	0.070	0.012
8	1.000	1.000	1.000	1.000	0.998	0.989	0.954	0.851	0.624	0.264	0.086
9	1.000	1.000	1.000	1.000	1.000	0.999	0.994	0.972	0.893	0.651	0.401
10	1.000	1.000	1.000	1.000	1.000	1.000	1.000	1.000	1.000	1.000	1.000

TABLE 1   BINOMIAL PROBABILITIES (continued)

n = 15   **n = 15**

The Tabulated Values Are the Cumulative Probabilities $P(x \le k)$ for $k = 0$ through 15.

k	p = .05	p = .10	p = .20	p = .30	p = .40	p = .50	p = .60	p = .70	p = .80	p = .90	p = .95
0	0.463	0.206	0.035	0.005	0.000	0.000	0.000	0.000	0.000	0.000	0.000
1	0.829	0.549	0.167	0.035	0.005	0.000	0.000	0.000	0.000	0.000	0.000
2	0.964	0.816	0.398	0.127	0.027	0.004	0.000	0.000	0.000	0.000	0.000
3	0.995	0.944	0.648	0.297	0.091	0.018	0.002	0.000	0.000	0.000	0.000
4	0.999	0.987	0.836	0.515	0.217	0.059	0.009	0.001	0.000	0.000	0.000
5	1.000	0.998	0.939	0.722	0.403	0.151	0.034	0.004	0.000	0.000	0.000
6	1.000	1.000	0.982	0.869	0.610	0.304	0.095	0.015	0.001	0.000	0.000
7	1.000	1.000	0.996	0.950	0.787	0.500	0.213	0.050	0.004	0.000	0.000
8	1.000	1.000	0.999	0.985	0.905	0.696	0.390	0.131	0.018	0.000	0.000
9	1.000	1.000	1.000	0.996	0.966	0.849	0.597	0.278	0.061	0.002	0.000
10	1.000	1.000	1.000	0.999	0.991	0.941	0.783	0.485	0.164	0.013	0.001
11	1.000	1.000	1.000	1.000	0.998	0.982	0.909	0.703	0.352	0.056	0.005
12	1.000	1.000	1.000	1.000	1.000	0.996	0.973	0.873	0.602	0.184	0.036
13	1.000	1.000	1.000	1.000	1.000	1.000	0.995	0.965	0.833	0.451	0.171
14	1.000	1.000	1.000	1.000	1.000	1.000	1.000	0.995	0.965	0.794	0.537
15	1.000	1.000	1.000	1.000	1.000	1.000	1.000	1.000	1.000	1.000	1.000

n = 20   **n = 20**

The Tabulated Values Are the Cumulative Probabilities $P(x \le k)$ for $k = 0$ through 20.

k	p = .05	p = .10	p = .20	p = .30	p = .40	p = .50	p = .60	p = .70	p = .80	p = .90	p = .95
0	0.358	0.122	0.012	0.001	0.000	0.000	0.000	0.000	0.000	0.000	0.000
1	0.736	0.392	0.069	0.008	0.001	0.000	0.000	0.000	0.000	0.000	0.000
2	0.925	0.677	0.206	0.035	0.004	0.000	0.000	0.000	0.000	0.000	0.000
3	0.984	0.867	0.411	0.107	0.016	0.001	0.000	0.000	0.000	0.000	0.000
4	0.997	0.957	0.630	0.238	0.051	0.006	0.000	0.000	0.000	0.000	0.000
5	1.000	0.989	0.804	0.416	0.126	0.021	0.002	0.000	0.000	0.000	0.000
6	1.000	0.998	0.913	0.608	0.250	0.058	0.006	0.000	0.000	0.000	0.000
7	1.000	1.000	0.968	0.772	0.416	0.132	0.021	0.001	0.000	0.000	0.000
8	1.000	1.000	0.990	0.887	0.596	0.252	0.057	0.005	0.000	0.000	0.000
9	1.000	1.000	0.997	0.952	0.755	0.412	0.128	0.017	0.001	0.000	0.000
10	1.000	1.000	0.999	0.983	0.872	0.588	0.245	0.048	0.003	0.000	0.000
11	1.000	1.000	1.000	0.995	0.943	0.748	0.404	0.113	0.010	0.000	0.000
12	1.000	1.000	1.000	0.999	0.979	0.868	0.584	0.228	0.032	0.000	0.000
13	1.000	1.000	1.000	1.000	0.994	0.942	0.750	0.392	0.087	0.002	0.000
14	1.000	1.000	1.000	1.000	0.998	0.979	0.874	0.584	0.196	0.011	0.000
15	1.000	1.000	1.000	1.000	1.000	0.994	0.949	0.762	0.370	0.043	0.003
16	1.000	1.000	1.000	1.000	1.000	0.999	0.984	0.893	0.589	0.133	0.016
17	1.000	1.000	1.000	1.000	1.000	1.000	0.996	0.965	0.794	0.323	0.075
18	1.000	1.000	1.000	1.000	1.000	1.000	0.999	0.992	0.931	0.608	0.264
19	1.000	1.000	1.000	1.000	1.000	1.000	1.000	0.999	0.988	0.878	0.642
20	1.000	1.000	1.000	1.000	1.000	1.000	1.000	1.000	1.000	1.000	1.000

TABLE 2   VALUES OF $e^{-\mu}$

μ	$e^{-\mu}$	μ	$e^{-\mu}$	μ	$e^{-\mu}$	μ	$e^{-\mu}$	μ	$e^{-\mu}$	μ	$e^{-\mu}$
0.00	1.000000	2.40	0.090718	4.80	0.008230	7.20	0.000747	9.60	0.000068		
0.05	0.951229	2.45	0.086294	4.85	0.007828	7.25	0.000710	9.65	0.000064		
0.10	0.904837	2.50	0.082085	4.90	0.007447	7.30	0.000676	9.70	0.000061		
0.15	0.860708	2.55	0.078082	4.95	0.007083	7.35	0.000643	9.75	0.000058		
0.20	0.818731	2.60	0.074274	5.00	0.006738	7.40	0.000611	9.80	0.000055		
0.25	0.778801	2.65	0.070651	5.05	0.006409	7.45	0.000581	9.85	0.000053		
0.30	0.740818	2.70	0.067206	5.10	0.006097	7.50	0.000553	9.90	0.000050		
0.35	0.704688	2.75	0.063928	5.15	0.005799	7.55	0.000526	9.95	0.000048		
0.40	0.670320	2.80	0.060810	5.20	0.005517	7.60	0.000500	10.00	0.000045		
0.45	0.637628	2.85	0.057844	5.25	0.005248	7.65	0.000476	10.05	0.000043		
0.50	0.606531	2.90	0.055023	5.30	0.004992	7.70	0.000453	10.10	0.000041		
0.55	0.576950	2.95	0.052340	5.35	0.004748	7.75	0.000431	10.15	0.000039		
0.60	0.548812	3.00	0.049787	5.40	0.004517	7.80	0.000410	10.20	0.000037		
0.65	0.522046	3.05	0.047359	5.45	0.004296	7.85	0.000390	10.25	0.000035		
0.70	0.496585	3.10	0.045049	5.50	0.004087	7.90	0.000371	10.30	0.000034		
0.75	0.472367	3.15	0.042852	5.55	0.003887	7.95	0.000353	10.35	0.000032		
0.80	0.449329	3.20	0.040762	5.60	0.003698	8.00	0.000335	10.40	0.000030		
0.85	0.427415	3.25	0.038774	5.65	0.003518	8.05	0.000319	10.45	0.000029		
0.90	0.406570	3.30	0.036883	5.70	0.003346	8.10	0.000304	10.50	0.000028		
0.95	0.386741	3.35	0.035084	5.75	0.003183	8.15	0.000289	10.55	0.000026		
1.00	0.367879	3.40	0.033373	5.80	0.003028	8.20	0.000275	10.60	0.000025		
1.05	0.349938	3.45	0.031746	5.85	0.002880	8.25	0.000261	10.65	0.000024		
1.10	0.332871	3.50	0.030197	5.90	0.002739	8.30	0.000249	10.70	0.000023		
1.15	0.316637	3.55	0.028725	5.95	0.002606	8.35	0.000236	10.75	0.000021		
1.20	0.301194	3.60	0.027324	6.00	0.002479	8.40	0.000225	10.80	0.000020		
1.25	0.286505	3.65	0.025991	6.05	0.002358	8.45	0.000214	10.85	0.000019		
1.30	0.272532	3.70	0.024724	6.10	0.002243	8 50	0.000203	10.90	0.000018		
1.35	0.259240	3.75	0.023518	6.15	0.002133	8.55	0.000194	10.95	0.000018		
1.40	0.246597	3.80	0.022371	6.20	0.002029	8.60	0.000184	11.00	0.000017		
1.45	0.234570	3.85	0.021280	6.25	0.001930	8.65	0.000175	11.05	0.000016		
1.50	0.223130	3.90	0.020242	6.30	0.001836	8.70	0.000167	11.10	0.000015		
1.55	0.212248	3.95	0.019255	6.35	0.001747	8.75	0.000158	11.15	0.000014		
1.60	0.201897	4.00	0.018316	6.40	0.001662	8.80	0.000151	11.20	0.000014		
1.65	0.192050	4.05	0.017422	6.45	0.001581	8.85	0.000143	11.25	0.000013		
1.70	0.182684	4.10	0.016573	6.50	0.001503	8.90	0.000136	11.30	0.000012		
1.75	0.173774	4.15	0.015764	6.55	0.001430	8.95	0.000130	11.35	0.000012		
1.80	0.165299	4.20	0.014996	6.60	0.001360	9.00	0.000123	11.40	0.000011		
1.85	0.157237	4.25	0.014264	6.65	0.001294	9.05	0.000117	11.45	0.000011		
1.90	0.149569	4.30	0.013569	6.70	0.001231	9.10	0.000112	11.50	0.000010		
1.95	0.142274	4.35	0.012907	6.75	0.001171	9.15	0.000106	11.55	0.000010		
2.00	0.135335	4.40	0.012277	6.80	0.001114	9.20	0.000101	11.60	0.000009		
2.05	0.128735	4.45	0.011679	6.85	0.001059	9.25	0.000096	11.65	0.000009		
2.10	0.122456	4.50	0.011109	6.90	0.001008	9.30	0.000091	11.70	0.000008		
2.15	0.116484	4.55	0.010567	6.95	0.000959	9.35	0.000087	11.75	0.000008		
2.20	0.110803	4.60	0.010052	7.00	0.000912	9.40	0.000083	11.80	0.000008		
2.25	0.105399	4.65	0.009562	7.05	0.000867	9.45	0.000079	11.85	0.000007		
2.30	0.100259	4.70	0.009095	7.10	0.000825	9.50	0.000075	11.90	0.000007		
2.35	0.095369	4.75	0.008652	7.15	0.000785	9.55	0.000071	11.95	0.000006		

TABLE 3A   STANDARD NORMAL CURVE AREAS (NEGATIVE z-VALUES)

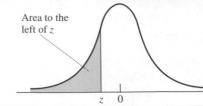

Area to the left of z

Each table value is the cumulative area to the left of the specified z-value.

z	.00	.01	.02	.03	.04	.05	.06	.07	.08	.09
The Second Decimal Digit of z										
−3.4	.0003	.0003	.0003	.0003	.0003	.0003	.0003	.0003	.0003	.0002
−3.3	.0005	.0005	.0005	.0004	.0004	.0004	.0004	.0004	.0004	.0003
−3.2	.0007	.0007	.0006	.0006	.0006	.0006	.0006	.0005	.0005	.0005
−3.1	.0010	.0009	.0009	.0009	.0008	.0008	.0008	.0008	.0007	.0007
−3.0	.0013	.0013	.0013	.0012	.0012	.0011	.0011	.0011	.0010	.0010
−2.9	.0019	.0018	.0018	.0017	.0016	.0016	.0015	.0015	.0014	.0014
−2.8	.0026	.0025	.0024	.0023	.0023	.0022	.0021	.0021	.0020	.0019
−2.7	.0035	.0034	.0033	.0032	.0031	.0030	.0029	.0028	.0027	.0026
−2.6	.0047	.0045	.0044	.0043	.0041	.0040	.0039	.0038	.0037	.0036
−2.5	.0062	.0060	.0059	.0057	.0055	.0054	.0052	.0051	.0049	.0048
−2.4	.0082	.0080	.0078	.0075	.0073	.0071	.0069	.0068	.0066	.0064
−2.3	.0107	.0104	.0102	.0099	.0096	.0094	.0091	.0089	.0087	.0084
−2.2	.0139	.0136	.0132	.0129	.0125	.0122	.0119	.0116	.0113	.0110
−2.1	.0179	.0174	.0170	.0166	.0162	.0158	.0154	.0150	.0146	.0143
−2.0	.0228	.0222	.0217	.0212	.0207	.0202	.0197	.0192	.0188	.0183
−1.9	.0287	.0281	.0274	.0268	.0262	.0256	.0250	.0244	.0239	.0233
−1.8	.0359	.0351	.0344	.0336	.0329	.0322	.0314	.0307	.0301	.0294
−1.7	.0446	.0436	.0427	.0418	.0409	.0401	.0392	.0384	.0375	.0367
−1.6	.0548	.0537	.0526	.0516	.0505	.0495	.0485	.0475	.0465	.0455
−1.5	.0668	.0655	.0643	.0630	.0618	.0606	.0594	.0582	.0571	.0559
−1.4	.0808	.0793	.0778	.0764	.0749	.0735	.0721	.0708	.0694	.0681
−1.3	.0968	.0951	.0934	.0918	.0901	.0885	.0869	.0853	.0838	.0823
−1.2	.1151	.1131	.1112	.1093	.1075	.1056	.1038	.1020	.1003	.0985
−1.1	.1357	.1335	.1314	.1292	.1271	.1251	.1230	.1210	.1190	.1170
−1.0	.1587	.1562	.1539	.1515	.1492	.1469	.1446	.1423	.1401	.1379
−0.9	.1841	.1814	.1788	.1762	.1736	.1711	.1685	.1660	.1635	.1611
−0.8	.2119	.2090	.2061	.2033	.2005	.1977	.1949	.1922	.1894	.1867
−0.7	.2420	.2389	.2358	.2327	.2296	.2266	.2236	.2206	.2177	.2148
−0.6	.2743	.2709	.2676	.2643	.2611	.2578	.2546	.2514	.2483	.2451
−0.5	.3085	.3050	.3015	.2981	.2946	.2912	.2877	.2843	.2810	.2776
−0.4	.3446	.3409	.3372	.3336	.3300	.3264	.3228	.3192	.3156	.3121
−0.3	.3821	.3783	.3745	.3707	.3669	.3632	.3594	.3557	.3520	.3483
−0.2	.4207	.4168	.4129	.4090	.4052	.4013	.3974	.3936	.3897	.3859
−0.1	.4602	.4562	.4522	.4483	.4443	.4404	.4364	.4325	.4286	.4247
−0.0	.5000	.4960	.4920	.4880	.4840	.4801	.4761	.4721	.4681	.4641

TABLE 3B STANDARD NORMAL CURVE AREAS (POSITIVE z-VALUES)

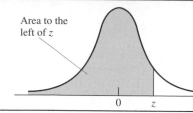

Area to the left of z

Each table value is the cumulative area to the left of the specified z-value.

The Second Decimal Digit of z

z	.00	.01	.02	.03	.04	.05	.06	.07	.08	.09
0.0	.5000	.5040	.5080	.5120	.5160	.5199	.5239	.5279	.5319	.5359
0.1	.5398	.5438	.5478	.5517	.5557	.5596	.5636	.5675	.5714	.5753
0.2	.5793	.5832	.5871	.5910	.5948	.5987	.6026	.6064	.6103	.6141
0.3	.6179	.6217	.6255	.6293	.6331	.6368	.6406	.6443	.6480	.6517
0.4	.6554	.6591	.6628	.6664	.6700	.6736	.6772	.6808	.6844	.6879
0.5	.6915	.6950	.6985	.7019	.7054	.7088	.7123	.7157	.7190	.7224
0.6	.7257	.7291	.7324	.7357	.7389	.7422	.7454	.7486	.7517	.7549
0.7	.7580	.7611	.7642	.7673	.7704	.7734	.7764	.7794	.7823	.7852
0.8	.7881	.7910	.7939	.7967	.7995	.8023	.8051	.8078	.8106	.8133
0.9	.8159	.8186	.8212	.8238	.8264	.8289	.8315	.8340	.8365	.8389
1.0	.8413	.8438	.8461	.8485	.8508	.8531	.8554	.8577	.8599	.8621
1.1	.8643	.8665	.8686	.8708	.8729	.8749	.8770	.8790	.8810	.8830
1.2	.8849	.8869	.8888	.8907	.8925	.8944	.8962	.8980	.8997	.9015
1.3	.9032	.9049	.9066	.9082	.9099	.9115	.9131	.9147	.9162	.9177
1.4	.9192	.9207	.9222	.9236	.9251	.9265	.9279	.9292	.9306	.9319
1.5	.9332	.9345	.9357	.9370	.9382	.9394	.9406	.9418	.9429	.9441
1.6	.9452	.9463	.9474	.9484	.9495	.9505	.9515	.9525	.9535	.9545
1.7	.9554	.9564	.9573	.9582	.9591	.9599	.9608	.9616	.9625	.9633
1.8	.9641	.9649	.9656	.9664	.9671	.9678	.9686	.9693	.9699	.9706
1.9	.9713	.9719	.9726	.9732	.9738	.9744	.9750	.9756	.9761	.9767
2.0	.9772	.9778	.9783	.9788	.9793	.9798	.9803	.9808	.9812	.9817
2.1	.9821	.9826	.9830	.9834	.9838	.9842	.9846	.9850	.9854	.9857
2.2	.9861	.9864	.9868	.9871	.9875	.9878	.9881	.9884	.9887	.9890
2.3	.9893	.9896	.9898	.9901	.9904	.9906	.9909	.9911	.9913	.9916
2.4	.9918	.9920	.9922	.9925	.9927	.9929	.9931	.9932	.9934	.9936
2.5	.9938	.9940	.9941	.9943	.9945	.9946	.9948	.9949	.9951	.9952
2.6	.9953	.9955	.9956	.9957	.9959	.9960	.9961	.9962	.9963	.9964
2.7	.9965	.9966	.9967	.9968	.9969	.9970	.9971	.9972	.9973	.9974
2.8	.9974	.9975	.9976	.9977	.9977	.9978	.9979	.9979	.9980	.9981
2.9	.9981	.9982	.9982	.9983	.9984	.9984	.9985	.9985	.9986	.9986
3.0	.9987	.9987	.9987	.9988	.9988	.9989	.9989	.9989	.9990	.9990
3.1	.9990	.9991	.9991	.9991	.9992	.9992	.9992	.9992	.9993	.9993
3.2	.9993	.9993	.9994	.9994	.9994	.9994	.9994	.9995	.9995	.9995
3.3	.9995	.9995	.9995	.9996	.9996	.9996	.9996	.9996	.9996	.9997
3.4	.9997	.9997	.9997	.9997	.9997	.9997	.9997	.9997	.9997	.9998

TABLE 4　STUDENT'S *t*-VALUES FOR SPECIFIED TAIL AREAS

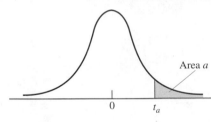

Each table entry is the *t*-value whose right-tail area equals the column heading, and whose degrees of freedom (*df*) equal the row number.

Degrees of Freedom (*df*)	Amount of Area in One Tail				
	.100	.050	.025	.010	.005
1	3.078	6.314	12.706	31.821	63.657
2	1.886	2.920	4.303	6.965	9.925
3	1.638	2.353	3.182	4.541	5.841
4	1.533	2.132	2.776	3.747	4.604
5	1.476	2.015	2.571	3.365	4.032
6	1.440	1.943	2.447	3.143	3.707
7	1.415	1.895	2.365	2.998	3.499
8	1.397	1.860	2.306	2.896	3.355
9	1.383	1.833	2.262	2.821	3.250
10	1.372	1.812	2.228	2.764	3.169
11	1.363	1.796	2.201	2.718	3.106
12	1.356	1.782	2.179	2.681	3.055
13	1.350	1.771	2.160	2.650	3.012
14	1.345	1.761	2.145	2.624	2.977
15	1.341	1.753	2.131	2.602	2.947
16	1.337	1.746	2.120	2.583	2.921
17	1.333	1.740	2.110	2.567	2.898
18	1.330	1.734	2.101	2.552	2.878
19	1.328	1.729	2.093	2.539	2.861
20	1.325	1.725	2.086	2.528	2.845
21	1.323	1.721	2.080	2.518	2.831
22	1.321	1.717	2.074	2.508	2.819
23	1.319	1.714	2.069	2.500	2.807
24	1.318	1.711	2.064	2.492	2.797
25	1.316	1.708	2.060	2.485	2.787
26	1.315	1.706	2.056	2.479	2.779
27	1.314	1.703	2.052	2.473	2.771
28	1.313	1.701	2.048	2.467	2.763
29	1.311	1.699	2.045	2.462	2.756
Infinity	1.282	1.645	1.960	2.326	2.576

TABLE 5 CHI-SQUARE VALUES FOR SPECIFIED AREAS

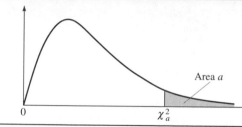

Each table entry is the χ^2-value whose area to the right equals the column heading, and whose degrees of freedom (*df*) equal the row number.

Amount of Area to the Right of the Table Value

df	.995	.990	.975	.950	.900	.100	.050	.025	.010	.005
1	0.00	0.00	0.00	0.00	0.02	2.71	3.84	5.02	6.63	7.88
2	0.01	0.02	0.05	0.10	0.21	4.61	5.99	7.38	9.21	10.60
3	0.07	0.11	0.22	0.35	0.58	6.25	7.81	9.35	11.34	12.84
4	0.21	0.30	0.48	0.71	1.06	7.78	9.49	11.14	13.28	14.86
5	0.41	0.55	0.83	1.15	1.61	9.24	11.07	12.83	15.09	16.75
6	0.68	0.87	1.24	1.64	2.20	10.64	12.59	14.45	16.81	18.55
7	0.99	1.24	1.69	2.17	2.83	12.02	14.07	16.01	18.48	20.28
8	1.34	1.65	2.18	2.73	3.49	13.36	15.51	17.53	20.09	21.95
9	1.73	2.09	2.70	3.33	4.17	14.68	16.92	19.02	21.67	23.59
10	2.16	2.56	3.25	3.94	4.87	15.99	18.31	20.48	23.21	25.19
11	2.60	3.05	3.82	4.57	5.58	17.28	19.68	21.92	24.72	26.76
12	3.07	3.57	4.40	5.23	6.30	18.55	21.03	23.34	26.22	28.30
13	3.57	4.11	5.01	5.89	7.04	19.81	22.36	24.74	27.69	29.82
14	4.07	4.66	5.63	6.57	7.79	21.06	23.68	26.12	29.14	31.32
15	4.60	5.23	6.26	7.26	8.55	22.31	25.00	27.49	30.58	32.80
16	5.14	5.81	6.91	7.96	9.31	23.54	26.30	28.85	32.00	34.27
17	5.70	6.41	7.56	8.67	10.09	24.77	27.59	30.19	33.41	35.72
18	6.26	7.01	8.23	9.39	10.86	25.99	28.87	31.53	34.81	37.16
19	6.84	7.63	8.91	10.12	11.65	27.20	30.14	32.85	36.19	38.58
20	7.43	8.26	9.59	10.85	12.44	28.41	31.41	34.17	37.57	40.00
21	8.03	8.90	10.28	11.59	13.24	29.62	32.67	35.48	38.93	41.40
22	8.64	9.54	10.98	12.34	14.04	30.81	33.92	36.78	40.29	42.80
23	9.26	10.20	11.69	13.09	14.85	32.01	35.17	38.08	41.64	44.18
24	9.89	10.86	12.40	13.85	15.66	33.20	36.42	39.36	42.98	45.56
25	10.52	11.52	13.12	14.61	16.47	34.38	37.65	40.65	44.31	46.93
26	11.16	12.20	13.84	15.38	17.29	35.56	38.89	41.92	45.64	48.29
27	11.81	12.88	14.57	16.15	18.11	36.74	40.11	43.19	46.96	49.65
28	12.46	13.56	15.31	16.93	18.94	37.92	41.34	44.46	48.28	50.99
29	13.12	14.26	16.05	17.71	19.77	39.09	42.56	45.72	49.59	52.34
30	13.79	14.95	16.79	18.49	20.60	40.26	43.77	46.98	50.89	53.67
40	20.71	22.16	24.43	26.51	29.05	51.81	55.76	59.34	63.69	66.77
50	27.99	29.71	32.36	34.76	37.69	63.17	67.50	71.42	76.15	79.49
60	35.53	37.48	40.48	43.19	46.46	74.40	79.08	83.30	88.38	91.95
70	43.28	45.44	48.76	51.74	55.33	85.53	90.53	95.02	100.42	104.21
80	51.17	53.54	57.15	60.39	64.28	96.58	101.88	106.63	112.33	116.32
90	59.20	61.75	65.65	69.13	73.29	107.56	113.14	118.14	124.11	128.30
100	67.33	70.06	74.22	77.93	82.36	118.50	124.34	129.56	135.81	140.18

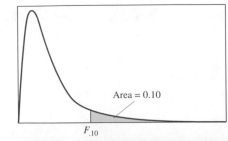

Each table entry is the F-value that has an area of 0.10 to the right and whose degrees of freedom are given by the column and row numbers.

TABLE 6A   *F-DISTRIBUTION**

Right-Hand Tail Area of 0.10

ddf	\multicolumn{9}{c}{Numerator *df (ndf)*}
	1	2	3	4	5	6	7	8	9
1	39.86	49.50	53.59	55.83	57.24	58.20	58.91	59.44	59.86
2	8.53	9.00	9.16	9.24	9.29	9.33	9.35	9.37	9.38
3	5.54	5.46	5.39	5.34	5.31	5.28	5.27	5.25	5.24
4	4.54	4.32	4.19	4.11	4.05	4.01	3.98	3.95	3.94
5	4.06	3.78	3.62	3.52	3.45	3.40	3.37	3.34	3.32
6	3.78	3.46	3.29	3.18	3.11	3.05	3.01	2.98	2.96
7	3.59	3.26	3.07	2.96	2.88	2.83	2.78	2.75	2.72
8	3.46	3.11	2.92	2.81	2.73	2.67	2.62	2.59	2.56
9	3.36	3.01	2.81	2.69	2.61	2.55	2.51	2.47	2.44
10	3.29	2.92	2.73	2.61	2.52	2.46	2.41	2.38	2.35
11	3.23	2.86	2.66	2.54	2.45	2.39	2.34	2.30	2.27
12	3.18	2.81	2.61	2.48	2.39	2.33	2.28	2.24	2.21
13	3.14	2.76	2.56	2.43	2.35	2.28	2.23	2.20	2.16
14	3.10	2.73	2.52	2.39	2.31	2.24	2.19	2.15	2.12
15	3.07	2.70	2.49	2.36	2.27	2.21	2.16	2.12	2.09
16	3.05	2.67	2.46	2.33	2.24	2.18	2.13	2.09	2.06
17	3.03	2.64	2.44	2.31	2.22	2.15	2.10	2.06	2.03
18	3.01	2.62	2.42	2.29	2.20	2.13	2.08	2.04	2.00
19	2.99	2.61	2.40	2.27	2.18	2.11	2.06	2.02	1.98
20	2.97	2.59	2.38	2.25	2.16	2.09	2.04	2.00	1.96
21	2.96	2.57	2.36	2.23	2.14	2.08	2.02	1.98	1.95
22	2.95	2.56	2.35	2.22	2.13	2.06	2.01	1.97	1.93
23	2.94	2.55	2.34	2.21	2.11	2.05	1.99	1.95	1.92
24	2.93	2.54	2.33	2.19	2.10	2.04	1.98	1.94	1.91
25	2.92	2.53	2.32	2.18	2.09	2.02	1.97	1.93	1.89
26	2.91	2.52	2.31	2.17	2.08	2.01	1.96	1.92	1.88
27	2.90	2.51	2.30	2.17	2.07	2.00	1.95	1.91	1.87
28	2.89	2.50	2.29	2.16	2.06	2.00	1.94	1.90	1.87
29	2.89	2.50	2.28	2.15	2.06	1.99	1.93	1.89	1.86
30	2.88	2.49	2.28	2.14	2.05	1.98	1.93	1.88	1.85
40	2.84	2.44	2.23	2.09	2.00	1.93	1.87	1.83	1.79
60	2.79	2.39	2.18	2.04	1.95	1.87	1.82	1.77	1.74
120	2.75	2.35	2.13	1.99	1.90	1.82	1.77	1.72	1.68
∞	2.71	2.30	2.08	1.94	1.85	1.77	1.72	1.67	1.63

* Reprinted, with permission of the *Biometrika* trustees, from Merrington, M. and C. M. Thompson. "Tables of Percentage Points of the Inverted Beta (*F*) Distribution," *Biometrika* 33(1943).

TABLE 6A   *F*-DISTRIBUTION (continued)

Right-Hand Tail Area of 0.10

	Numerator *df (ndf)*									
ddf	10	12	15	20	24	30	40	60	120	∞
1	60.19	60.71	61.22	61.74	62.00	62.26	62.53	62.79	63.06	63.33
2	9.39	9.41	9.42	9.44	9.45	9.46	9.47	9.47	9.48	9.49
3	5.23	5.22	5.20	5.18	5.18	5.17	5.16	5.15	5.14	5.13
4	3.92	3.90	3.87	3.84	3.83	3.82	3.80	3.79	3.78	3.76
5	3.30	3.27	3.24	3.21	3.19	3.17	3.16	3.14	3.12	3.10
6	2.94	2.90	2.87	2.84	2.82	2.80	2.78	2.76	2.74	2.72
7	2.70	2.67	2.63	2.59	2.58	2.56	2.54	2.51	2.49	2.47
8	2.54	2.50	2.46	2.42	2.40	2.38	2.36	2.34	2.32	2.29
9	2.42	2.38	2.34	2.30	2.28	2.25	2.23	2.21	2.18	2.16
10	2.32	2.28	2.24	2.20	2.18	2.16	2.13	2.11	2.08	2.06
11	2.25	2.21	2.17	2.12	2.10	2.08	2.05	2.03	2.00	1.97
12	2.19	2.15	2.10	2.06	2.04	2.01	1.99	1.96	1.93	1.90
13	2.14	2.10	2.05	2.01	1.98	1.96	1.93	1.90	1.88	1.85
14	2.10	2.05	2.01	1.96	1.94	1.91	1.89	1.86	1.83	1.80
15	2.06	2.02	1.97	1.92	1.90	1.87	1.85	1.82	1.79	1.76
16	2.03	1.99	1.94	1.89	1.87	1.84	1.81	1.78	1.75	1.72
17	2.00	1.96	1.91	1.86	1.84	1.81	1.78	1.75	1.72	1.69
18	1.98	1.93	1.89	1.84	1.81	1.78	1.75	1.72	1.69	1.66
19	1.96	1.91	1.86	1.81	1.79	1.76	1.73	1.70	1.67	1.63
20	1.94	1.89	1.84	1.79	1.77	1.74	1.71	1.68	1.64	1.61
21	1.92	1.87	1.83	1.78	1.75	1.72	1.69	1.66	1.62	1.59
22	1.90	1.86	1.81	1.76	1.73	1.70	1.67	1.64	1.60	1.57
23	1.89	1.84	1.80	1.74	1.72	1.69	1.66	1.62	1.59	1.55
24	1.88	1.83	1.78	1.73	1.70	1.67	1.64	1.61	1.57	1.53
25	1.87	1.82	1.77	1.72	1.69	1.66	1.63	1.59	1.56	1.52
26	1.86	1.81	1.76	1.71	1.68	1.65	1.61	1.58	1.54	1.50
27	1.85	1.80	1.75	1.70	1.67	1.64	1.60	1.57	1.53	1.49
28	1.84	1.79	1.74	1.69	1.66	1.63	1.59	1.56	1.52	1.48
29	1.83	1.78	1.73	1.68	1.65	1.62	1.58	1.55	1.51	1.47
30	1.82	1.77	1.72	1.67	1.64	1.61	1.57	1.54	1.50	1.46
40	1.76	1.71	1.66	1.61	1.57	1.54	1.51	1.47	1.42	1.38
60	1.71	1.66	1.60	1.54	1.51	1.48	1.44	1.40	1.35	1.29
120	1.65	1.60	1.55	1.48	1.45	1.41	1.37	1.32	1.26	1.19
∞	1.60	1.55	1.49	1.42	1.38	1.34	1.30	1.24	1.17	1.00

Denominator *df*

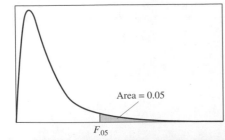

Each table entry is the F-value that has an area of 0.05 to the right and whose degrees of freedom are given by the column and row numbers.

Area = 0.05

$F_{.05}$

TABLE 6B F-DISTRIBUTION

Right-Hand Tail Area of 0.05

					Numerator *df (ndf)*				
ddf	1	2	3	4	5	6	7	8	9
1	161.4	199.5	215.7	224.6	230.2	234.0	236.8	238.9	240.5
2	18.51	19.00	19.16	19.25	19.30	19.33	19.35	19.37	19.38
3	10.13	9.55	9.28	9.12	9.01	8.94	8.89	8.85	8.81
4	7.71	6.94	6.59	6.39	6.26	6.16	6.09	6.04	6.00
5	6.61	5.79	5.41	5.19	5.05	4.95	4.88	4.82	4.77
6	5.99	5.14	4.76	4.53	4.39	4.28	4.21	4.15	4.10
7	5.59	4.74	4.35	4.12	3.97	3.87	3.79	3.73	3.68
8	5.32	4.46	4.07	3.84	3.69	3.58	3.50	3.44	3.39
9	5.12	4.26	3.86	3.63	3.48	3.37	3.29	3.23	3.18
10	4.96	4.10	3.71	3.48	3.33	3.22	3.14	3.07	3.02
11	4.84	3.98	3.59	3.36	3.20	3.09	3.01	2.95	2.90
12	4.75	3.89	3.49	3.26	3.11	3.00	2.91	2.85	2.80
13	4.67	3.81	3.41	3.18	3.03	2.92	2.83	2.77	2.71
14	4.60	3.74	3.34	3.11	2.96	2.85	2.76	2.70	2.65
15	4.54	3.68	3.29	3.06	2.90	2.79	2.71	2.64	2.59
16	4.49	3.63	3.24	3.01	2.85	2.74	2.66	2.59	2.54
17	4.45	3.59	3.20	2.96	2.81	2.70	2.61	2.55	2.49
18	4.41	3.55	3.16	2.93	2.77	2.66	2.58	2.51	2.46
19	4.38	3.52	3.13	2.90	2.74	2.63	2.54	2.48	2.42
20	4.35	3.49	3.10	2.87	2.71	2.60	2.51	2.45	2.39
21	4.32	3.47	3.07	2.84	2.68	2.57	2.49	2.42	2.37
22	4.30	3.44	3.05	2.82	2.66	2.55	2.46	2.40	2.34
23	4.28	3.42	3.03	2.80	2.64	2.53	2.44	2.37	2.32
24	4.26	3.40	3.01	2.78	2.62	2.51	2.42	2.36	2.30
25	4.24	3.39	2.99	2.76	2.60	2.49	2.40	2.34	2.28
26	4.23	3.37	2.98	2.74	2.59	2.47	2.39	2.32	2.27
27	4.21	3.35	2.96	2.73	2.57	2.46	2.37	2.31	2.25
28	4.20	3.34	2.95	2.71	2.56	2.45	2.36	2.29	2.24
29	4.18	3.33	2.93	2.70	2.55	2.43	2.35	2.28	2.22
30	4.17	3.32	2.92	2.69	2.53	2.42	2.33	2.27	2.21
40	4.08	3.23	2.84	2.61	2.45	2.34	2.25	2.18	2.12
60	4.00	3.15	2.76	2.53	2.37	2.25	2.17	2.10	2.04
120	3.92	3.07	2.68	2.45	2.29	2.17	2.09	2.02	1.96
∞	3.84	3.00	2.60	2.37	2.21	2.10	2.01	1.94	1.88

Denominator *df*

TABLE 6B   *F*-DISTRIBUTION (continued)

Right-Hand Tail Area of 0.05

ddf	Numerator *df (ndf)*									
	10	12	15	20	24	30	40	60	120	∞
1	241.9	243.9	245.9	248.0	249.1	250.1	251.1	252.2	253.3	254.3
2	19.40	19.41	19.43	19.45	19.45	19.46	19.47	19.48	19.49	19.50
3	8.79	8.74	8.70	8.66	8.64	8.62	8.59	8.57	8.55	8.53
4	5.96	5.91	5.86	5.80	5.77	5.75	5.72	5.69	5.66	5.63
5	4.74	4.68	4.62	4.56	4.53	4.50	4.46	4.43	4.40	4.36
6	4.06	4.00	3.94	3.87	3.84	3.81	3.77	3.74	3.70	3.67
7	3.64	3.57	3.51	3.44	3.41	3.38	3.34	3.30	3.27	3.23
8	3.35	3.28	3.22	3.15	3.12	3.08	3.04	3.01	2.97	2.93
9	3.14	3.07	3.01	2.94	2.90	2.86	2.83	2.79	2.75	2.71
10	2.98	2.91	2.85	2.77	2.74	2.70	2.66	2.62	2.58	2.54
11	2.85	2.79	2.72	2.65	2.61	2.57	2.53	2.49	2.45	2.40
12	2.75	2.69	2.62	2.54	2.51	2.47	2.43	2.38	2.34	2.30
13	2.67	2.60	2.53	2.46	2.42	2.38	2.34	2.30	2.25	2.21
14	2.60	2.53	2.46	2.39	2.35	2.31	2.27	2.22	2.18	2.13
15	2.54	2.48	2.40	2.33	2.29	2.25	2.20	2.16	2.11	2.07
16	2.49	2.42	2.35	2.28	2.24	2.19	2.15	2.11	2.06	2.01
17	2.45	2.38	2.31	2.23	2.19	2.15	2.10	2.06	2.01	1.96
18	2.41	2.34	2.27	2.19	2.15	2.11	2.06	2.02	1.97	1.92
19	2.38	2.31	2.23	2.16	2.11	2.07	2.03	1.98	1.93	1.88
20	2.35	2.28	2.20	2.12	2.08	2.04	1.99	1.95	1.90	1.84
21	2.32	2.25	2.18	2.10	2.05	2.01	1.96	1.92	1.87	1.81
22	2.30	2.23	2.15	2.07	2.03	1.98	1.94	1.89	1.84	1.78
23	2.27	2.20	2.13	2.05	2.01	1.96	1.91	1.86	1.81	1.76
24	2.25	2.18	2.11	2.03	1.98	1.94	1.89	1.84	1.79	1.73
25	2.24	2.16	2.09	2.01	1.96	1.92	1.87	1.82	1.77	1.71
26	2.22	2.15	2.07	1.99	1.95	1.90	1.85	1.80	1.75	1.69
27	2.20	2.13	2.06	1.97	1.93	1.88	1.84	1.79	1.73	1.67
28	2.19	2.12	2.04	1.96	1.91	1.87	1.82	1.77	1.71	1.65
29	2.18	2.10	2.03	1.94	1.90	1.85	1.81	1.75	1.70	1.64
30	2.16	2.09	2.01	1.93	1.89	1.84	1.79	1.74	1.68	1.62
40	2.08	2.00	1.92	1.84	1.79	1.74	1.69	1.64	1.58	1.51
60	1.99	1.92	1.84	1.75	1.70	1.65	1.59	1.53	1.47	1.39
120	1.91	1.83	1.75	1.66	1.61	1.55	1.50	1.43	1.35	1.25
∞	1.83	1.75	1.67	1.57	1.52	1.46	1.39	1.32	1.22	1.00

(Denominator df labels the row at left: D e n o m i n a t o r   d f)

Adapted from Newmark J. *Statistics and Probability in Modern Life*, 5th ed., Saunders College Publishing, Philadelphia, 1992, p. A.14.

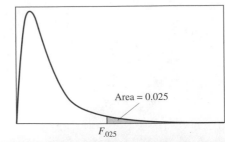

Each table entry is the F-value that has an area of 0.025 to the right and whose degrees of freedom are given by the column and row numbers.

Area = 0.025

$F_{.025}$

TABLE 6C   F-DISTRIBUTION

Right-Hand Tail Area of 0.025

ddf	Numerator df (ndf)								
	1	2	3	4	5	6	7	8	9
1	647.8	799.5	864.2	899.6	921.8	937.1	948.2	956.7	963.3
2	38.51	39.00	39.17	39.25	39.30	39.33	39.36	39.37	39.39
3	17.44	16.04	15.44	15.10	14.88	14.73	14.62	14.54	14.47
4	12.22	10.65	9.98	9.60	9.36	9.20	9.07	8.98	8.90
5	10.01	8.43	7.76	7.39	7.15	6.98	6.85	6.76	6.68
6	8.81	7.26	6.60	6.23	5.99	5.82	5.70	5.60	5.52
7	8.07	6.54	5.89	5.52	5.29	5.12	4.99	4.90	4.82
8	7.57	6.06	5.42	5.05	4.82	4.65	4.53	4.43	4.36
9	7.21	5.71	5.08	4.72	4.48	4.32	4.20	4.10	4.03
10	6.94	5.46	4.83	4.47	4.24	4.07	3.95	3.85	3.78
11	6.72	5.26	4.63	4.28	4.04	3.88	3.76	3.66	3.59
12	6.55	5.10	4.47	4.12	3.89	3.73	3.61	3.51	3.44
13	6.41	4.97	4.35	4.00	3.77	3.60	3.48	3.39	3.31
14	6.30	4.86	4.24	3.89	3.66	3.50	3.38	3.29	3.21
15	6.20	4.77	4.15	3.80	3.58	3.41	3.29	3.20	3.12
16	6.12	4.69	4.08	3.73	3.50	3.34	3.22	3.12	3.05
17	6.04	4.62	4.01	3.66	3.44	3.28	3.16	3.06	2.98
18	5.98	4.56	3.95	3.61	3.38	3.22	3.10	3.01	2.93
19	5.92	4.51	3.90	3.56	3.33	3.17	3.05	2.96	2.88
20	5.87	4.46	3.86	3.51	3.29	3.13	3.01	2.91	2.84
21	5.83	4.42	3.82	3.48	3.25	3.09	2.97	2.87	2.80
22	5.79	4.38	3.78	3.44	3.22	3.05	2.93	2.84	2.76
23	5.75	4.35	3.75	3.41	3.18	3.02	2.90	2.81	2.73
24	5.72	4.32	3.72	3.38	3.15	2.99	2.87	2.78	2.70
25	5.69	4.29	3.69	3.35	3.13	2.97	2.85	2.75	2.68
26	5.66	4.27	3.67	3.33	3.10	2.94	2.82	2.73	2.65
27	5.63	4.24	3.65	3.31	3.08	2.92	2.80	2.71	2.63
28	5.61	4.22	3.63	3.29	3.06	2.90	2.78	2.69	2.61
29	5.59	4.20	3.61	3.27	3.04	2.88	2.76	2.67	2.59
30	5.57	4.18	3.59	3.25	3.03	2.87	2.75	2.65	2.57
40	5.42	4.05	3.46	3.13	2.90	2.74	2.62	2.53	2.45
60	5.29	3.93	3.34	3.01	2.79	2.63	2.51	2.41	2.33
120	5.15	3.80	3.23	2.89	2.67	2.52	2.39	2.30	2.22
∞	5.02	3.69	3.12	2.79	2.57	2.41	2.29	2.19	2.11

Denominator df

TABLE 6C   F-DISTRIBUTION (continued)

Right-Hand Tail Area of 0.025

ddf	Numerator df (ndf)									
	10	12	15	20	24	30	40	60	120	∞
1	968.6	976.7	984.9	993.1	997.2	1001	1006	1010	1014	1018
2	39.40	39.41	39.43	39.45	39.46	39.46	39.47	39.48	39.49	39.50
3	14.42	14.34	14.25	14.17	14.12	14.08	14.04	13.99	13.95	13.90
4	8.84	8.75	8.66	8.56	8.51	8.46	8.41	8.36	8.31	8.26
5	6.62	6.52	6.43	6.33	6.28	6.23	6.18	6.12	6.07	6.02
6	5.46	5.37	5.27	5.17	5.12	5.07	5.01	4.96	4.90	4.85
7	4.76	4.67	4.57	4.47	4.42	4.36	4.31	4.25	4.20	4.14
8	4.30	4.20	4.10	4.00	3.95	3.89	3.84	3.78	3.73	3.67
9	3.96	3.87	3.77	3.67	3.61	3.56	3.51	3.45	3.39	3.33
10	3.72	3.62	3.52	3.42	3.37	3.31	3.26	3.20	3.14	3.08
11	3.53	3.43	3.33	3.23	3.17	3.12	3.06	3.00	2.94	2.88
12	3.37	3.28	3.18	3.07	3.02	2.96	2.91	2.85	2.79	2.72
13	3.25	3.15	3.05	2.95	2.89	2.84	2.78	2.72	2.66	2.60
14	3.15	3.05	2.95	2.84	2.79	2.73	2.67	2.61	2.55	2.49
15	3.06	2.96	2.86	2.76	2.70	2.64	2.59	2.52	2.46	2.40
16	2.99	2.89	2.79	2.68	2.63	2.57	2.51	2.45	2.38	2.32
17	2.92	2.82	2.72	2.62	2.56	2.50	2.44	2.38	2.32	2.25
18	2.87	2.77	2.67	2.56	2.50	2.44	2.38	2.32	2.26	2.19
19	2.82	2.72	2.62	2.51	2.45	2.39	2.33	2.27	2.20	2.13
20	2.77	2.68	2.57	2.46	2.41	2.35	2.29	2.22	2.16	2.09
21	2.73	2.64	2.53	2.42	2.37	2.31	2.25	2.18	2.11	2.04
22	2.70	2.60	2.50	2.39	2.33	2.27	2.21	2.14	2.08	2.00
23	2.67	2.57	2.47	2.36	2.30	2.24	2.18	2.11	2.04	1.97
24	2.64	2.54	2.44	2.33	2.27	2.21	2.15	2.08	2.01	1.94
25	2.61	2.51	2.41	2.30	2.24	2.18	2.12	2.05	1.98	1.91
26	2.59	2.49	2.39	2.28	2.22	2.16	2.09	2.03	1.95	1.88
27	2.57	2.47	2.36	2.25	2.19	2.13	2.07	2.00	1.93	1.85
28	2.55	2.45	2.34	2.23	2.17	2.11	2.05	1.98	1.91	1.83
29	2.53	2.43	2.32	2.21	2.15	2.09	2.03	1.96	1.89	1.81
30	2.51	2.41	2.31	2.20	2.14	2.07	2.01	1.94	1.87	1.79
40	2.39	2.29	2.18	2.07	2.01	1.94	1.88	1.80	1.72	1.64
60	2.27	2.17	2.06	1.94	1.88	1.82	1.74	1.67	1.58	1.48
120	2.16	2.05	1.94	1.82	1.76	1.69	1.61	1.53	1.43	1.31
∞	2.05	1.94	1.83	1.71	1.64	1.57	1.48	1.39	1.27	1.00

Denominator df

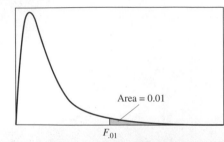

Each table entry is the F-value that has an area of 0.01 to the right and whose degrees of freedom are given by the column and row numbers.

Area = 0.01

$F_{.01}$

TABLE 6D	F-DISTRIBUTION

Right-Hand Tail Area of 0.01

Numerator df (ndf)

ddf	1	2	3	4	5	6	7	8	9
1	4052	4999.5	5403	5625	5764	5859	5928	5982	6022
2	98.50	99.00	99.17	99.25	99.30	99.33	99.36	99.37	99.39
3	34.12	30.82	29.46	28.71	28.24	27.91	27.67	27.49	27.35
4	21.20	18.00	16.69	15.98	15.52	15.21	14.98	14.80	14.66
5	16.26	13.27	12.06	11.39	10.97	10.67	10.46	10.29	10.16
6	13.75	10.92	9.78	9.15	8.75	8.47	8.26	8.10	7.98
7	12.25	9.55	8.45	7.85	7.46	7.19	6.99	6.84	6.72
8	11.26	8.65	7.59	7.01	6.63	6.37	6.18	6.03	5.91
9	10.56	8.02	6.99	6.42	6.06	5.80	5.61	5.47	5.35
10	10.04	7.56	6.55	5.99	5.64	5.39	5.20	5.06	4.94
11	9.65	7.21	6.22	5.67	5.32	5.07	4.89	4.74	4.63
12	9.33	6.93	5.95	5.41	5.06	4.82	4.64	4.50	4.39
13	9.07	6.70	5.74	5.21	4.86	4.62	4.44	4.30	4.19
14	8.86	6.51	5.56	5.04	4.69	4.46	4.28	4.14	4.03
15	8.68	6.36	5.42	4.89	4.56	4.32	4.14	4.00	3.89
16	8.53	6.23	5.29	4.77	4.44	4.20	4.03	3.89	3.78
17	8.40	6.11	5.18	4.67	4.34	4.10	3.93	3.79	3.68
18	8.29	6.01	5.09	4.58	4.25	4.01	3.84	3.71	3.60
19	8.18	5.93	5.01	4.50	4.17	3.94	3.77	3.63	3.52
20	8.10	5.85	4.94	4.43	4.10	3.87	3.70	3.56	3.46
21	8.02	5.78	4.87	4.37	4.04	3.81	3.64	3.51	3.40
22	7.95	5.72	4.82	4.31	3.99	3.76	3.59	3.45	3.35
23	7.88	5.66	4.76	4.26	3.94	3.71	3.54	3.41	3.30
24	7.82	5.61	4.72	4.22	3.90	3.67	3.50	3.36	3.26
25	7.77	5.57	4.68	4.18	3.85	3.63	3.46	3.32	3.22
26	7.72	5.53	4.64	4.14	3.82	3.59	3.42	3.29	3.18
27	7.68	5.49	4.60	4.11	3.78	3.56	3.39	3.26	3.15
28	7.64	5.45	4.57	4.07	3.75	3.53	3.36	3.23	3.12
29	7.60	5.42	4.54	4.04	3.73	3.50	3.33	3.20	3.09
30	7.56	5.39	4.51	4.02	3.70	3.47	3.30	3.17	3.07
40	7.31	5.18	4.31	3.83	3.51	3.29	3.12	2.99	2.89
60	7.08	4.98	4.13	3.65	3.34	3.12	2.95	2.82	2.72
120	6.85	4.79	3.95	3.48	3.17	2.96	2.79	2.66	2.56
∞	6.63	4.61	3.78	3.32	3.02	2.80	2.64	2.51	2.41

Denominator df

TABLE 6D *F*-DISTRIBUTION (continued)

Right-Hand Tail Area of 0.01

ddf	\multicolumn{10}{c}{Numerator *df (ndf)*}
	10	12	15	20	24	30	40	60	120	∞
1	6056	6106	6157	6209	6235	6261	6287	6313	6339	6366
2	99.40	99.42	99.43	99.45	99.46	99.47	99.47	99.48	99.49	99.50
3	27.23	27.05	26.87	26.69	26.60	26.50	26.41	26.32	26.22	26.13
4	14.55	14.37	14.20	14.02	13.93	13.84	13.75	13.65	13.56	13.46
5	10.05	9.89	9.72	9.55	9.47	9.38	9.29	9.20	9.11	9.02
6	7.87	7.72	7.56	7.40	7.31	7.23	7.14	7.06	6.97	6.88
7	6.62	6.47	6.31	6.16	6.07	5.99	5.91	5.82	5.74	5.65
8	5.81	5.67	5.52	5.36	5.28	5.20	5.12	5.03	4.95	4.86
9	5.26	5.11	4.96	4.81	4.73	4.65	4.57	4.48	4.40	4.31
10	4.85	4.71	4.56	4.41	4.33	4.25	4.17	4.08	4.00	3.91
11	4.54	4.40	4.25	4.10	4.02	3.94	3.86	3.78	3.69	3.60
12	4.30	4.16	4.01	3.86	3.78	3.70	3.62	3.54	3.45	3.36
13	4.10	3.96	3.82	3.66	3.59	3.51	3.43	3.34	3.25	3.17
14	3.94	3.80	3.66	3.51	3.43	3.35	3.27	3.18	3.09	3.00
15	3.80	3.67	3.52	3.37	3.29	3.21	3.13	3.05	2.96	2.87
16	3.69	3.55	3.41	3.26	3.18	3.10	3.02	2.93	2.84	2.75
17	3.59	3.46	3.31	3.16	3.08	3.00	2.92	2.83	2.75	2.65
18	3.51	3.37	3.23	3.08	3.00	2.92	2.84	2.75	2.66	2.57
19	3.43	3.30	3.15	3.00	2.92	2.84	2.76	2.67	2.58	2.49
20	3.37	3.23	3.09	2.94	2.86	2.78	2.69	2.61	2.52	2.42
21	3.31	3.17	3.03	2.88	2.80	2.72	2.64	2.55	2.46	2.36
22	3.26	3.12	2.98	2.83	2.75	2.67	2.58	2.50	2.40	2.31
23	3.21	3.07	2.93	2.78	2.70	2.62	2.54	2.45	2.35	2.26
24	3.17	3.03	2.89	2.74	2.66	2.58	2.49	2.40	2.31	2.21
25	3.13	2.99	2.85	2.70	2.62	2.54	2.45	2.36	2.27	2.17
26	3.09	2.96	2.81	2.66	2.58	2.50	2.42	2.33	2.23	2.13
27	3.06	2.93	2.78	2.63	2.55	2.47	2.38	2.29	2.20	2.10
28	3.03	2.90	2.75	2.60	2.52	2.44	2.35	2.26	2.17	2.06
29	3.00	2.87	2.73	2.57	2.49	2.41	2.33	2.23	2.14	2.03
30	2.98	2.84	2.70	2.55	2.47	2.39	2.30	2.21	2.11	2.01
40	2.80	2.66	2.52	2.37	2.29	2.20	2.11	2.02	1.92	1.80
60	2.63	2.50	2.35	2.20	2.12	2.03	1.94	1.84	1.73	1.60
120	2.47	2.34	2.19	2.03	1.95	1.86	1.76	1.66	1.53	1.38
∞	2.32	2.18	2.04	1.88	1.79	1.70	1.59	1.47	1.32	1.00

Denominator df

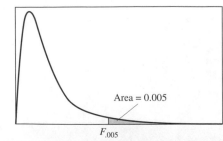

Each table entry is the F-value that has an area of 0.005 to the right and whose degrees of freedom are given by the column and row numbers.

Area = 0.005

$F_{.005}$

TABLE 6E   *F*-DISTRIBUTION

Right-Hand Tail Area of 0.005

ddf	\multicolumn{9}{c}{Numerator *df (ndf)*}								
	1	2	3	4	5	6	7	8	9
1	16211	20000	21615	22500	23056	23437	23715	23925	24091
2	198.5	199.0	199.2	199.2	199.3	199.3	199.4	199.4	199.4
3	55.55	49.80	47.47	46.19	45.39	44.84	44.43	44.13	43.88
4	31.33	26.28	24.26	23.15	22.46	21.97	21.62	21.35	21.14
5	22.78	18.31	16.53	15.56	14.94	14.51	14.20	13.96	13.77
6	18.63	14.54	12.92	12.03	11.46	11.07	10.79	10.57	10.39
7	16.24	12.40	10.88	10.05	9.52	9.16	8.89	8.68	8.51
8	14.69	11.04	9.60	8.81	8.30	7.95	7.69	7.50	7.34
9	13.61	10.11	8.72	7.96	7.47	7.13	6.88	6.69	6.54
10	12.83	9.43	8.08	7.34	6.87	6.54	6.30	6.12	5.97
11	12.23	8.91	7.60	6.88	6.42	6.10	5.86	5.68	5.54
12	11.75	8.51	7.23	6.52	6.07	5.76	5.52	5.35	5.20
13	11.37	8.19	6.93	6.23	5.79	5.48	5.25	5.08	4.94
14	11.06	7.92	6.68	6.00	5.56	5.26	5.03	4.86	4.72
15	10.80	7.70	6.48	5.80	5.37	5.07	4.85	4.67	4.54
16	10.58	7.51	6.30	5.64	5.21	4.91	4.69	4.52	4.38
17	10.38	7.35	6.16	5.50	5.07	4.78	4.56	4.39	4.25
18	10.22	7.21	6.03	5.37	4.96	4.66	4.44	4.28	4.14
19	10.17	7.09	5.92	5.27	4.85	4.56	4.34	4.18	4.04
20	9.94	6.99	5.82	5.17	4.76	4.47	4.26	4.09	3.96
21	9.83	6.89	5.73	5.09	4.68	4.39	4.18	4.01	3.88
22	9.73	6.81	5.65	5.02	4.61	4.32	4.11	3.94	3.81
23	9.63	6.73	5.58	4.95	4.54	4.26	4.05	3.88	3.75
24	9.55	6.66	5.52	4.89	4.49	4.20	3.99	3.83	3.69
25	9.48	6.60	5.46	4.84	4.43	4.15	3.94	3.78	3.64
26	9.41	6.54	5.41	4.79	4.38	4.10	3.89	3.73	3.60
27	9.34	6.49	5.36	4.74	4.34	4.06	3.85	3.69	3.56
28	9.28	6.44	5.32	4.70	4.30	4.02	3.81	3.65	3.52
29	9.23	6.40	5.28	4.66	4.26	3.98	3.77	3.61	3.48
30	9.18	6.35	5.24	4.62	4.23	3.95	3.74	3.58	3.45
40	8.83	6.07	4.98	4.37	3.99	3.71	3.51	3.35	3.22
60	8.49	5.79	4.73	4.14	3.76	3.49	3.29	3.13	3.01
120	8.18	5.54	4.50	3.92	3.55	3.28	3.09	2.93	2.81
∞	7.88	5.30	4.28	3.72	3.35	3.09	2.90	2.74	2.62

D
e
n
o
m
i
n
a
t
o
r

d
f

TABLE 6E **F-DISTRIBUTION (continued)**

Right-Hand Tail Area of 0.005

					Numerator *df (ndf)*					
ddf	10	12	15	20	24	30	40	60	120	∞
1	24224	24426	24630	24836	24940	25044	25148	25253	25359	25465
2	199.4	199.4	199.4	199.4	199.5	199.5	199.5	199.5	199.5	199.5
3	43.69	43.39	43.08	42.78	42.62	42.47	42.31	42.15	41.99	41.83
4	20.97	20.70	20.44	20.17	20.03	19.89	19.75	19.61	19.47	19.32
5	13.62	13.38	13.15	12.90	12.78	12.66	12.53	12.40	12.27	12.14
6	10.25	10.03	9.81	9.59	9.47	9.36	9.24	9.12	9.00	8.88
7	8.38	8.18	7.97	7.75	7.65	7.53	7.42	7.31	7.19	7.08
8	7.21	7.01	6.81	6.61	6.50	6.40	6.29	6.18	6.06	5.95
9	6.42	6.23	6.03	5.83	5.73	5.62	5.52	5.41	5.30	5.19
10	5.85	5.66	5.47	5.27	5.17	5.07	4.97	4.86	4.75	4.64
11	5.42	5.24	5.05	4.86	4.76	4.65	4.55	4.44	4.34	4.23
12	5.09	4.91	4.72	4.53	4.43	4.33	4.23	4.12	4.01	3.90
13	4.82	4.64	4.46	4.27	4.17	4.07	3.97	3.87	3.76	3.65
14	4.60	4.43	4.25	4.06	3.96	3.86	3.76	3.66	3.55	3.44
15	4.42	4.25	4.07	3.88	3.79	3.69	3.58	3.48	3.37	3.26
16	4.27	4.10	3.92	3.73	3.64	3.54	3.44	3.33	3.22	3.11
17	4.14	3.97	3.79	3.61	3.51	3.41	3.31	3.21	3.10	2.98
18	4.03	3.86	3.68	3.50	3.40	3.30	3.20	3.10	2.99	2.87
19	3.93	3.76	3.59	3.40	3.31	3.21	3.11	3.00	2.89	2.78
20	3.85	3.68	3.50	3.32	3.22	3.12	3.02	2.92	2.81	2.69
21	3.77	3.60	3.43	3.24	3.15	3.05	2.95	2.84	2.73	2.61
22	3.70	3.54	3.36	3.18	3.08	2.98	2.88	2.77	2.66	2.55
23	3.64	3.47	3.30	3.12	3.02	2.92	2.82	2.71	2.60	2.48
24	3.59	3.42	3.25	3.06	2.97	2.87	2.77	2.66	2.55	2.43
25	3.54	3.37	3.20	3.01	2.92	2.82	2.72	2.61	2.50	2.38
26	3.49	3.33	3.15	2.97	2.87	2.77	2.67	2.56	2.45	2.33
27	3.45	3.28	3.11	2.93	2.83	2.73	2.63	2.52	2.41	2.29
28	3.41	3.25	3.07	2.89	2.79	2.69	2.59	2.48	2.37	2.25
29	3.38	3.21	3.04	2.86	2.76	2.66	2.56	2.45	2.33	2.21
30	3.34	3.18	3.01	2.82	2.73	2.63	2.52	2.42	2.30	2.18
40	3.12	2.95	2.78	2.60	2.50	2.40	2.30	2.18	2.06	1.93
60	2.90	2.74	2.57	2.39	2.29	2.19	2.08	1.96	1.83	1.69
120	2.71	2.54	2.37	2.19	2.09	1.98	1.87	1.75	1.61	1.43
∞	2.52	2.36	2.19	2.00	1.90	1.79	1.67	1.53	1.36	1.00

Denominator *df* (left-side label)

TABLE 7   CRITICAL VALUES FOR THE NUMBER OF RUNS IN A TWO-TAILED TEST WITH $\alpha = 0.05$
EACH CELL CONTAINS THE CRITICAL VALUES $L_{.025}$ AND $R_{.025}$

Each cell shows $L_{.025}$ (top) and $R_{.025}$ (bottom), written here as L/R.

The Smaller of n_1 and n_2	\\ The Larger of n_1 and n_2 → 5	6	7	8	9	10	11	12	13	14	15	16	17	18	19	20
2								2/6	2/6	2/6	2/6	2/6	2/6	2/6	2/6	2/6
3		2/8	2/8	2/8	2/8	2/8	2/8	2/8	2/8	2/8	3/8	3/8	3/8	3/8	3/8	3/8
4	2/9	2/9	2/10	3/10	3/10	3/10	3/10	3/10	3/10	3/10	3/10	4/10	4/10	4/10	4/10	4/10
5	2/10	3/10	3/11	3/11	3/12	3/12	4/12	4/12	4/12	4/12	4/12	4/12	4/12	5/12	5/12	5/12
6		3/11	3/12	3/12	4/13	4/13	4/13	4/13	5/14	5/14	5/14	5/14	5/14	5/14	6/14	6/14
7			3/13	4/13	4/14	5/14	5/14	5/14	5/15	5/15	5/15	6/16	6/16	6/16	6/16	6/16
8				4/14	5/14	5/15	5/15	6/16	6/16	6/16	6/16	6/17	7/17	7/17	7/17	7/17
9					5/15	5/16	6/16	6/16	6/17	7/17	7/18	7/18	7/18	8/18	8/18	8/18
10						6/16	6/17	7/17	7/18	7/18	7/18	8/19	8/19	8/19	8/20	9/20
11							7/17	7/18	7/19	8/19	8/19	8/20	9/20	9/20	9/21	9/21
12								7/19	8/19	8/20	8/20	9/21	9/21	9/21	10/22	10/22
13									8/20	9/20	9/21	9/21	10/22	10/22	10/23	10/23
14										9/21	9/22	10/22	10/23	10/23	11/23	11/24
15											10/22	10/23	11/23	11/24	11/24	12/25
16												11/23	11/24	11/25	12/25	12/25
17													11/25	12/25	12/26	13/26
18														12/26	13/26	13/27
19															13/27	13/27
20																14/28

Table values are for a two-tailed test with $\alpha = 0.05$.

From C. Eisenhart and F. Swed, "Tables for Testing Randomness of Grouping in a Sequence of Alternatives," *The Annals of Statistics,* 14(1943), pp. 66–87. Reprinted by permission.

TABLE 8   CRITICAL VALUES OF SPEARMAN'S RANK CORRELATION COEFFICIENT

Sample Size n	Amount of Probability in One Tail			
	0.05	0.025	0.01	0.005
5	0.900	—	—	—
6	0.829	0.886	0.943	—
7	0.714	0.786	0.893	0.929
8	0.643	0.738	0.833	0.881
9	0.600	0.683	0.783	0.833
10	0.564	0.648	0.745	0.794
11	0.523	0.623	0.736	0.818
12	0.497	0.591	0.703	0.780
13	0.475	0.566	0.673	0.745
14	0.457	0.545	0.646	0.716
15	0.441	0.525	0.623	0.689
16	0.425	0.507	0.601	0.666
17	0.412	0.490	0.582	0.645
18	0.399	0.476	0.564	0.625
19	0.388	0.462	0.549	0.608
20	0.377	0.450	0.534	0.591
21	0.368	0.438	0.521	0.576
22	0.359	0.428	0.508	0.562
23	0.351	0.418	0.496	0.549
24	0.343	0.409	0.485	0.537
25	0.336	0.400	0.475	0.526
26	0.329	0.392	0.465	0.515
27	0.323	0.385	0.456	0.505
28	0.317	0.377	0.448	0.496
29	0.311	0.370	0.440	0.487
30	0.305	0.364	0.432	0.478

Adapted from Newmark J *Statistics and Probability in Modern Life,* 5th ed. Saunders College Publishing, Philadelphia, 1992, p. A.19.

APPENDIX B   Data Disk

A computer diskette is included with the purchase of this book. The disk contains the data sets for 550 exercises, including all data sets for the MINITAB problems. Data sets are stored as ASCII files.

All data files are stored on the diskette in a directory called DATA. The name of each data file identifies the chapter and the problem number. A file name is of the form **CxPy.DAT,** where x is the chapter number and y is the problem number. For example, the data for Problem 11.124 (Problem 124 in Chapter 11) is stored with the file name C11P124.DAT. The data files can be imported into MINITAB by using the **READ** command. To illustrate, consider Problem 11.124 that involves two samples. To store samples 1 and 2 in columns C1 and C2, insert the data disk in your disk drive A and type the following at the MINITAB prompt MTB >.

```
READ 'A:\DATA\C11P124' C1 C2
```

If drive B is used, replace A in the above command with B. Because each file includes MINITAB's default file extension DAT, it does not have to be specified in the **READ** command.

The **READ** command can also be used for data files that pertain to only one sample. For example, Problem 2.19 involves a single sample. To store the values of the sample in column C1, type the following command.

```
READ 'A:\DATA\C2P19' C1
```

The data disk includes the data files listed below and a text file named README.TXT that contains these instructions for importing the files into MINITAB.

Chapter 1:

C1P15.DAT	C1P16.DAT	C1P17.DAT	C1P18.DAT	C1P23.DAT	C1P24.DAT
C1P25.DAT	C1P26.DAT	C1P27.DAT	C1P28.DAT	C1P29.DAT	C1P30.DAT
C1P31.DAT	C1P32.DAT	C1P33.DAT	C1P34.DAT		

Chapter 2:

C2P1.DAT	C2P2.DAT	C2P3.DAT	C2P4.DAT	C2P14.DAT	C2P15.DAT
C2P16.DAT	C2P17.DAT	C2P18.DAT	C2P19.DAT	C2P20.DAT	C2P21.DAT
C2P22.DAT	C2P23.DAT	C2P24.DAT	C2P29.DAT	C2P30.DAT	C2P31.DAT
C2P32.DAT	C2P33.DAT	C2P34.DAT	C2P35.DAT	C2P36.DAT	C2P37.DAT
C2P38.DAT	C2P39.DAT	C2P40.DAT	C2P41.DAT	C2P42.DAT	C2P43.DAT
C2P50.DAT	C2P51.DAT	C2P58.DAT	C2P59.DAT	C2P68.DAT	C2P79.DAT
C2P80.DAT	C2P81.DAT	C2P82.DAT	C2P83.DAT	C2P84.DAT	C2P95.DAT
C2P96.DAT	C2P97.DAT	C2P98.DAT	C2P99.DAT	C2P109.DAT	C2P110.DAT
C2P111.DAT	C2P119.DAT	C2P126.DAT	C2P127.DAT	C2P128.DAT	C2P129.DAT
C2P130.DAT	C2P131.DAT	C2P138.DAT	C2P139.DAT	C2P143.DAT	C2P144.DAT
C2P145.DAT	C2P148.DAT	C2P149.DAT	C2P150.DAT	C2P151.DAT	C2P152.DAT
C2P153.DAT	C2P154.DAT				

Chapter 3:

C3P21.DAT	C3P23.DAT	C3P24.DAT	C3P25.DAT	C3P26.DAT	C3P28.DAT
C3P30.DAT	C3P31.DAT	C3P32.DAT	C3P33.DAT	C3P34.DAT	C3P35.DAT
C3P36.DAT	C3P37.DAT	C3P47.DAT	C3P48.DAT	C3P49.DAT	C3P50.DAT
C3P51.DAT	C3P52.DAT	C3P53.DAT	C3P54.DAT	C3P55.DAT	C3P56.DAT
C3P57.DAT	C3P66.DAT	C3P67.DAT	C3P68.DAT	C3P69.DAT	C3P70.DAT
C3P71.DAT	C3P72.DAT	C3P73.DAT			

Chapter 4:
C4P95.DAT

Chapter 5:

C5P46.DAT	C5P49.DAT	C5P50.DAT	C5P52.DAT	C5P53.DAT	C5P73.DAT
C5P75.DAT	C5P107.DAT	C5P108.DAT			

Chapter 9:

C9P20.DAT	C9P21.DAT	C9P22.DAT	C9P23.DAT	C9P24.DAT	C9P25.DAT
C9P71.DAT	C9P77.DAT	C9P80.DAT	C9P81.DAT	C9P82.DAT	C9P83.DAT
C9P84.DAT	C9P85.DAT	C9P154.DAT	C9P155.DAT	C9P162.DAT	C9P163.DAT
C9P170.DAT	C9P171.DAT	C9P182.DAT	C9P183.DAT	C9P184.DAT	C9P190.DAT
C9P191.DAT	C9P192.DAT	C9P193.DAT	C9P194.DAT		

Chapter 10:

C10P27.DAT	C10P32.DAT	C10P35.DAT	C10P36.DAT	C10P37.DAT	C10P38.DAT
C10P42.DAT	C10P43.DAT	C10P50.DAT	C10P51.DAT	C10P54.DAT	C10P55.DAT
C10P56.DAT	C10P57.DAT	C10P79.DAT	C10P83.DAT	C10P95.DAT	C10P96.DAT
C10P100.DAT	C10P101.DAT	C10P102.DAT	C10P119.DAT	C10P120.DAT	C10P130.DAT
C10P131.DAT	C10P132.DAT	C10P136.DAT			

Chapter 11:

C11P22.DAT	C11P24.DAT	C11P30.DAT	C11P32.DAT	C11P34.DAT	C11P37.DAT
C11P38.DAT	C11P39.DAT	C11P40.DAT	C11P41.DAT	C11P42.DAT	C11P45.DAT
C11P46.DAT	C11P47.DAT	C11P48.DAT	C11P49.DAT	C11P50.DAT	C11P51.DAT
C11P52.DAT	C11P54.DAT	C11P55.DAT	C11P56.DAT	C11P57.DAT	C11P58.DAT
C11P59.DAT	C11P60.DAT	C11P61.DAT	C11P62.DAT	C11P63.DAT	C11P122.DAT
C11P124.DAT	C11P129.DAT	C11P130.DAT	C11P140.DAT	C11P141.DAT	C11P154.DAT
C11P155.DAT	C11P156.DAT	C11P157.DAT	C11P158.DAT	C11P159.DAT	C11P162.DAT

Chapter 12:

C12P21.DAT	C12P22.DAT	C12P23.DAT	C12P24.DAT	C12P25.DAT	C12P26.DAT
C12P27.DAT	C12P28.DAT	C12P29.DAT	C12P30.DAT	C12P31.DAT	C12P32.DAT
C12P33.DAT	C12P34.DAT	C12P35.DAT	C12P36.DAT	C12P37.DAT	C12P38.DAT
C12P39.DAT	C12P40.DAT	C12P41.DAT	C12P44.DAT	C12P51.DAT	C12P52.DAT
C12P53.DAT	C12P54.DAT	C12P55.DAT	C12P56.DAT	C12P57.DAT	C12P58.DAT
C12P59.DAT	C12P60.DAT	C12P61.DAT	C12P62.DAT	C12P63.DAT	C12P64.DAT
C12P69.DAT	C12P70.DAT	C12P73.DAT	C12P74.DAT	C12P81.DAT	C12P82.DAT
C12P83.DAT	C12P85.DAT				

Chapter 13:

C13P7.DAT	C13P8.DAT	C13P9.DAT	C13P10.DAT	C13P11.DAT	C13P14.DAT
C13P15.DAT	C13P16.DAT	C13P17.DAT	C13P18.DAT	C13P19.DAT	C13P20.DAT
C13P21.DAT	C13P22.DAT	C13P23.DAT	C13P24.DAT	C13P25.DAT	C13P26.DAT
C13P27.DAT	C13P28.DAT	C13P29.DAT	C13P30.DAT	C13P31.DAT	C13P32.DAT
C13P33.DAT	C13P34.DAT	C13P35.DAT	C13P36.DAT	C13P37.DAT	C13P40.DAT
C13P41.DAT	C13P42.DAT	C13P43.DAT	C13P44.DAT	C13P45.DAT	C13P54.DAT
C13P55.DAT	C13P56.DAT	C13P57.DAT	C13P58.DAT	C13P59.DAT	C13P60.DAT
C13P61.DAT	C13P62.DAT	C13P63.DAT	C13P64.DAT	C13P65.DAT	C13P66.DAT
C13P67.DAT	C13P68.DAT	C13P69.DAT	C13P70.DAT	C13P71.DAT	C13P72.DAT
C13P73.DAT	C13P74.DAT	C13P75.DAT	C13P81.DAT	C13P83.DAT	C13P84.DAT
C13P85.DAT	C13P86.DAT	C13P87.DAT	C13P88.DAT	C13P89.DAT	C13P90.DAT
C13P91.DAT	C13P92.DAT	C13P93.DAT	C13P97.DAT	C13P98.DAT	C13P99.DAT
C13P100.DAT	C13P101.DAT	C13P102.DAT	C13P103.DAT	C13P104.DAT	C13P105.DAT
C13P106.DAT	C13P107.DAT	C13P108.DAT	C13P109.DAT		

Chapter 14:

C14P6.DAT	C14P7.DAT	C14P8.DAT	C14P9.DAT	C14P12.DAT	C14P13.DAT
C14P14.DAT	C14P15.DAT	C14P16.DAT	C14P17.DAT	C14P18.DAT	C14P19.DAT
C14P20.DAT	C14P21.DAT	C14P23.DAT	C14P25.DAT	C14P26.DAT	C14P27.DAT
C14P28.DAT	C14P29.DAT	C14P30.DAT	C14P31.DAT	C14P32.DAT	C14P33.DAT
C14P34.DAT	C14P35.DAT	C14P39.DAT	C14P40.DAT	C14P41.DAT	C14P42.DAT
C14P43.DAT	C14P44.DAT	C14P45.DAT	C14P46.DAT	C14P47.DAT	C14P48.DAT
C14P49.DAT	C14P50.DAT	C14P51.DAT	C14P52.DAT	C14P55.DAT	C14P56.DAT
C14P57.DAT	C14P58.DAT	C14P59.DAT	C14P60.DAT	C14P61.DAT	C14P62.DAT
C14P63.DAT	C14P64.DAT	C14P65.DAT	C14P66.DAT	C14P67.DAT	

Chapter 15:

C15P1.DAT	C15P2.DAT	C15P3.DAT	C15P4.DAT	C15P5.DAT	C15P6.DAT
C15P7.DAT	C15P8.DAT	C15P9.DAT	C15P10.DAT	C15P11.DAT	C15P12.DAT
C15P13.DAT	C15P14.DAT	C15P15.DAT	C15P16.DAT	C15P17.DAT	C15P18.DAT
C15P19.DAT	C15P20.DAT	C15P21.DAT	C15P22.DAT	C15P23.DAT	C15P24.DAT
C15P25.DAT	C15P26.DAT	C15P27.DAT	C15P28.DAT	C15P29.DAT	C15P30.DAT
C15P31.DAT	C15P32.DAT	C15P33.DAT	C15P34.DAT	C15P35.DAT	C15P36.DAT
C15P37.DAT	C15P38.DAT	C15P39.DAT	C15P40.DAT	C15P41.DAT	C15P42.DAT
C15P43.DAT	C15P44.DAT	C15P45.DAT	C15P46.DAT	C15P47.DAT	C15P48.DAT
C15P49.DAT	C15P50.DAT	C15P51.DAT	C15P52.DAT	C15P53.DAT	C15P54.DAT
C15P55.DAT	C15P56.DAT	C15P57.DAT	C15P58.DAT	C15P59.DAT	C15P60.DAT
C15P61.DAT	C15P62.DAT	C15P63.DAT	C15P64.DAT	C15P65.DAT	C15P67.DAT
C15P70.DAT	C15P77.DAT	C15P78.DAT	C15P79.DAT	C15P80.DAT	C15P84.DAT
C15P85.DAT	C15P86.DAT	C15P87.DAT	C15P88.DAT	C15P89.DAT	C15P90.DAT
C15P91.DAT	C15P92.DAT	C15P93.DAT	C15P94.DAT	C15P95.DAT	C15P98.DAT
C15P99.DAT	C15P100.DAT	C15P101.DAT	C15P102.DAT	C15P103.DAT	C15P104.DAT
C15P105.DAT	C15P106.DAT	C15P107.DAT	C15P108.DAT	C15P109.DAT	C15P110.DAT
C15P111.DAT	C15P112.DAT	C15P113.DAT	C15P114.DAT	C15P115.DAT	C15P116.DAT
C15P117.DAT	C15P118.DAT	C15P119.DAT	C15P120.DAT	C15P121.DAT	C15P122.DAT

Chapter 16:

C16P13.DAT	C16P14.DAT	C16P15.DAT	C16P16.DAT	C16P17.DAT	C16P33.DAT
C16P34.DAT	C16P35.DAT	C16P36.DAT	C16P37.DAT	C16P38.DAT	C16P53.DAT
C16P54.DAT	C16P55.DAT	C16P56.DAT	C16P57.DAT	C16P58.DAT	C16P61.DAT
C16P62.DAT	C16P69.DAT	C16P70.DAT	C16P71.DAT	C16P76.DAT	C16P77.DAT
C16P78.DAT	C16P79.DAT	C16P80.DAT	C16P81.DAT	C16P100.DAT	C16P101.DAT
C16P102.DAT	C16P111.DAT	C16P112.DAT	C16P113.DAT	C16P114.DAT	C16P115.DAT
C16P116.DAT	C16P117.DAT	C16P118.DAT			

APPENDIX C Bibliography

Bell, E. T. *Men of Mathematics.* Simon and Schuster, New York, 1937.

Cochran, W. G. *Sampling Techniques.* Wiley, New York, 1963.

Cochran, W. G. and Cox, G. M. *Experimental Designs,* 2d ed. Wiley, New York, 1957.

Devore, J. and Peck, R. *Statistics: The Exploration and Analysis of Data,* West, St. Paul, MN, 1986.

Draper, N. R. and Smith, H. *Applied Regression Analysis,* 2d ed. Wiley, New York, 1981.

Feller, W. *An Introduction to Probability Theory and Its Applications,* Vol. 1, 3d ed. Wiley, New York, 1968.

Freedman et al. *Statistics,* 2d ed. Norton, New York, 1991.

Freund, J. E. *Mathematical Statistics,* 5th ed. Prentice Hall, Englewood Cliffs, NJ, 1992.

Freund, J. E. and Simon, G. A. *Statistics: A First Course,* 5th ed. Prentice Hall, Englewood Cliffs, NJ, 1991.

Goldhaber, M. and Lehman, J. *Crisis Evacuation During the Three Mile Island Nuclear Accident: The TMI Population Registry,* Pennsylvania Department of Health, August 1982.

Hacking, I. *The Emergence of Probability,* Cambridge University Press, New York, 1975.

Hald, A. *A History of Probability and Statistics and Their Applications Before 1750,* Wiley, New York, 1990.

Hicks, C. R. *Fundamental Concepts in the Design of Experiments,* 3d ed. Holt, Rinehart, and Winston, New York, 1982.

Hogg, R. V. and Craig, A. T. *Introduction to Mathematical Statistics,* 4th ed. Macmillan, New York, 1986.

Johnson, R. *Elementary Statistics,* 5th ed. PWS-KENT, Boston, 1988.

Khazanie, R. *Elementary Statistics in a World of Applications,* 3d ed. Scott, Foresman, Glenview, IL, 1990.

Kish, L. *Survey Sampling.* Wiley, New York, 1965.

Larsen, R. J. and Marx, M. L. *Statistics,* Prentice Hall, Englewood Cliffs, NJ, 1990.

McClave, J. T. and Dietrich, F. H. *Statistics,* 5th ed. Dellen, San Francisco, 1991.

Mendenhall, W. and Beaver, R. *Introduction to Probability and Statistics,* 8th ed. PWS-KENT, Boston, 1991.

Mendenhall, W., Wackerly, D., and Scheaffer, R. L. *Mathematical Statistics with Applications,* 4th ed. PWS-KENT, Boston, 1990.

Meyers, R. H. *Classical and Modern Regression with Applications,* 2d ed. PWS-KENT, Boston, 1990.

Minitab Inc., *MINITAB Reference Manual,* Release 8, PC Version, Minitab, Inc., State College, PA, 1991.

Minitab Inc., *MINITAB Reference Manual,* Release 11 for Windows, Minitab, Inc., State College, PA, 1996.

Mood, A. M., Graybill, F. A., and Boes, D. C. *Introduction to the Theory of Statistics,* 3d ed. McGraw-Hill, New York, 1974.

Moore, D. S. and McCabe, G. P. *Introduction to the Practice of Statistics,* Freeman, New York, 1989.

Mosteller, F. and Rourke, R. E. *Sturdy Statistics,* Addison-Wesley, Reading, MA, 1973.

National Bureau of Standards Handbook 91, *Experimental Statistics,* U.S. Government Printing Office, Washington, D.C., 1963.

Neter, J., Wasserman, W., and Whitmore, G. A. *Applied Statistics,* 3d ed. Allyn and Bacon, Boston, 1987.

Newmark, J. *Statistics and Probability in Modern Life,* 5th ed. Saunders College Publishing, Philadelphia, 1992.

Ott, L. *An Introduction to Statistical Methods and Data Analysis,* 3d ed. PWS-KENT, Boston, 1988.

Salvia, A. A. *Introduction to Statistics,* Saunders College Publishing, Philadelphia, 1990.

Scheaffer, R. L. and Farber, E. *The Student Edition of MINITAB, Release 8,* Addison-Wesley, Reading, MA, 1992.

Scheaffer, R. L., Mendenhall, W., and Ott, L. *Elementary Survey Sampling,* 4th ed. PWS-KENT, Boston, 1990.

Snedecor, G. W. and Cochran, W. G. *Statistical Methods,* 8th ed. Iowa State University Press, Ames, IA, 1989.

Stigler, S. M. *The History of Statistics: The Measurement of Uncertainty Before 1900,* Belkamp Press of Harvard University, Cambridge, MA, 1986.

Tanur, J. M. et al. eds. *Statistics: A Guide to the Unknown,* 3d ed. Wadsworth/Brooks-Cole, Belmont, CA, 1989.

Triola, M. F. *Elementary Statistics,* 4th ed. Addison-Wesley, Reading, MA, 1989.

Tukey, J. W. *Exploratory Data Analysis,* Addison-Wesley, Reading, MA, 1977.

U.S. Bureau of the Census, *Statistical Abstract of the United States: 1995,* U.S. Government Printing Office, Washington, D.C., 1995.

Velleman, P. F. and Hoaglin, D. C. *Applications, Basics, and Computing of Exploratory Data Analysis,* PWS-KENT, Boston, 1981.

Walpole, R. E. and Myers, R. H. *Probability and Statistics for Engineers and Scientists,* 3d ed. Macmillan, New York, 1985.

Weiss, N. A. *Elementary Statistics,* Addison-Wesley, Reading, MA, 1989.

Weiss, N. A. and Hassett, M. J. *Introductory Statistics,* 3d ed. Addison-Wesley, Reading, MA, 1991.

Williams, B. *A Sampler on Sampling,* Wiley, New York, 1978.

ANSWERS TO ODD-NUMBERED SECTION EXERCISES AND ALL REVIEW EXERCISES

CHAPTER 13
Sections 13.1 and 13.2

13.1 42.5714

13.3 37.48

13.5 a. 11.6
b. 1,012.8
c. 16.7647
d.

Source	df	SS	MS	F
Treatments	2	1,012.8	506.4	30.21
Error	17	285.0	16.7647	
Total	19	1,297.8		

13.7 a. 336
b. 336
c. 46
d. 46
e.

Source	df	SS	MS	F
Treatments	2	336	168	32.87
Error	9	46	5.111	
Total	11	382		

Sections 13.3 and 13.4

13.15 a.

The sample means appear to differ significantly.

b. $H_0: \mu_1 = \mu_2 = \mu_3$
H_a: Not all the means are equal.

Source	df	SS	MS	F
Treatments	2	146.92	73.46	15.74
Error	9	42.00	4.667	
Total	11	188.92		

RR: $F > 4.26$
Reject H_0

13.9

Source	df	SS	MS	F
Treatments	2	157.11	78.55	9.67
Error	10	81.20	8.12	
Total	12	238.31		

13.11

Source	df	SS	MS	F
Treatments	4	752.58	188.15	9.91
Error	19	360.75	18.99	
Total	23	1,113.33		

13.13 a.

Source	df	SS	MS	F
Treatments	7	21.7	3.1	1.82
Error	28	47.6	1.7	
Total	35	69.3		

b. 8
c. 36

13.17 a. 18 ± 2.19
　　　b. 5 ± 3.57
13.19 p-value < 0.005
13.21 $H_0: \mu_1 = \mu_2$
　　　$H_a: \mu_1 \neq \mu_2$

Source	df	SS	MS	F
Treatments	1	59.10	59.10	10.82
Error	20	109.23	5.462	
Total	21	168.33		

　　RR: $F > 4.35$
　　Reject H_0
13.23 $\$3.29 \pm \$2.85$
13.25 p-value < 0.005
13.27 2.475 ± 0.852
13.29 4.2 ± 0.644
13.31 $H_0: \mu_1 = \mu_2 = \mu_3$
　　　$H_a:$ Not all the means are equal.

Source	df	SS	MS	F
Treatments	2	11.66	5.83	2.37
Error	20	49.21	2.46	
Total	22	60.87		

　　RR: $F > 3.49$
　　Fail to reject H_0 (no)

M▶ 13.41 MTB > # FROM THE MINITAB OUTPUT, MEAN (2) = 87.60, MEAN(4) = 79.60,
　　　MTB > # MSE = 12.9 WITH DF = 36. FROM THESE RESULTS, WE OBTAIN:
　　　MTB > # 95% CI FOR MU(2) – MU(4) IS 8.00 +/– 3.15

M▶ 13.43 ANALYSIS OF VARIANCE

SOURCE	DF	SS	MS	F	p
FACTOR	3	51.123	17.041	29.82	0.000
ERROR	20	11.430	0.572		
TOTAL	23	62.553			

M▶ 13.45 ANALYSIS OF VARIANCE

SOURCE	DF	SS	MS	F	p
FACTOR	3	14.149	4.716	12.68	0.000
ERROR	19	7.065	0.372		
TOTAL	22	21.215			

13.33 -1.68 ± 1.40
13.35 p-value < 0.005
13.37 $H_0: \mu_1 = \mu_3 = \mu_4$
　　　$H_a:$ Not all the means are equal.

Source	df	SS	MS	F
Treatments	2	3.210	1.605	3.53
Error	14	6.365	0.455	
Total	16	9.575		

　　RR: $F > 3.74$
　　Fail to reject H_0 (no)
13.39 $-\$1,554 \pm \$858$

Section 13.5

13.47 The randomized block design would be more appropriate, because the analysis accounts for variation due to differences between the blocks (cars).

13.49 a. 6
b. 9
c.

Source	df	SS	MS	F
Treatments	5	15.65	3.13	1.98
Blocks	8	23.12	2.89	1.83
Error	40	63.20	1.58	
Total	53	101.97		

13.51 H_0: $\mu_A = \mu_B = \mu_C = \mu_D = \mu_E = \mu_F$
H_a: Not all the treatment means are equal.
$F = 1.98$ from the ANOVA table
RR: $F > 2.45$
Fail to reject H_0 (no)

13.53 H_0: $\mu_1 = \mu_2 = \mu_3 = \mu_4 = \mu_5 = \mu_6 = \mu_7 = \mu_8 = \mu_9$
H_a: Not all the block means are equal.
$F = 1.83$ from the ANOVA table
RR: $F > 2.18$
Fail to reject H_0 (no)

13.55 H_0: $\mu_A = \mu_B = \mu_C = \mu_D$
H_a: Not all the treatment means are equal.

Source	df	SS	MS	F
Treatments	3	15.79375	5.26458	101.32
Blocks	7	5.09375	0.72768	14.00
Error	21	1.09125	0.05196	
Total	31	21.97875		

$F = 101.32$ from the ANOVA table
RR: $F > 3.07$
Reject H_0 (yes)

13.57 H_0: $\mu_1 = \mu_2 = \mu_3 = \mu_4 = \mu_5 = \mu_6 = \mu_7 = \mu_8$
H_a: Not all the block means are equal.
$F = 14.00$ from the ANOVA table
RR: $F > 2.02$
Reject H_0 (yes)

13.59 0.65 ± 0.196

13.61 p-value < 0.005; p-value < 0.005

13.63 p-value < 0.005

13.65 p-value > 0.10

 13.67
```
Analysis of Variance for C1
Source      DF        SS         MS        F      P
C2           3    45.282     15.094    11.44  0.001
C3           4   161.063     40.266    30.52  0.000
Error       12    15.833      1.319
Total       19   222.178
MTB > # REJECT Ho SINCE P-VALUE = 0.001 < 0.01.
MTB > # THERE IS SUFFICIENT EVIDENCE OF A DIFFERENCE.
```

13.69

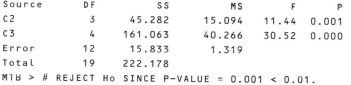

Main Effects Plot - Means for Dry Time

13.71
```
Test of μ =   0.0 vs μ not =   0.0
Variable     N      Mean    StDev   SE Mean        T    P-Value
C6          29     196.0    280.3      52.1     3.77     0.0008
MTB > # REJECT Ho SINCE P-VALUE = 0.0008 < 0.05.
MTB > # THERE IS A SIGNIFICANT DIFFERENCE.
MTB > # T-VALUE EQUALS SQUARE ROOT OF F-VALUE.
```

Review Exercises

13.72 $H_0: \mu_1 = \mu_2 = \mu_3$
H_a: Not all the means are equal.

Source	df	SS	MS	F
Treatments	2	5,942	2,971	3.74
Error	18	14,286.57	793.70	
Total	20	20,228.57		

RR: $F > 6.01$
Fail to reject H_0

13.73 $0.025 < p$-value < 0.05

13.74 $\$934.43 \pm \$18.46$

13.75 $-\$38.71 \pm \$26.11$

13.76 19.13

13.77 If $n_1 = n_2 = \cdots = n_k = n$, then

$$MSE = \frac{(n-1)s_1^2 + (n-1)s_2^2 + \cdots + (n-1)s_k^2}{(n-1) + (n-1) + \cdots + (n-1)}$$

$$= \frac{(n-1)(s_1^2 + s_2^2 + \cdots + s_k^2)}{k(n-1)}$$

$$= \frac{s_1^2 + s_2^2 + \cdots + s_k^2}{k}$$

13.81 a.

There appears to be a significant difference in the sample means.

b. $H_0: \mu_1 = \mu_2 = \mu_3 = \mu_4$
H_a: Not all the means are equal.

Source	df	SS	MS	F
Treatments	3	75.05	25.017	8.55
Error	14	40.95	2.925	
Total	17	116		

RR: $F > 3.34$
Reject H_0

13.78 a. 991.2
b. 21.423
c.

Source	df	SS	MS	F
Treatments	3	991.2	330.4	15.42
Error	26	557	21.423	
Total	29	1,548.2		

d. $H_0: \mu_1 = \mu_2 = \mu_3 = \mu_4$
H_a: Not all the means are equal.
$F = 15.42$
RR: $F > 2.98$
Reject H_0

13.79 39 ± 2.497

13.80 12 ± 4.404

13.82 $H_0: \mu_1 = \mu_2 = \mu_3 = \mu_4 = \mu_5 = \mu_6 = \mu_7 = \mu_8$
H_a: Not all the means are equal.

Source	df	SS	MS	F
Treatments	7	493	70.43	6.80
Error	22	228	10.364	
Total	29	721		

RR: $F > 3.59$
Reject H_0

13.83 $H_0: \mu_1 = \mu_2 = \mu_3$
H_a: Not all the means are equal.

Source	df	SS	MS	F
Treatments	2	244.33	122.17	17.99
Error	21	142.63	6.792	
Total	23	386.96		

RR: $F > 3.47$
Reject H_0 (yes)

13.84 p-value < 0.005

13.85 $65.875 + 1.917$

13.86 7.75 ± 2.710

13.87 $H_0: \mu_1 = \mu_2 = \mu_3 = \mu_4$
H_a: Not all the means are equal.

Source	df	SS	MS	F
Treatments	3	107.28	35.76	2.92
Error	17	208.53	12.266	
Total	20	315.81		

RR: $F > 3.20$
Fail to reject H_0 (no)

13.88 $0.05 < p$-value < 0.10

13.89 6.4 ± 3.85

13.90 $H_0: \mu_1 = \mu_2 = \mu_3$
H_a: Not all the means are equal.

Source	df	SS	MS	F
Treatments	2	6.729	3.365	7.90
Error	12	5.108	0.426	
Total	14	11.837		

RR: $F > 3.89$
Reject H_0 (yes)

13.91 $H_0: \mu_1 = \mu_2 = \mu_3 = \mu_4$
H_a: Not all the treatment means are equal.

Source	df	SS	MS	F
Treatments	3	389.361	129.787	3.40
Blocks	3	71,627.442	23,875.814	625.94
Error	9	343.300	38.144	
Total	15	72,360.103		

$F = 3.40$ from the ANOVA table
RR: $F > 2.81$
Reject H_0 (yes)

13.92 $0.05 < p$-value < 0.10

13.93 $\$11.54 \pm \$8.00$

13.94 a. 4
b. 10
c.

Source	df	SS	MS	F
Treatments	3	13.71	4.57	2.51
Blocks	9	37.08	4.12	2.26
Error	27	49.14	1.82	
Total	39	99.93		

13.95 $H_0: \mu_1 = \mu_2 = \mu_3 = \mu_4$
H_a: Not all the treatment means are equal.
$F = 2.51$ from the ANOVA table
RR: $F > 2.96$
Fail to reject H_0 (no)

13.96 $H_0: \mu_1 = \mu_2 = \mu_3 = \cdots = \mu_{10}$
H_a: Not all the block means are equal.
$F = 2.26$ from the ANOVA table
RR: $F > 1.87$
Reject H_0 (yes)

13.97 $H_0: \mu_A = \mu_B = \mu_C = \mu_D$
H_a: Not all the treatment means are equal.

Source	df	SS	MS	F
Treatments	3	132.55	44.1833	8.46
Blocks	4	27.3	6.825	1.31
Error	12	62.7	5.225	
Total	19	222.55		

$F = 8.46$ from the ANOVA table
RR: $F > 3.49$
Reject H_0 (yes)

13.98 $H_0: \mu_1 = \mu_2 = \mu_3 = \mu_4 = \mu_5$
H_a: Not all the block means are equal.
$F = 1.31$ from the ANOVA table
RR: $F > 2.48$
Fail to reject H_0 (no)

13.99 $H_0: \mu_A = \mu_B = \mu_C$
H_a: Not all the treatment means are equal.

Source	df	SS	MS	F
Treatments	2	2,435.167	1,217.584	4.54
Blocks	3	7,070.917	2,356.972	8.78
Error	6	1,610.833	268.472	
Total	11	11,116.917		

$F = 4.54$ from the ANOVA table
RR: $F > 5.14$
Fail to reject H_0 (no)

13.100 $0.05 < p\text{-value} < 0.10$

13.101 2.75 ± 22.51

M▶ 13.102 ANALYSIS OF VARIANCE

```
SOURCE     DF       SS        MS        F        p
FACTOR      2    206.10    103.05    19.39    0.000
ERROR      30    159.41      5.31
TOTAL      32    365.52
```

INDIVIDUAL 95 PCT CI'S FOR MEAN
BASED ON POOLED STDEV

```
LEVEL    N     MEAN     STDEV   ------+---------+---------+---------+
C1      12    38.292    1.827                    (----*-----)
C2      10    35.000    2.838   (-----*-----)
C3      11    41.273    2.240                              (-----*-----)
                                ------+---------+---------+---------+
POOLED STDEV =    2.305         35.0      37.5      40.0      42.5
```

MTB > # REJECT Ho SINCE P-VALUE < 0.01. SUFF. EVIDENCE OF A DIFFERENCE.
MTB > # PART B: IT APPEARS THAT MU(2) DIFFERS FROM MU(1) AND MU(3).
MTB > # ALSO MU(1) MAY DIFFER FROM MU(3).

M▶ 13.103 MTB > # FROM THE MINITAB OUTPUT, MEAN(1) = 38.29, MEAN(3) = 41.27,
MTB > # MSE = 5.31 WITH DF = 30. FROM THESE RESULTS, WE OBTAIN:
MTB > # 95% CI FOR MU(1) - MU(3) IS -2.98 +/- 1.89

M▶ 13.104 ANALYSIS OF VARIANCE

```
SOURCE     DF      SS        MS        F        p
FACTOR      2    2556.8    1278.4    29.90    0.000
ERROR      45    1924.2      42.8
TOTAL      47    4481.0
```

INDIVIDUAL 95 PCT CI'S FOR MEAN
BASED ON POOLED STDEV

```
LEVEL    N     MEAN     STDEV   -----+---------+---------+---------+-
C1      16    79.187    7.626                   (----*----)
C2      16    70.000    6.673   (----*----)
C3      16    87.875    5.058                             (----*---)
                                -----+---------+---------+---------+-
POOLED STDEV =    6.539         70.0      77.0      84.0      91.0
```

MTB > # REJECT Ho SINCE P-VALUE < 0.05. SUFF. EVIDENCE OF A DIFFERENCE.

M▶ 13.105 ANALYSIS OF VARIANCE

SOURCE	DF	SS	MS	F	p
FACTOR	1	603.8	603.8	14.42	0.001
ERROR	30	1256.2	41.9		
TOTAL	31	1860.0			

INDIVIDUAL 95 PCT CI'S FOR MEAN
BASED ON POOLED STDEV

```
 LEVEL     N      MEAN     STDEV  ---------+---------+---------+-------
 C1        16    79.187    7.626   (-----*------)
 C3        16    87.875    5.058                    (------*-----)
                                   ---------+---------+---------+-------
 POOLED STDEV =    6.471                    80.0      85.0      90.0
```

MTB > # REJECT Ho SINCE P-VALUE < 0.05. SUFF. EVIDENCE OF A DIFFERENCE.

M▶ 13.106 Analysis of Variance for C1

Source	DF	SS	MS	F	P
C2	3	0.3335	0.1112	20.18	0.000
C3	19	63.7624	3.3559	609.00	0.000
Error	57	0.3141	0.0055		
Total	79	64.4101			

MTB > # REJECT Ho SINCE P-VALUE = 0.000 < 0.05.
MTB > # THERE IS SUFFICIENT EVIDENCE OF A DIFFERENCE.

M▶ 13.107 Analysis of Variance for C1

Source	DF	SS	MS	F	P
C2	2	0.3162	0.1581	23.25	0.000
C3	19	48.6459	2.5603	376.53	0.000
Error	38	0.2584	0.0068		
Total	59	49.2205			

MTB > # REJECT Ho SINCE P-VALUE = 0.000 < 0.05.
MTB > # THERE IS SUFFICIENT EVIDENCE OF A DIFFERENCE.

M▶ 13.108 Analysis of Variance for C1

Source	DF	SS	MS	F	P
C2	1	84.05	84.05	16.64	0.003
C3	9	3273.05	363.67	72.01	0.000
Error	9	45.45	5.05		
Total	19	3402.55			

MTB > # REJECT Ho SINCE P-VALUE = 0.003 < 0.01.
MTB > # THERE IS SUFFICIENT EVIDENCE OF A DIFFERENCE.

M▶ 13.109

Main Effects Plot - Means for Grade

CHAPTER 14
Sections 14.1 and 14.2

14.1 A deterministic model assumes that the phenomenon of interest can be predicted precisely, while a probabilistic model allows for variation in the predictions.

14.3 A probabilistic model produces different values of y for the same x because it contains a random error component. A deterministic model has no random component and thus gives the same y-value for a given value of x.

14.5 $\hat{y} = 50 + 45x$
Deterministic model

14.7

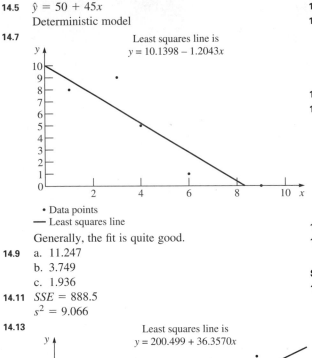

Least squares line is
$y = 10.1398 - 1.2043x$

• Data points
— Least squares line

Generally, the fit is quite good.

14.9 a. 11.247
  b. 3.749
  c. 1.936

14.11 $SSE = 888.5$
$s^2 = 9.066$

14.13

Least squares line is
$y = 200.499 + 36.3570x$

• Data points
— Least squares line

The least squares line suggests a linear model.

14.15 $s = 17.60; 2s = 35.2$

Sections 14.3 and 14.4

14.17 a. -0.910
  b. 0.827
  c. The least squares line accounts for 82.7 percent of the total variation in the y-values.

14.19 $-0.934, 0.872$

14.21 $H_0: \beta_1 = 0$
$H_a: \beta_1 < 0$
$t = -7.37$
RR: $t < -2.896$
Reject H_0 (yes)

14.23 p-value < 0.005

14.25 a. -0.872
  b. y tends to decrease as x increases.
  c. 0.760
  d. The least squares line accounts for 76.0 percent of the total variation in the y-values.

14.27 -0.611 ± 0.519

14.29 a. 0.847
  The least squares line accounts for 84.7 percent of the total variation in the gold prices.
  b. $H_0: \beta_1 = 0$
  $H_a: \beta_1 \neq 0$
  $t = 6.23$
  RR: $t < -3.499, t > 3.499$
  Reject H_0 (yes)

14.31 p-value < 0.01

14.33 -20.995 ± 8.318

Section 14.5

14.35 a.

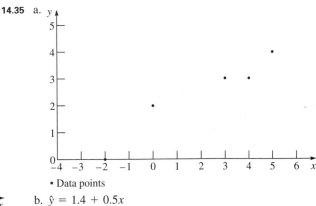

• Data points

  b. $\hat{y} = 1.4 + 0.5x$
  c. 0.924
  The least squares line accounts for 92.4 percent of the total variation in the y-values.

d. $H_0: \beta_1 = 0$
$H_a: \beta_1 \neq 0$
$t = 6.04$
RR: $t < -3.182, t > 3.182$
Reject H_0 (yes)
e. 2.9 ± 0.736
f. 2.9 ± 1.704
g. Widths are 1.472 and 3.408

14.43   a.

• Data points

b. $\hat{y} = 40.678 - 1.396x$
c. 0.654
The least squares line accounts for 65.4 percent of the total variation in the y-values.
d. $H_0: \beta_1 = 0$
$H_a: \beta_1 \neq 0$
$t = -3.89$
RR: $t < -2.306, t > 2.306$
Reject H_0
e. 33.70 ± 7.51

14.37   a. 291.7 ± 2.967
b. Narrower because 21 is closer to $\bar{x}$.
c. 20

14.39 73.19 ± 3.90

14.41 73.19 ± 15.01

Section 14.6

M▶ **14.45**   a. $\hat{y} = 4.57 - 0.0503x$
b. 0.898
The least squares line accounts for 89.8 percent of the total variation in the y-values.
c. $t = -7.86$
p-value $= 0.000$
Yes
d. For $x = 50$, we are 95 percent confident that the long run average years of added life is between 1.8318 and 2.2763.

M▶ **14.47**   a. $r^2 = 3.7901/4.22 = 0.898$
b. $s = 0.2478$; approximately 95 percent of the data points are expected to be within $2s = 0.4956$ of the least squares line.
c. $SSE = 0.4298$; no other line has a smaller value.

M▶ **14.49**   a. $r^2 = 84.7$ percent
The least squares line accounts for 84.7 percent of the total variation in the y-values.
b. $r^2 = 12,034/14,202 = 0.847$

M► **14.51** MTB > # PART A

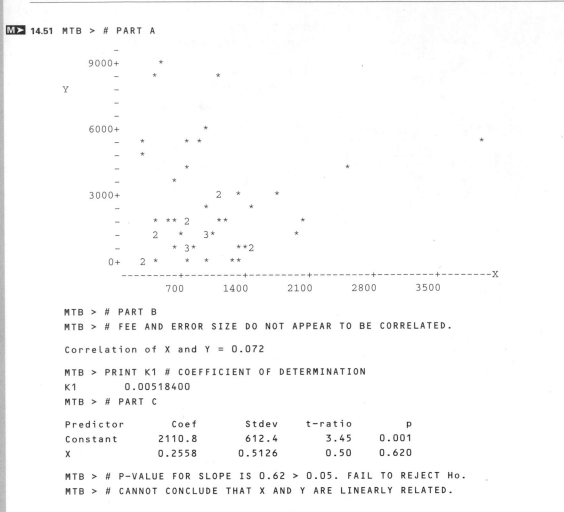

```
MTB > # PART B
MTB > # FEE AND ERROR SIZE DO NOT APPEAR TO BE CORRELATED.

Correlation of X and Y = 0.072

MTB > PRINT K1 # COEFFICIENT OF DETERMINATION
K1        0.00518400
MTB > # PART C

Predictor       Coef        Stdev       t-ratio        p
Constant        2110.8       612.4        3.45       0.001
X               0.2558       0.5126       0.50       0.620

MTB > # P-VALUE FOR SLOPE IS 0.62 > 0.05. FAIL TO REJECT Ho.
MTB > # CANNOT CONCLUDE THAT X AND Y ARE LINEARLY RELATED.
```

Review Exercises

14.52 a.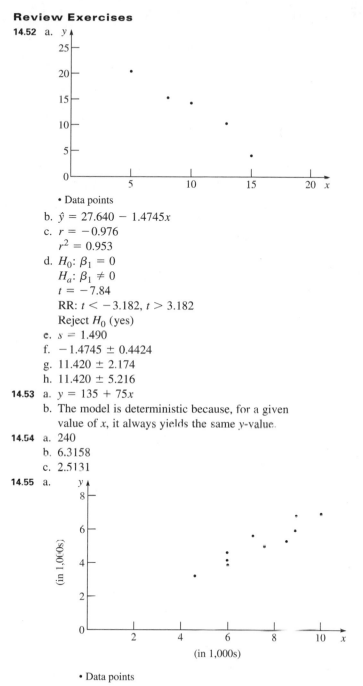

• Data points

b. $\hat{y} = 27.640 - 1.4745x$

c. $r = -0.976$
$r^2 = 0.953$

d. $H_0: \beta_1 = 0$
$H_a: \beta_1 \neq 0$
$t = -7.84$
RR: $t < -3.182$, $t > 3.182$
Reject H_0 (yes)

e. $s = 1.490$

f. -1.4745 ± 0.4424

g. 11.420 ± 2.174

h. 11.420 ± 5.216

14.53 a. $y = 135 + 75x$

b. The model is deterministic because, for a given
value of x, it always yields the same y-value.

14.54 a. 240

b. 6.3158

c. 2.5131

14.55 a.

• Data points

b. $\hat{y} = 179.26 + 0.69367x$

c. 0.879
The least squares line accounts for 87.9 percent of the total variation in the y-values.

14.56 $H_0: \beta_1 = 0$
$H_a: \beta_1 \neq 0$
$t = 7.62$
RR: $t < -2.306, t > 2.306$
Reject H_0
p-value < 0.01

14.57 $4{,}341 \pm 420$

14.58 $4{,}341 \pm 1{,}138$

14.59 a.

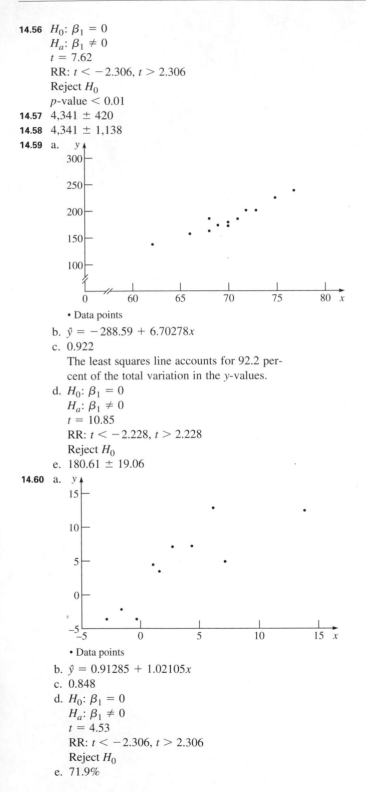

• Data points

b. $\hat{y} = -288.59 + 6.70278x$

c. 0.922
The least squares line accounts for 92.2 percent of the total variation in the y-values.

d. $H_0: \beta_1 = 0$
$H_a: \beta_1 \neq 0$
$t = 10.85$
RR: $t < -2.228, t > 2.228$
Reject H_0

e. 180.61 ± 19.06

14.60 a.

• Data points

b. $\hat{y} = 0.91285 + 1.02105x$

c. 0.848

d. $H_0: \beta_1 = 0$
$H_a: \beta_1 \neq 0$
$t = 4.53$
RR: $t < -2.306, t > 2.306$
Reject H_0

e. 71.9%

14.61 $2.95\% \pm 2.53\%$

14.62 $2.95\% \pm 8.15\%$

M▶ **14.63**   MTB > # PART A

```
y -
  -                                                              *
  -
  -
90+
  -                                              *
  -
  -
  -                                    ***
75+                      *
  -                 *
  -            *                        *
  -               *   *
  -
60+   *
  -        *                       *
  -
    --+---------+  -------+---------+---------+---------+----x
    0.2480    0.2560    0.2640    0.2720    0.2800    0.2880
```

MTB > # PART B
The regression equation is
y = - 114 + 690 x

Predictor	Coef	StDev	T	P
Constant	-114.43	61.54	-1.86	0.088
x	690.4	227.6	3.03	0.010

S = 9.389   R-Sq = 43.4%   R-Sq(adj) = 38.7%

MTB > # PART C: R-SQ = 43.4%.
MTB > # PART D: P-VALUE FOR SLOPE = 0.01 < 0.05. THUS,
MTB > # SUFFICIENT EVIDENCE THAT MODEL PROVIDES USEFUL INFO.

M▶ **14.64**   a. $\hat{y} = -108 + 0.0233x$

b. $r^2 = 85.8$ percent

The least squares line accounts for 85.8 percent of the total variation in sales.

M▶ **14.65** $H_0: \beta_1 = 0$

$H_a: \beta_1 \neq 0$

$t = 12.99$

p-value = 0.000

Reject H_0 (yes)

M▶ **14.66**   For weeks when $x = 7{,}600$, we are 95 percent confident that the long run average of weekly sales is between \$66,626 and \$72,398.

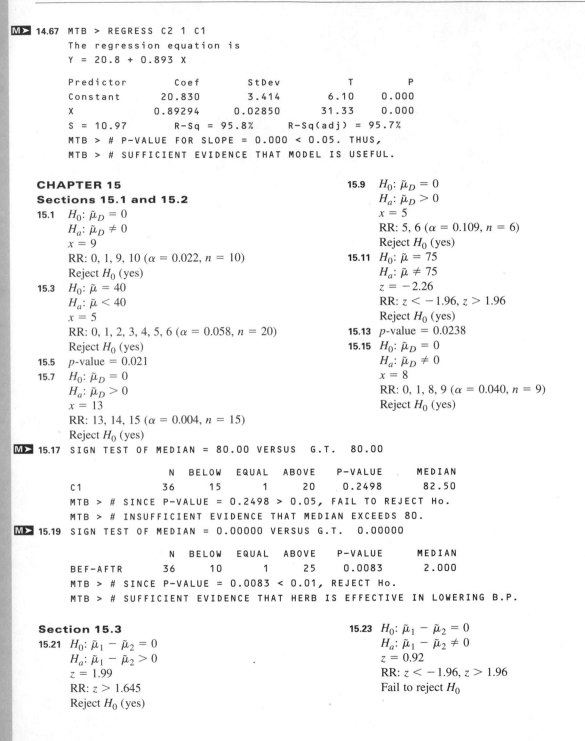

14.67 MTB > REGRESS C2 1 C1

The regression equation is

Y = 20.8 + 0.893 X

Predictor	Coef	StDev	T	P
Constant	20.830	3.414	6.10	0.000
X	0.89294	0.02850	31.33	0.000

S = 10.97   R-Sq = 95.8%   R-Sq(adj) = 95.7%

MTB > # P-VALUE FOR SLOPE = 0.000 < 0.05. THUS,

MTB > # SUFFICIENT EVIDENCE THAT MODEL IS USEFUL.

CHAPTER 15
Sections 15.1 and 15.2

15.1 H_0: $\tilde{\mu}_D = 0$
H_a: $\tilde{\mu}_D \neq 0$
$x = 9$
RR: 0, 1, 9, 10 ($\alpha = 0.022$, $n = 10$)
Reject H_0 (yes)

15.3 H_0: $\tilde{\mu} = 40$
H_a: $\tilde{\mu} < 40$
$x = 5$
RR: 0, 1, 2, 3, 4, 5, 6 ($\alpha = 0.058$, $n = 20$)
Reject H_0 (yes)

15.5 p-value = 0.021

15.7 H_0: $\tilde{\mu}_D = 0$
H_a: $\tilde{\mu}_D > 0$
$x = 13$
RR: 13, 14, 15 ($\alpha = 0.004$, $n = 15$)
Reject H_0 (yes)

15.9 H_0: $\tilde{\mu}_D = 0$
H_a: $\tilde{\mu}_D > 0$
$x = 5$
RR: 5, 6 ($\alpha = 0.109$, $n = 6$)
Reject H_0 (yes)

15.11 H_0: $\tilde{\mu} = 75$
H_a: $\tilde{\mu} \neq 75$
$z = -2.26$
RR: $z < -1.96$, $z > 1.96$
Reject H_0 (yes)

15.13 p-value = 0.0238

15.15 H_0: $\tilde{\mu}_D = 0$
H_a: $\tilde{\mu}_D \neq 0$
$x = 8$
RR: 0, 1, 8, 9 ($\alpha = 0.040$, $n = 9$)
Reject H_0 (yes)

15.17 SIGN TEST OF MEDIAN = 80.00 VERSUS G.T. 80.00

	N	BELOW	EQUAL	ABOVE	P-VALUE	MEDIAN
C1	36	15	1	20	0.2498	82.50

MTB > # SINCE P-VALUE = 0.2498 > 0.05, FAIL TO REJECT Ho.

MTB > # INSUFFICIENT EVIDENCE THAT MEDIAN EXCEEDS 80.

15.19 SIGN TEST OF MEDIAN = 0.00000 VERSUS G.T. 0.00000

	N	BELOW	EQUAL	ABOVE	P-VALUE	MEDIAN
BEF-AFTR	36	10	1	25	0.0083	2.000

MTB > # SINCE P-VALUE = 0.0083 < 0.01, REJECT Ho.

MTB > # SUFFICIENT EVIDENCE THAT HERB IS EFFECTIVE IN LOWERING B.P.

Section 15.3

15.21 H_0: $\tilde{\mu}_1 - \tilde{\mu}_2 = 0$
H_a: $\tilde{\mu}_1 - \tilde{\mu}_2 > 0$
$z = 1.99$
RR: $z > 1.645$
Reject H_0 (yes)

15.23 H_0: $\tilde{\mu}_1 - \tilde{\mu}_2 = 0$
H_a: $\tilde{\mu}_1 - \tilde{\mu}_2 \neq 0$
$z = 0.92$
RR: $z < -1.96$, $z > 1.96$
Fail to reject H_0

15.25 H_0: $\tilde{\mu}_1 - \tilde{\mu}_2 = 0$
H_a: $\tilde{\mu}_1 - \tilde{\mu}_2 \neq 0$
$z = 2.33$
RR: $z < -1.96$, $z > 1.96$
Reject H_0 (yes)
15.27 p-value $= 0.0198$

15.29 H_0: $\tilde{\mu}_1 - \tilde{\mu}_2 = 0$
H_a: $\tilde{\mu}_1 - \tilde{\mu}_2 > 0$
$z = 2.77$
RR: $z > 2.326$
Reject H_0 (yes)
15.31 p-value $= 0.0028$

M▶ 15.33 Mann-Whitney Confidence Interval and Test

```
C1            N =  11     Median =      28.000
C2            N =  14     Median =      22.300
Point estimate for ETA1-ETA2 is       5.350
95.4 pct c.i. for ETA1-ETA2 is (3.501,6.699)
W = 217.0
Test of ETA1 = ETA2  vs.  ETA1  n.e. ETA2 is significant at 0.0001
The test is significant at 0.0001 (adjusted for ties)

MTB > # SINCE P-VALUE = 0.0001 < 0.05, REJECT Ho.
MTB > # SUFFICIENT EVIDENCE THAT THE MEDIANS DIFFER.
```

M▶ 15.35 Mann-Whitney Confidence Interval and Test
```
C1            N =  16     Median =      184.00
C2            N =  14     Median =      147.00
Point estimate for ETA1-ETA2 is       30.00
95.2 Percent CI for ETA1-ETA2 is (4.99,62.99)
W = 304.0
Test of ETA1 = ETA2  vs  ETA1 not = ETA2 is significant at 0.0210
The test is significant at 0.0208 (adjusted for ties)

MTB > # SINCE P-VALUE = 0.0208 < 0.05, REJECT Ho.
MTB > # SUFFICIENT EVIDENCE THAT THE MEDIANS DIFFER.
```

Section 15.4

15.37 H_0: $\tilde{\mu}_1 = \tilde{\mu}_2 = \tilde{\mu}_3$
H_a: At least 2 populations have different medians.
$H = 13.11$
RR: $\chi^2 > 4.61$
Reject H_0
15.39 H_0: $\tilde{\mu}_1 = \tilde{\mu}_2$
H_a: $\tilde{\mu}_1 \neq \tilde{\mu}_2$
$H = 16.41$
RR: $\chi^2 > 3.84$
Reject H_0
15.41 p-value < 0.005

15.43 H_0: $\tilde{\mu}_1 = \tilde{\mu}_2 = \mu_3 = \tilde{\mu}_4$
H_a: At least 2 populations have different medians.
$H = 8.18$
RR: $\chi^2 > 7.81$
Reject H_0 (yes)
15.45 H_0: $\tilde{\mu}_1 = \tilde{\mu}_2 = \mu_3$
H_a: At least 2 populations have different medians.
$H = 15.53$
RR: $\chi^2 > 5.99$
Reject H_0 (yes)
15.47 p-value < 0.005

M▶ 15.49
```
LEVEL      NOBS     MEDIAN   AVE. RANK   Z VALUE
    1        16      78.00      24.4      -0.02
    2        16      71.50      11.6      -4.52
    3        16      87.00      37.5       4.54
OVERALL      48                 24.5

H = 27.33  d.f. = 2   p = 0.000
H = 27.41  d.f. = 2   p = 0.000 (adj. for ties)

MTB > # SINCE P-VALUE = 0.000 < 0.05, REJECT Ho.
MTB > # THERE IS SUFFICIENT EVIDENCE THAT A DIFFERENCE EXISTS.
```

M▶ **15.51**

LEVEL	N	Median	Ave Rank	Z
1	5	21.00	4.6	-2.64
2	6	26.00	12.9	0.90
3	5	25.00	15.2	1.73
4	5	24.00	10.9	-0.04
Overall	21		11.0	

H = 8.18 DF = 3 P = 0.042
H = 8.33 DF = 3 P = 0.040 (adjusted for ties)

MTB > # SINCE P-VALUE = 0.042 < 0.05, REJECT Ho.
MTB > # SUFFICIENT EVIDENCE TO INDICATE DIFFERENCES.

Section 15.5

15.53 H_0: $\tilde{\mu}_A = \tilde{\mu}_B = \tilde{\mu}_C = \tilde{\mu}_D$
H_a: At least 2 treatments have different medians.
$F_r = 19.61$
RR: $\chi^2 > 7.81$
Reject H_0 (yes)

15.55 H_0: $\tilde{\mu}_1 = \tilde{\mu}_2$
H_a: $\tilde{\mu}_1 \neq \tilde{\mu}_2$
$F_r = 4.08$
RR: $\chi^2 > 2.71$
Reject H_0 (yes)

15.57 $0.025 < p\text{-value} < 0.05$

15.59 H_0: $\tilde{\mu}_1 = \tilde{\mu}_2 = \tilde{\mu}_3 = \tilde{\mu}_4$
H_a: At least 2 treatments have different medians.
$F_r = 11.35$
RR: $\chi^2 > 6.25$
Reject H_0 (yes)

15.61 $0.005 < p\text{-value} < 0.01$

M▶ **15.63** Friedman test of C1 by C2 blocked by C3
S = 7.76 d.f. = 1 p = 0.005

C2	N	Est. Median	Sum of RANKS
1	29	558.0	51.0
2	29	308.3	36.0

Grand median = 433.1
MTB > # REJECT Ho SINCE P-VALUE = 0.005 < 0.05.
MTB > # THERE IS A SIGNIFICANT DIFFERENCE.

M▶ **15.65** Friedman test of C1 by C2 blocked by C3
S = 17.73 d.f. = 2 p = 0.000
S = 18.91 d.f. = 2 p = 0.000 (adjusted for ties)

C2	N	Est. Median	Sum of RANKS
2	20	1.3958	52.5
3	20	1.3442	41.5
4	20	1.2125	26.0

Grand median = 1.3175
MTB > # REJECT Ho SINCE P-VALUE = 0.000 < 0.05.
MTB > # A SIGNIFICANT DIFFERENCE DOES EXIST.

Section 15.6

15.67 7

15.69 1, 2, 26, 3, 4, 25, 27, 28, 29, 30, 31, 5, 6
(Other answers are possible.)

15.71 H_0: The sequence of observations is random.
H_a: The sequence is not random.
$x = 15$
RR: $x \leq 8$, $x \geq 19$
Fail to reject H_0 (no)

15.73 H_0: The sequence of observations is random.
H_a: The sequence is not random.
$z = -0.74$
RR: $z < -2.576$, $z > 2.576$
Fail to reject H_0 (no)

15.75 H_0: The sequence of observations is random.
H_a: The sequence is not random.
$z = -2.56$
RR: $z < -1.96$, $z > 1.96$
Reject H_0 (yes)

15.77 H_0: The sequence of observations is random.
H_a: The sequence is not random.
$z = -1.14$
RR: $z < -1.96$, $z > 1.96$
Fail to reject H_0 (no)

M▶ 15.79 THE OBSERVED NO. OF RUNS = 10
THE EXPECTED NO. OF RUNS = 7.8571
6 OBSERVATIONS ABOVE K   8 BELOW

Section 15.7

15.81 $r_s < -0.409$, $r_s > 0.409$

15.83 $r_s < -0.475$

15.85 $r_s = 0.905$

15.87 $r_s = -0.866$

15.89 H_0: $\rho_s = 0$
H_a: $\rho_s < 0$
$r_s = -0.866$
RR: $r_s < -0.564$
Reject H_0 (yes)

15.91 H_0: $\rho_s = 0$
H_a: $\rho_s \neq 0$
$r_s = -0.779$
RR: $r_s < -0.689$, $r_s > 0.689$
Reject H_0 (yes)

M▶ 15.93 Correlation of C3 and C4 = -0.779

Review Exercises

15.95 H_0: $\tilde{\mu}_1 - \tilde{\mu}_2 = 0$
H_a: $\tilde{\mu}_1 - \tilde{\mu}_2 \neq 0$
$z = 2.57$
RR: $z < -1.96$, $z > 1.96$
Reject H_0 (yes)

15.96 H_0: The sequence of observations is random.
H_a: The sequence is not random.
$x = 12$
RR: $x \leq 9$, $x \geq 21$
Fail to reject H_0

15.97 H_0: The sequence of observations is random.
H_a: The sequence is not random.
$z = -1.13$
RR: $z < -1.96$, $z > 1.96$
Fail to reject H_0

15.98 H_0: $\tilde{\mu}_D = 0$
H_a: $\tilde{\mu}_D > 0$
$x = 9$
RR: 8, 9, 10 ($\alpha = 0.055$, $n = 10$)
Reject H_0 (yes)

15.99 p-value = 0.011

15.100 H_0: $\tilde{\mu} = 300$
H_a: $\tilde{\mu} < 300$
$z = -3.21$
RR: $z < -2.326$
Reject H_0 (yes)

15.101 p-value = 0.0007

15.102 H_0: $\rho_s = 0$
H_a: $\rho_s > 0$
$r_s = 0.685$
RR: $r_s > 0.745$
Fail to reject H_0 (no)

15.103 $0.01 < p$-value < 0.025

15.104 H_0: $\tilde{\mu}_1 = \tilde{\mu}_2 = \tilde{\mu}_3 = \tilde{\mu}_4$
H_a: At least 2 populations have different medians.
$H = 17.33$
RR: $\chi^2 > 11.34$
Reject H_0 (yes)

15.105 H_0: The sequence of observations is random.
H_a: The sequence is not random.
$x = 14$
RR: $x \leq 10$, $x \geq 22$
Fail to reject H_0

15.106 H_0: The sequence of observations is random.
H_a: The sequence is not random.
$z = -0.74$
RR: $z < -1.96$, $z > 1.96$
Fail to reject H_0

15.107 H_0: $\tilde{\mu}_1 = \tilde{\mu}_2 = \tilde{\mu}_3 = \tilde{\mu}_4$
H_a: At least 2 populations have different medians.
$H = 15.21$
RR: $\chi^2 > 7.81$
Reject H_0 (yes)

15.108 p-value < 0.005

15.109 H_0: $\tilde{\mu}_1 = \tilde{\mu}_2 = \tilde{\mu}_3$
H_a: At least 2 treatments have different medians.
$F_r = 9.25$
RR: $\chi^2 > 5.99$
Reject H_0

15.110 H_0: $\tilde{\mu}_1 = \tilde{\mu}_2 = \tilde{\mu}_3 = \tilde{\mu}_4 = \tilde{\mu}_5$
H_a: At least 2 treatments have different medians.
$F_r = 32.26$
RR: $\chi^2 > 13.28$
Reject H_0 (yes)

15.111 H_0: $\tilde{\mu}_1 = \tilde{\mu}_2 = \tilde{\mu}_3 = \tilde{\mu}_4$
H_a: At least 2 treatments have different medians.
$F_r = 10.31$
RR: $\chi^2 > 7.81$
Reject H_0 (yes)

15.112 $0.01 < p\text{-value} < 0.025$

15.113 H_0: $\tilde{\mu}_1 = \tilde{\mu}_2$
H_a: $\tilde{\mu}_1 \neq \tilde{\mu}_2$
$F_r = 0.50$
RR: $\chi^2 > 3.84$
Fail to reject H_0 (no)

M▶ **15.114** SIGN TEST OF MEDIAN = 300.0 VERSUS L.T.   300.0
```
                    N   BELOW  EQUAL  ABOVE   P-VALUE    MEDIAN
C1                  36    27     1      8     0.0009     294.5
MTB > # SINCE P-VALUE = 0.0009 < 0.05, REJECT Ho.
MTB > # SUFFICIENT EVIDENCE THAT MEDIAN IS LESS THAN 5 MINUTES.
```

M▶ **15.115** Correlation of C3 and C4 = 0.685

M▶ **15.116** MEDIAN =   485.00
```
     K =     485.0000

     THE OBSERVED NO. OF RUNS =   14
     THE EXPECTED NO. OF RUNS =   16.0000
     15 OBSERVATIONS ABOVE K    15 BELOW
```

M▶ **15.117**
```
LEVEL     NOBS     MEDIAN   AVE. RANK    Z VALUE
   1       10      81.40      11.4       -2.83
   2       10      86.75      25.0        1.42
   3       10      96.85      35.0        4.51
   4       10      78.10      10.6       -3.11
OVERALL    40                 20.5

H = 30.03   d.f. = 3   p = 0.000
H = 30.04   d.f. = 3   p = 0.000 (adj. for ties)
MTB > # SINCE P-VALUE = 0.000 < 0.05, REJECT Ho.
MTB > # SUFFICIENT EVIDENCE THAT THE MEDIANS ARE NOT ALL EQUAL.
```

M▶ **15.118** Mann-Whitney Confidence Interval and Test
```
C1           N =  10      Median =      81.400
C3           N =  10      Median =      96.850
Point estimate for ETA1-ETA2 is     -15.500
95.5 pct c.i. for ETA1-ETA2 is (-17.000,-12.301)
W = 55.0
Test of ETA1 = ETA2  vs.  ETA1 n.e. ETA2 is significant at 0.0002
The test is significant at 0.0002 (adjusted for ties)
MTB > # SINCE P-VALUE = 0.0002 < 0.01, REJECT Ho.
MTB > # SUFFICIENT EVIDENCE THAT MEDIANS 1 AND 3 DIFFER.
```

15.119 SIGN TEST OF MEDIAN = 0.00000 VERSUS N.E. 0.00000

	N	BELOW	EQUAL	ABOVE	P-VALUE	MEDIAN
C3	33	4	3	26	0.0001	0.06000

MTB > # SINCE P-VALUE = 0.0001 < 0.01, REJECT Ho.
MTB > # SUFFICIENT EVIDENCE THAT THE AVERAGE PRICES DIFFER.

15.120 Friedman test of C1 by C2 blocked by C3
$S = 14.67$ d.f. = 1 p = 0.000
$S = 16.13$ d.f. = 1 p = 0.000 (adjusted for ties)

C2	N	Est. Median	Sum of RANKS
1	33	3.7800	60.5
2	33	3.7200	38.5

Grand median = 3.7500
MTB > # REJECT Ho SINCE P-VALUE = 0.000 < 0.01.
MTB > # THERE IS A SIGNIFICANT DIFFERENCE.

15.121 Friedman test of C1 by C2 blocked by C3
$S = 32.26$ d.f. = 4 p = 0.000
$S = 32.92$ d.f. = 4 p = 0.000(adjusted for ties)

C2	N	Est. Median	Sum of RANKS
1	10	211.55	33.0
2	10	210.58	22.0
3	10	212.47	48.5
4	10	211.71	35.5
5	10	210.34	11.0

Grand median = 211.33
MTB > # REJECT Ho SINCE P-VALUE = 0.000 < 0.01.
MTB > # THERE IS A SIGNIFICANT DIFFERENCE.

15.122 Friedman test of C1 by C2 blocked by C3
$S = 10.31$ d.f. = 3 p = 0.016
$S = 10.44$ d.f. = 3 p = 0.015 (adjusted for ties)

C2	N	Est. Median	Sum of RANKS
1	8	81.813	17.5
2	8	80.062	15.0
3	8	92.313	30.0
4	8	80.563	17.5

Grand median = 83.688
MTB > # REJECT Ho SINCE P-VALUE = 0.016 < 0.05.
MTB > # YES, THERE IS A SIGNIFICANT DIFFERENCE.

CHAPTER 16
Sections 16.1–16.2

16.1 For a particular pair of x_1 and x_2 values, the model always gives the same value of y. It does not contain a random error component.

16.3 a. The estimate of the partial slope β_1 is 47.6. The mean of y increases by 47.6 for each 1-unit increase in x_1 while x_2 is held fixed. The partial slope β_2 is estimated by $\hat{\beta}_2 = 9.8$. The mean of y increases by 9.8 for each 1-unit increase in x_2 while x_1 is held fixed.

b. $\hat{y} = 190.4$

c. $H_0: \beta_1 = \beta_2 = 0$
H_a: At least one $\beta_i \neq 0$
$F = 3.68$
RR: $F > 3.89$
Fail to reject H_0

16.5 $0.05 < p\text{-value} < 0.10$

16.7 $R^2 = 0.782$
The estimated model accounts for 78.2 percent of the total variation in the y-values.

16.9 $H_0: \beta_1 = \beta_2 = \beta_3 = \beta_4 = 0$
H_a: At least one $\beta_i \neq 0$
$F = 23.51$
RR: $F > 2.76$
Reject H_0

16.11 $H_0: \beta_1 = \beta_2 = \cdots = \beta_{12} = 0$
H_a: At least one $\beta_i \neq 0$
$F = 4.00$
RR: $F > 19.41$
Fail to reject H_0

16.13 a. $\hat{y} = 1,548 + 1.10x_1 + 0.738x_2$

b. $\hat{\beta}_1 = 1.10$; the mean of y increases by \$1.10 for each \$1 increase in x_1 while x_2 is held fixed.

c. $R^2 = 69.1$ percent
The estimated model accounts for 69.1 percent of the total variation in the y-values.

d. $\hat{\sigma} = s = 737.6$

e. $H_0: \beta_1 = \beta_2 = 0$
H_a: At least one $\beta_i \neq 0$
$F = 13.42$
$p\text{-value} = 0.001$
Reject H_0

M▶ 16.15 a. $\hat{y} = 58.8 - 19.0x_1 + 323x_2 + 0.181x_3$

b. $R^2 = 85.4$ percent
The estimated model accounts for 85.4 percent of the total variation in the y-values.

c. $H_0: \beta_1 = \beta_2 = \beta_3 = 0$
H_a: At least one $\beta_i \neq 0$
$F = 19.47$
$p\text{-value} = 0.000$
Reject H_0

M▶ 16.17 $H_0: \beta_1 = \beta_2 = 0$
H_a: At least one $\beta_i \neq 0$
$F = 104.86$
$p\text{-value} = 0.000$
Reject H_0 (yes)

Sections 16.3–16.4

16.19 $\mu_y = \beta_0 + \beta_1 x_1 + \beta_2 x_2$　　where

$$x_1 = \begin{cases} 1 & \text{if 2.0 liter} \\ 0 & \text{if not} \end{cases} \qquad x_2 = \begin{cases} 1 & \text{if 2.5 liter} \\ 0 & \text{if not} \end{cases}$$

16.21 $\mu_y = \beta_0 + \beta_1 x_1 + \beta_2 x_2 + \beta_3 x_3$　　where

$$x_1 = \begin{cases} 1 & \text{if 2.0 liter} \\ 0 & \text{if not} \end{cases} \qquad x_2 = \begin{cases} 1 & \text{if 2.5 liter} \\ 0 & \text{if not} \end{cases}$$

$$x_3 = \begin{cases} 1 & \text{if 5-speed} \\ 0 & \text{if not} \end{cases}$$

16.23 a. β_0

b. $\beta_0 + \beta_1$

c. $\beta_0 = $ (mean number of afternoon calls)
$\beta_1 = $ (mean number of morning calls) $-$ (mean number of afternoon calls)
$\beta_2 = $ (mean number of evening calls) $-$ (mean number of afternoon calls)

16.25 $H_0: \beta_1 = \beta_2 = \beta_3 = \beta_4 = 0$
H_a: At least one $\beta_i \neq 0$
$F = 36.23$
RR: $F > 2.61$
Reject H_0

16.27 $\mu_y = \beta_0 + \beta_1 x_1 + \beta_2 x_2 + \beta_3 x_3 + \beta_4 x_4 + \beta_5 x_1 x_2 + \beta_6 x_1 x_3 + \beta_7 x_1 x_4 + \beta_8 x_2 x_3 + \beta_9 x_2 x_4 + \beta_{10} x_3 x_4 + \beta_{11} x_1^2 + \beta_{12} x_2^2 + \beta_{13} x_3^2 + \beta_{14} x_4^2$

16.29 a. Response surface is a plane in 3 dimensions.

b. μ_y increases by 5 units.

c. μ_y increases by 5 units.

d. Add to the model an interaction term $x_1 x_2$.

16.31 $y = \beta_0 + \beta_1 x + \beta_2 x^2 + \epsilon$ where $\beta_2 > 0$

16.33 a. $\hat{y} = 67.1 - 1.41x + 0.0111x^2$

b. $H_0: \beta_1 = \beta_2 = 0$
H_a: At least one $\beta_i \neq 0$
$F = 25.84$
$p\text{-value} = 0.000$
Reject H_0

M▶ 16.35 $\hat{y} = -86.0 + 0.0185x_1 + 3.79x_2 + 1.98x_3$

M▶ 16.37 $H_0: \beta_1 = \beta_2 = \beta_3 = 0$
H_a: At least one $\beta_i \neq 0$
$F = 70.33$
$p\text{-value} = 0.000$
Reject H_0 (yes)

Sections 16.5–16.6

16.39 H_0: $\beta_1 = 0$
H_a: $\beta_1 > 0$
$t = 2.74$
RR: $t > 2.508$
Reject H_0

16.41 $0.005 < p\text{-value} < 0.01$

16.43 9.8 ± 5.185

16.45 a. H_0: $\beta_1 = \beta_2 = 0$
H_a: At least one $\beta_i \neq 0$
$F = 38.16$
$p\text{-value} = 0.000$
Reject H_0
b. H_0: $\beta_2 = 0$
H_a: $\beta_2 \neq 0$
$t = -2.10$
$p\text{-value} = 0.044$
Reject H_0 (yes)
c. 0.2488 ± 0.0785

16.47 H_0: $\beta_5 = 0$
H_a: $\beta_5 \neq 0$
$t = 1.84$
RR: $t < -1.96, t > 1.96$
Fail to reject H_0

16.49 a. H_0: $\beta_1 = \beta_2 = \beta_3 = 0$
H_a: At least one $\beta_i \neq 0$
b. H_0: $\beta_4 = 0$
H_a: $\beta_4 \neq 0$

16.51 H_0: $\beta_4 = 0$
H_a: $\beta_4 \neq 0$
$F = 14.54$
RR: $F > 3.92$
Reject H_0 (yes)

16.53 H_0: $\beta_2 = \beta_3 = 0$
H_a: At least one $\beta_i \neq 0$
$F = 4.72$
RR: $F > 4.10$
Reject H_0 (yes)

M▶ 16.55

Obs.	X	Y	Fit	Residual
1	82.0	33.000	26.427	6.573
2	43.0	30.000	27.093	2.907
3	73.0	23.000	23.572	-0.572
4	35.0	32.000	31.418	0.582
5	29.0	33.000	35.598	-2.598
6	60.0	25.000	22.635	2.365
7	75.0	19.000	24.051	-5.051
8	61.0	23.000	22.574	0.426
9	49.0	26.000	24.784	1.216
10	57.0	23.000	22.954	0.046
11	94.0	33.000	33.041	-0.041
12	77.0	22.000	24.618	-2.618
13	87.0	31.000	28.793	2.207
14	46.0	27.000	25.838	1.162
15	72.0	23.000	23.367	-0.367
16	33.0	33.000	32.722	0.278
17	30.0	31.000	34.846	-3.846
18	62.0	23.000	22.534	0.466
19	78.0	18.000	24.935	-6.935
20	61.0	21.000	22.574	-1.574
21	45.0	26.000	26.234	-0.234
22	55.0	27.000	23.278	3.722
23	88.0	34.000	29.333	4.667
24	77.0	24.000	24.618	-0.618
25	81.0	31.000	26.021	4.979
26	44.0	30.000	26.652	3.348
27	70.0	21.000	23.022	-2.022
28	38.0	32.000	29.629	2.371
29	31.0	32.000	34.116	-2.116
30	61.0	23.000	22.574	0.426
31	76.0	18.000	24.323	-6.323
32	60.0	20.000	22.635	-2.635
33	50.0	28.000	24.477	3.523
34	53.0	23.000	23.691	-0.691
35	92.0	34.000	31.716	2.284
36	84.0	22.000	27.306	-5.306

M▶ 16.57 a. $\hat{y} = -86.0 + 0.0185x_1 + 3.79x_2 + 1.98x_3$
b. $\hat{y} = -108 + 0.0233x_1$
c. H_0: $\beta_2 = \beta_3 = 0$
H_a: At least one $\beta_i \neq 0$
$F = 3.86$
RR: $F > 3.37$
Reject H_0 (yes)

Sections 16.7–16.8

16.59 For each predictor variable x_i, construct a plot of the residuals against the corresponding values of x_i. If curvature is exhibited, try a higher-order term such as βx_i^2 in the model.

16.61 Yes; try adding a higher-order term such as $\beta_2 x^2$ to the model.

16.63 There is 1 outlier with a fit of about 140 and a standardized residual of approximately 3.6.

16.65 Multicollinearity refers to the existence of correlations among some of the predictor variables. When some of the correlations are large, variables contain redundant information, the estimated coefficients are unstable with large standard deviations, and they are difficult to interpret.

16.67 There is 1 outlier with a fit of approximately 240 and a standardized residual of about 3.6.

16.69 The stem-and-leaf display is approximately mound shaped, suggesting that the normality assumption appears reasonable.

16.71

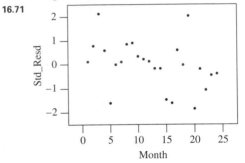

The sequences of 4 positives, 7 positives, 4 negatives, and 5 negatives cast some doubt on the independence of the random error components.

16.73 Some of the estimated standard deviations of the regression coefficients are large relative to the coefficients.

16.75 a. x_2, x_{10}, x_6, and x_5.
  b. $0.02 < p\text{-value} < 0.05$
  c. R^2 changes from 78.30 to 79.34 percent, for an increase of 1.04 percent.
  d. $\hat{y} = 62.34 + 2.41x_2 + 9.26x_{10} + 4.72x_6 + 0.85x_5$
  e. Model: $\mu_y = \beta_0 + \beta_1 x_2 + \beta_2 x_{10} + \beta_3 x_6 + \beta_4 x_5$
  $H_0: \beta_1 = \beta_2 = \beta_3 = \beta_4 = 0$
  $H_a:$ At least one $\beta_i \neq 0$
  $F = 24.96$
  RR: $F > 2.74$
  Reject H_0

M▶ 16.77 a. The regression equation is

Y = −3.59 + 0.886X1 + 5.84X2 −0.941X2−SQ

```
STD_RESD
-0.39936   1.23170   1.00426   -1.45196
-1.69412   0.78891   0.12155
 0.19344  -0.35159   1.14516
```

b. A plot of the standardized residuals against the values of x_2 appears below.

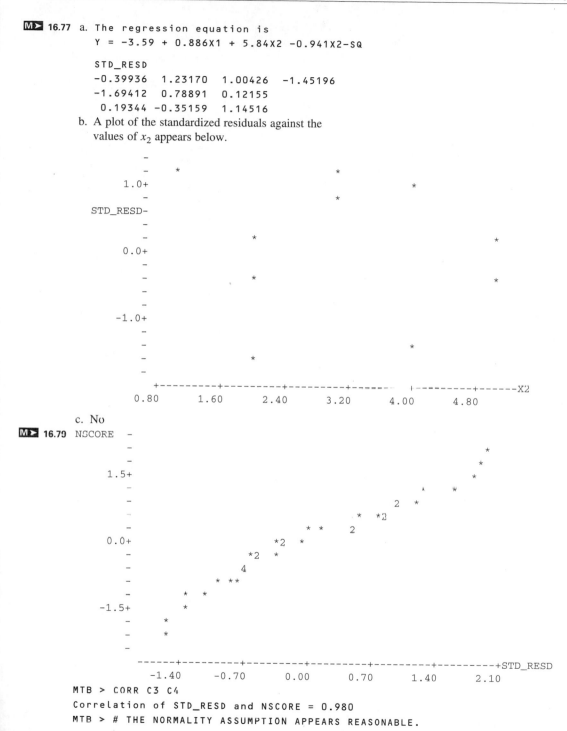

c. No

M▶ 16.79

```
MTB > CORR C3 C4
Correlation of STD_RESD and NSCORE = 0.980
MTB > # THE NORMALITY ASSUMPTION APPEARS REASONABLE.
```

M▶ 16.81 Response is Y on 5 predictors,
with N = 14

Step	1	2
Constant	171.24	80.05
X1	-21.0	-15.0
T-Value	-5.50	-4.65
X4		0.086
T-Value		3.50
S	6.65	4.78
R-Sq	71.59	86.57

The estimated model is
$\hat{y} = 80.05 - 15.0x_1 + 0.086x_4$
$R^2 = 86.57$ percent

Review Exercises

16.82 a. $\hat{y} = -5.43 + 3.64x_1 + 0.733x_2 - 0.228x_1x_2$
b. 2
c. 4.62
d. $\hat{\sigma} = s = 0.2914$
e. $R^2 = 93.4$ percent
The estimated model accounts for 93.4 percent of the total variation in the y-values.

16.83 $H_0: \beta_1 = \beta_2 = \beta_3 = 0$
H_a: At least one $\beta_i \neq 0$
$F = 122.66$
p-value $= 0.000$
Reject H_0

16.84 3.6422 ± 1.61663

16.85 $H_0: \beta_3 = 0$
$H_a: \beta_3 \neq 0$
$t = -1.64$
p-value $= 0.114$
Fail to reject H_0

16.86 a. Qualitative variable
b. 3 levels
c. $\mu_y = \beta_0 + \beta_1x_1 + \beta_2x_2$ where

$x_1 = \begin{cases} 1 & \text{if priority} \\ 0 & \text{if not} \end{cases}$ $x_2 = \begin{cases} 1 & \text{if standard} \\ 0 & \text{if not} \end{cases}$

16.87 a. $\beta_0 + \beta_4$
b. β_0
c. $\beta_0 =$ (mean number of M1000 sales)
$\beta_1 =$ (mean number of M500 sales) $-$
     (mean number of M1000 sales)
$\beta_2 =$ (mean number of M550 sales) $-$
     (mean number of M1000 sales)
$\beta_3 =$ (mean number of M700 sales) $-$
     (mean number of M1000 sales)
$\beta_4 =$ (mean number of M950 sales) $-$
     (mean number of M1000 sales)

16.88 For each combination of values of the predictor variables, the random error component ϵ has a normal distribution with 0 mean and constant variance σ^2. Also, for each pair of observations y_i and y_j, the associated error components are independent.

16.89 a. The partial slope β_1 is estimated by $\hat{\beta}_1 = 2.1$. The mean of y increases by 2.1 for each 1-unit increase in x_1 while x_2 and x_3 are held fixed. The partial slope β_2 is estimated by $\hat{\beta}_2 = -4.8$. The mean of y decreases by 4.8 for each 1-unit increase in x_2 while x_1 and x_3 are held fixed. The partial slope β_3 is estimated by $\hat{\beta}_3 = 10$. The mean of y increases by 10 for each 1-unit increase in x_3 while x_1 and x_2 are held fixed.
b. 54.6
c. $H_0: \beta_1 = \beta_2 = \beta_3 = 0$
H_a: At least one $\beta_i \neq 0$
$F = 14.17$
RR: $F > 3.10$
Reject H_0

16.90 p-value < 0.005

16.91 $H_0: \beta_3 = 0$
$H_a: \beta_3 \neq 0$
$t = 2.12$
RR: $t < -2.086, t > 2.086$
Reject H_0 (yes)

16.92 10.0 ± 9.83

16.93 A confidence interval should be used since a mean is to be estimated.

16.94 $R^2 = 0.684$
The estimated model accounts for 68.4 percent of the total variation in the y-values.

16.95 $H_0: \beta_1 = \beta_2 = 0$
H_a: At least one $\beta_i \neq 0$
$F = 11.91$
RR: $F > 7.21$
Reject H_0

16.96 a. 1
b. $y = (\beta_0 + \beta_3) + (\beta_1 + \beta_5)x_1 + (\beta_2 + \beta_6)x_2 + (\beta_4 + \beta_7)x_1x_2 + \epsilon$
c. $y = \beta_0 + \beta_1x_1 + \beta_2x_2 + \beta_4x_1x_2 + \epsilon$
d. $H_0: \beta_3 = \beta_5 = \beta_6 = \beta_7 = 0$
H_a: At least one $\beta_i \neq 0$

16.97 $H_0: \beta_3 = \beta_5 = \beta_6 = \beta_7 = 0$
H_a: At least one $\beta_i \neq 0$
$F = 0.75$
RR: $F > 2.78$
Fail to reject H_0

16.98 $y = \beta_0 + \beta_1x + \beta_2x^2 + \epsilon$ where $\beta_2 < 0$

16.99 e. $\beta_2 < 0$

16.100 a. $\hat{y} = -3.45 + 0.0743x_1 - 0.0074x_2 - 0.000069x_1x_2$

b. $H_0: \beta_1 = \beta_2 = \beta_3 = 0$

H_a: At least one $\beta_i \neq 0$

$F = 80.75$

p-value $= 0.000$

Reject H_0 (yes)

c. $\$4.072 < y < \$8.926$

16.101 $H_0: \beta_3 = 0$

$H_a: \beta_3 \neq 0$

$t = -3.00$

p-value $= 0.011$

Reject H_0

16.102 $H_0: \beta_2 = \beta_3 = 0$

H_a: At least one $\beta_i \neq 0$

$F = 56.56$

RR: $F > 3.89$

Reject H_0 (yes)

16.103 a. 2

b. The change in μ_y from a 1-unit increase in x_1 is dependent on the level at which x_2 is held fixed.

c. Remove the term $\beta_3x_1x_2$.

16.104 Yes; a higher-order term such as β_2x^2 should be added to the model.

16.105 There is 1 outlier with a fit of approximately 500 and a standardized residual of about -3.2.

16.106 No; the standardized residuals appear to fluctuate in a random manner.

16.107 The mound shape of the histogram suggests that the normality assumption appears reasonable.

16.108 Large (in absolute value) correlations between some of the predictor variables; regression coefficients with relatively large estimated standard deviations.

16.109 a. x_5

b. p-value < 0.01

c. $\hat{y} = 19,765 + 924x_7 + 534x_2 + 5,233x_5$

d. Model: $\mu_y = \beta_0 + \beta_1x_7 + \beta_2x_2 + \beta_3x_5$

$H_0: \beta_1 = \beta_2 = \beta_3 = 0$

H_a: At least one $\beta_i \neq 0$

$F = 76.30$

RR: $F > 2.98$

Reject H_0

16.110 A condition of multicollinearity exists where x_1 and x_2 are highly correlated.

M▶ 16.111 a. $R^2 = 95.3$ percent

The estimated model accounts for 95.3 percent of the total variation in the y-values.

b. $H_0: \beta_1 = \beta_2 = \beta_3 = 0$

H_a: At least one $\beta_i \neq 0$

$F = 80.75$

p-value $= 0.000$

Reject H_0

c. $H_0: \beta_3 = 0$

$H_a: \beta_3 \neq 0$

$t = -3.00$

p-value $= 0.011$

Reject H_0 (yes)

M▶ 16.112 $\$3.701 < y < \$8.587$

M▶ 16.113 $H_0: \beta_2 = \beta_3 = 0$

H_a: At least one $\beta_i \neq 0$

$F = 56.56$

RR: $F > 3.89$

Reject H_0 (yes)

M▶ 16.114 a. $\hat{y} = -86.0 + 0.0185x_1 + 3.79x_2 + 1.98x_3$

Row	FIT	STD_RESD
1	54.6966	0.85954
2	63.0401	-0.03372
3	45.1585	0.27123
.	....	....
.	....	....
28	52.6628	-1.63263
29	58.3933	-0.80249
30	71.3674	-0.28659

b.
```
STD_RESD-
    2.0+                                          *
       -                              *
       -
       -                      *
       -               *   *       *   *      *   *
       -      *   *        *   *                        *   *
    0.0+           *                     *   *              *
       -                                                       *
       -         *   2                 *                       *
       -
       -      *                    *
    -2.0+                                                    *
       -
       -         *
       -
       -
    -4.0+
       --------+---------+---------+---------+---------+--------X1
            6300      6600      6900      7200      7500
```

c. No; it does not appear that a higher-order term is needed.

M▶ 16.115 Histogram of STD_RESD   N = 30

Midpoint	Count	
-3.5	1	*
-3.0	0	
-2.5	0	
-2.0	1	*
-1.5	2	**
-1.0	3	***
-0.5	4	****
0.0	4	****
0.5	8	********
1.0	4	****
1.5	2	**
2.0	1	*

No; the normality assumption appears to be reasonable.

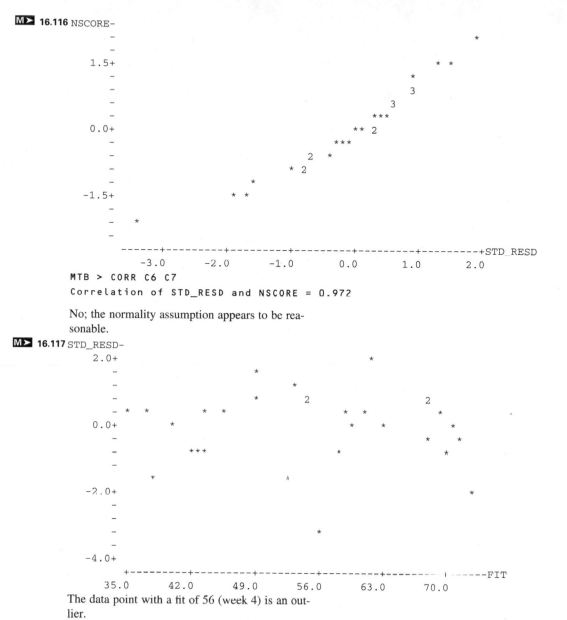

M▶ 16.116 NSCORE-

```
       -
       -                                                              *
  1.5+                                                      *  *
       -                                               *
       -                                              3
       -                                          3
       -                                      ***
  0.0+                                    **  2
       -                               ***
       -                          2   *
       -                     *   2
       -                *
 -1.5+              *   *
       -
       -       *
       -
       ------+---------+---------+---------+---------+---------+STD_RESD
           -3.0      -2.0      -1.0       0.0       1.0       2.0
```

MTB > CORR C6 C7
Correlation of STD_RESD and NSCORE = 0.972

No; the normality assumption appears to be reasonable.

M▶ 16.117 STD_RESD-

```
  2.0+                                          *
       -                        *
       -                            *
       -                   *      2                    2
       -   *   *        *   *           *   *        *
  0.0+          *                     *      *           *
       -                                          *      *
       -            ***                *           *
       -       *                  A
 -2.0+                                                  *
       -
       -
       -                        *
       -
 -4.0+
       +---------+---------+---------+---------+--------|------FIT
        35.0      42.0      49.0      56.0      63.0      70.0
```

The data point with a fit of 56 (week 4) is an outlier.

The variation in the standardized residuals tends to be less for smaller and larger fits. However, in light of the limited number of data points, there does not appear to be a serious violation in the assumption of constant variance.

M▶ 16.118 a.

```
MTB > STEPWISE C6 C1-C5

STEPWISE REGRESSION OF Y ON 5 PREDICTORS, WITH N = 12

       STEP        1        2        3
    CONSTANT    200.66    94.09  -731.40

    X5          -0.190   -0.155   -0.134
    T-RATIO      -3.37    -4.51    -4.35

    X4                    0.134    0.110
    T-RATIO               4.45     3.92

    X3                             844
    T-RATIO                        2.10

    S            7.65     4.51     3.84
    R-SQ        53.21    85.38    90.56
```

The selected variables are x_5, x_4, and x_3.

b. $\hat{y} = -731.40 - 0.134x_5 + 0.110x_4 + 844x_3$

c. $R^2 = 90.56$ percent

INDEX